THE LEPIDOPTERISTS' SOCIETY

COMMEMORATIVE VOLUME

1945 — 1973

THE LEPIDOPTERISTS' SOCIETY
COMMEMORATIVE VOLUME

1945 — 1973

Compiled by ROY O. KENDALL

Edited by HARRY K. CLENCH & THEODORE D. SARGENT

With contributions by PETER F. BELLINGER,
HARRY K. CLENCH & PAUL A. OPLER

Published by

THE LEPIDOPTERISTS' SOCIETY

1977

Library of Congress Catalog Card Number: 77-24451

ISBN: 0-930282-01-9

Printed in the United States of America

CONTENTS

FOREWORD *by Harry K. Clench* vii

INTRODUCTION *by Roy O. Kendall* xv

EARLY ORGANIZATION 1

CONSTITUTION & BY-LAWS 3

MEMBERSHIP
A Backward Glance - A Glimpse Ahead
by Harry K. Clench & Theodore D. Sargent . . . 15
Membership Table 20

FINANCIAL GROWTH 27

PUBLICATIONS 29

ANNUAL MEETINGS 31

ANNUAL FIELD SUMMARIES *by Paul A. Opler* 37

LIBRARY *by Peter F. Bellinger* 45

DISTINGUISHED SERVICE
Honorary Life Memberships 49
Karl Jordan Medal 51

PERSONNEL
Officers & Editors 55
Charter Members 58
Obituaries/Biographies/Bibliographies 64
Disposition of Collections 70

BIOGRAPHICAL SKETCHES 75

FINANCIAL SPONSORS
Members 115
Entomological Suppliers 121

INDICES 125
Author 127
Subject 141
Scientific Names 190

FOREWORD

Sheldon Remington and William Clench were boyhood friends in Boston in the years just before the first World War, and they shared an active interest in collecting butterflies. In the due course of time they grew up, went their separate ways, married, had children. Though it was often pushed to the background, that interest in butterflies was never

P. Sheldon Remington and son Charles L., out for a walk in St. Louis, Missouri, 1924.

Harry K. Clench, at age 1, surveying the area for lycaenids.

wholly lost. Years later Sheldon passed it on to his son, Charles; and William passed it on to his son, Harry (me).

Charles grew up in St. Louis, Missouri, and took his undergraduate work at the Principia College in Elsa, Illinois, while I grew up in Boston. Our mutual interests drew us into correspondence. Eventually - right after World War II - Charles and his wife, Jeanne, arrived in Cambridge, so that Charles could begin graduate studies at Harvard. I was then in my second undergraduate year, at Cambridge Junior College.

We lost no time getting together, and soon found that we had a lot more in common than fathers who had been boyhood companions. We were both deeply interested in butterflies. We both wanted to make the study of Lepidoptera our profession. We each had a wide circle of friends and correspondents many of whom, like ourselves, were amateur lepidopterists. Most of them were older and wiser than we, and much of what we knew we had learned from them. Yet at the same time we felt that some amateurs, with no guidance in what was scientifically useful, were misdirecting their efforts. In short, we believed that amateur and professional each had much to teach, and much to learn from, the other, if only some way could be found to bring them together.

The idea of a society of lepidopterists we had each mulled over a bit before we had met. Once together our bull sessions, whether in the Clench kitchen on Avon Hill Street or in the Remington office in Harvard's Biological Laboratories, tended more and more to that topic. We found it easy to agree, for our thoughts were much the same: we wanted an organization that would appeal to both beginner and seasoned veteran, to both amateur and professional; we wanted it to set high standards by example (nor were we averse to an occasional bit of preaching); and we wanted it to be a means for exchanging ideas and information among lepidopterists of all degrees of experience and expertise.

Our first concerted step in forming a society was to draft an "Aims of Organization," to summarize, for our own guidance, our basic policy and the content of our hoped-for journal. Then, in March 1947, we prepared and mailed to many Lepidopterists a letter announcing our new organization and soliciting members:

Cambridge, Massachusetts
24 March 1947

Dear Fellow Lepidopterist:

This letter is our announcement of the organization of a new association which will have the title The Lepidopterists' Union. Although no effort will be made at the outset to contact foreign entomologists, membership in the Union is open to anyone interested in the study of Lepidoptera, which certainly includes you.

You will want to know something about the Union, and we can begin by quoting the "Aims of Organization" from the initial draft of the plan of establishment:

> "To promote the scientifically sound and progressive study of Lepidoptera by: (1) distributing a periodical on Lepidoptera; (2) facilitating the exchange of specimens and ideas by both the professional worker and the interested amateur in the field."

For the present the periodical will be mimeographed and distributed monthly. The organization is strictly non-profit and the number of pages sent out per year will depend entirely on the amount of membership dues, which will be spent for paper, postage, envelopes, and other supplies and a small sum for the labor employed to do the mimeographing and addressing. The periodical will contain reviews of all pertinent papers on Lepidoptera, especially North American Rhopalocera, notes submitted by you on collecting, rearing, and handling of specimens, an exchange notice column, and articles on special subjects of interest to lepidopterists written by specialists, at first principally the two editors. Papers belonging in published journals will not be included. Although the periodical will begin in mimeographed form, it will be presented in a dignified style, and "newsy chatter" and the so-called "amateurish approach" will be avoided. We feel that writing of this sort has materially decreased the value of previous attempts at a periodical for Lepidopterists in the United States.

The first number will be mailed early in May and

another number will follow each month. For this first year the dues will be only one dollar ($1.00), for which you will receive one number for each of the last eight months of 1947, beginning with May. In addition, a list of names, addresses, and specialties of all members will be sent out to members probably late in the year.

We cordially invite your participation by enrolling as a charter member now. To do so, merely return the enclosed card with your name, address, and special interest in entomology, with the $1.00 dues. We know that we have missed the names of many would-be members, so please list on the reverse side of the card the names and addresses of Lepidopterists whom you suggest for membership. We also invite comments and queries from you concerning any phase of the Lepidopterists' Union. One dollar is the minimum and your generosity in sending more will permit more pages for circulation. Checks should be made out to "Charles L. Remington, Treasurer."

Charles L. Remington
and Harry K. Clench

The Lepidopterists' Union
48 Avon Hill Street
Cambridge 38, Mass.

(We had proposed to call the organization "The Lepidopterists' Union," in frank imitation of the British and the American Ornithologists' Unions and the American Malacological Union. So many of the recipients of our letter commented unfavorably on the term "Union," as generally inappropriate in the light of its present-day connotations, that we changed the name to "The Lepidopterists' Society.")

Response to the letter was gratifying. The society, to coin a phrase, was an idea whose time had come. In May the first number, a modest operation indeed, rolled off the presses. The masthead had been printed for us by Howard Allegeier of the Harvard Botanical Museum press; the printing was done on the mimeograph machine of the Harvard Biological Laboratory; Jeanne Remington did the careful stencil

---

← *Biological Laboratories, Harvard University, Cambridge, Mass., birthplace of* The Lepidopterists' News/Journal.

typing; Charles and I wrote the copy.

We continued to work together until the end of the summer. Then I had to leave for Ann Arbor to attend the University of Michigan. From that time my particiaption was reduced to writing an occasional article, and the active conduct of the society and its journal were entirely in the hands of Jeanne and Charles.

Jeanne and Charles Remington enjoying the hospitality of Don and Mary Jane Eff at their home in Boulder, Colorado, in July 1949.

Right here let me point out something important. Organizations of lepidopterists have come and gone over the years. Why has the Lepidopterists' Society succeeded where others failed? The reasons are probably many, but two in particular deserve notice.

First, our Society's Journal throughout has lived, and grown, within its means. It began inexpensively mimeographed; it graduated to photolithography only when revenue -- and that means membership -- grew enough to warrant the change; and from there to regular printing, again only when membership and income made this possible.

Second, our Society has always had a plan, or direction, for growth, keeping the needs of both beginner and veteran, of both amateur and professional, always in mind.

Charlie Remington knew what he wanted and how to go about getting it. Lepidopterists, evidently, have agreed with him. We, the members of the Society he largely shaped, have much indeed to thank him for.

Harry K. Clench

INTRODUCTION

The thought of a special publication to commemorate the Society's first 25 years came to mind during the June 1968 meetings in Washington, D. C.; the Executive Council had just tentatively granted my request to host the 1972 meetings in San Antonio. In 1969, during the meetings at Michigan State University, East Lansing, John Downey, then Secretary, presented, in my absence, my thoughts regarding publication of this Commemorative volume. While the consensus was affirmative, no positive action was taken by the Council at that time. At the Carson City meetings in June 1970, the Council not only voted affirmatively on the proposal but provided limited guidance for promoting the project (5:4/70).

Following this action by the Council, an *ad hoc* Finance Committee was formed, funds were solicited and deposited with the Treasurer, and data were gathered and compiled for the Commemorative volume. It was hoped that initial distribution of the volume would be made at the 1972 Anniversary meetings; however, data from various parts of the world and at home were slow coming and the manuscript was not completed until 1973 when it was made available for editing during the June meetings in Sarasota.

Editing has taken longer than anyone originally visualized. However, in the tradition of the Society founders, compilation and editing have been accomplished by individuals working on their own time and at their own expense. It is both unfortunate and regrettable that some of the workers had less personal time to devote to the project than others. While 1972 was the original cut-off date for inclusion of data, and although it is neither desirable nor feasible to bring current all of the materials, the time required for editing has made it desirable to up-date some of the sections as of the final editing (1976).

Although the information in this volume is generally self indexing, the reader will find it both convenient and helpful to refer to the CONTENTS page to locate specific sections of special interest. Basically this volume includes

information about the Society, information about people in the Society, and a consolidated index of authors, subjects, and scientific names (ca. 7,000) appearing in the *News/ Journal* during the first 25 years of publication.

Because of popular interest, it was intended in the beginning to publish a consolidated index to the Lepidoptera recorded in the Annual Field Summaries. These data were accumulated and organized by state, location within each state, and date of capture. Unfortunately, insufficient funds were available for publication of this information which would require a second volume. Perhaps if there is continuing interest, and funds become available, these consolidations can be published at a future date.

To facilitate referring to the various Society publications in the briefest and clearest way, the following abbreviations are used throughout this volume. Only the first page of a reference is given.

References to the consecutively numbered volumes of the *Lepidopterists' News*, the Society's principal publication from 1947 (vol. 1) through 1958 (vol. 12) or to its continuation, the *Journal of the Lepidopterists' Society*, from 1959 (vol. 13) to date, give volume and page numbers only, such as "7:33" or "17:100."

References to the *News of the Lepidopterists' Society*, the Society's newsletter for items of less permanent interest, beginning in 1959, give the issue number, page, and year published, such as "4:3/59" or "1:2/62."

The letter S preceding a citation refers to *Supplements* to the *Journal of the Lepidopterists' Society*. When first begun in 1969 these were identified with the particular volume number of the *Journal* with which they appeared. Later (1971) (4:1-2/71), this policy was changed to sequential numbering: S1 (=vol. 23, suppl. 1), S2 (=vol. 23, suppl. 2), S3 (=vol. 23, suppl. 3), S4 (=vol. 25, suppl. 1), S5 (=vol. 25, suppl. 2), S6 (=vol. 25, suppl. 3). As used in the cumulative index: "S1:22" means *J. Lepid. Soc.*, vol. 23, suppl. 1, p. 22.

I am most grateful to the following for their help in the preparation of this volume: Alfred Scherer and Daniel P. Rodriguez of Trinity University, San Antonio, Texas, for translating some of the biographical information; Cyril F. dos Passos and Sidney A. Hessel for serving with me as an *ad hoc* finance committee to raise funds for publication of the volume; Peter F. Bellinger, Harry K. Clench & Theodore D. Sargent, and Paul A. Opler for their written contributions; Glenn Y. Belyea, E. M. Kinch, W. W. McGuire and Susie Rickard for help in preparing the cumulative index; Harry K. Clench for several suggestions on content and arrangement; and the News Office of Harvard University, Cambridge, Massachusetts, for providing the photograph of the Biological Laboratories.

Publication of this volume was made possible through the efforts of many who gave generously of their time and money. A list of members and commercial firms who gave financial support to this project may be found under the title Financial Sponsors. The Society is most grateful to all who gave, and while the Society cannot guarantee the announcements of the entomological suppliers, the reader's attention is invited to their ads; they were all made in good faith.

Lastly, this volume is dedicated to my fellow lepidopterists throughout the International Lepidopterists' community. I hope that all recipients will find the contents interesting, factual, and useful.

Roy O. Kendall

EARLY ORGANIZATION

In the first issue of *The Lepidopterists' News* the founders proposed a set of Articles of Organization (1:2). These brief articles set forth the Society name, its non-profit educational aims, objectives, and management policies. No meetings or elected officials were contemplated at that time. By 1950, however, it became obvious that formalization was in the best interest of all. Cyril F. dos Passos was asked to prepare a draft Constitution and By-Laws (4:1).

On 31 July 1950, Charles Remington appointed an *Organization Committee*: Cyril F. dos Passos, Chairman; Don B. Stallings, Secretary; Axel Andersen (Denmark), Jean Bourgogne (France), Harry K. Clench, John G. Franclemont, T. N. Freeman (Canada), Lloyd M. Martin, Yoshio Okada (Japan), Lauro Pereira Travassos (Brazil), and C. B. Williams (England), Members (4:38).

On 1 October 1950 the Organization Committee appointed temporary officers to act until the membership could vote on its proposed slate of nominees who would serve as the first elected officials. Temporary appointments included: James H. McDunnough, President; Frederick H. Rindge, Secretary; and J. Benjamin Ziegler, Treasurer (4:38).

The Society's first elected officials who took office in January 1951 were: James H. McDunnough, President; Austin H. Clark, Senior Vice-President; Walter Forster (Germany), Vice-President; Kenneth J. Hayward (Argentina), Vice-President; Frederick H. Rindge, Secretary; J. Benjamin Ziegler, Treasurer; Thomas N. Freeman (Canada), Henri Stempffer (France), Lloyd M. Martin, Norman D. Riley (England), John G. Franclemont, and Takashi Shirozu (Japan), Members-at-Large (5:2).

The Organization Committee also adopted a Constitution and By-Laws which were presented to the membership for

approval and ratification at the Society's first annual meeting 29-30 December 1950 at the American Museum of Natural History (5:1). The newly adopted Constitution provided for an Executive Council of thirteen to control the affairs of the Society. These were the President, senior Vice-President, two Vice-Presidents, Secretary, Treasurer, six Members-at-Large, and the Editor-in-Chief (4:38). Later (1967) the Executive Council was enlarged to include the President-elect, and nine, rather than six, Members-at-Large *(Sec. 2, Art. IV, Constitution).*

CONSTITUTION & BY-LAWS

On the first of October 1950 the Organization Committee submitted a proposed Constitution and By-Laws to the Society for adoption. At the Society's first Annual Meeting, 30 December 1950, the membership ratified the proposed Constitution and By-Laws. As then adopted, the Constitution provided for a thirteen-member Executive Council to control the affairs of the Society (4:38).

The following year several amendments were proposed and all were passed at the third Annual Meeting (5:111, 6:53). Probably the most significant of these amendments was one requiring all Life Membership fees to be placed in a Permanent Publications Fund. In 1954, still other amendments were proposed, all of which were passed by mail ballot (8:48). These amendments increased the tenure of the Secretary and the Treasurer from two to three years with the provision that they could succeed themselves twice. They also established three categories of membership: Active, Sustaining, and Life (8:48, 9:35).

By early 1958, two more major amendments were proposed which would allow the Society to function more efficiently: 1) instead of one publication, the *Lepidopterists' News*, there would be three - the *Journal*, the *News*, and the *Memoirs*; and 2) an Editorial Board was established with an Editor-in-Chief as chairman, appointed by the Executive Council for a term of three years, with the provision that he could be re-appointed (12:157). In 1963, this Editor-in-Chief was discontinued. From that time forward each of the three editors functioned more or less independently but under the broad publication policies of the Editorial Board. The functions and composition of the Board were also revised in 1963 (17:63).

The Executive Council was increased from six to nine members in 1966 (7:4/66, 2:1/67), and in 1967, Article XIII was passed which provided for the disposition of Society assets should it be dissolved. This and Article XIV, General Prohibitions, were necessary for the Society to qualify under the Internal Revenue Code as a non-profit, scientific and educational organization (7:2/67, 3:1/68). The most recent amendments (29:67) changed the inauguration

date of the President, Vice-Presidents, and Members-at-Large of the Executive Council from 1 January to the date of the annual meeting.

The following is the present Constitution and By-Laws as amended January 1974 (29:67).

WHEREAS, The Lepidopterists' Society was formed on May 4, 1947, to promote the scientifically sound and progressive study of Lepidoptera by —

1. publishing a periodical on Lepidoptera.
2. facilitating the exchange of specimens and notes by both the professional worker and the amateur in the field,

AND WHEREAS, it is now proposed to organize said Society in a more formal manner, the following Constitution and By-Laws are hereby adopted by the duly appointed Organization Committee. [1 October 1950]

CONSTITUTION

ARTICLE I. NAME

Section 1. The organization shall be known as The Lepidopterists' Society.

ARTICLE II. OBJECT

Section 1. The Lepidopterists' Society is a non-profit educational and scientific organization. It shall be the purpose of the Society to promote internationally the science of lepidopterology in all its branches; to further the scientifically sound and progressive study of Lepidoptera; to publish periodicals and other publications on Lepidoptera; to facilitate the exchange of specimens and ideas by both the professional worker and the amateur in the field; and to secure cooperation in all measures tending to that end.

ARTICLE III. MEMBERSHIP

Section 1. All persons interested in lepidopterology shall be eligible for membership.

Section 2. All individual subscribers to the *Journal* and the *News of the Lepidopterists' Society*, who have paid their current dues, shall be deemed members of the Society.

Section 3. The membership of the Society shall consist of five classes——Active, Student, Sustaining, Life, and Honorary Life members. All persons who joined the Society before January 1, 1948, shall be designated Charter members.

Section 4. Application for Active, Student, Sustaining, and Life membership in the Society, received by the Secretary or Treasurer and accompanied by the appropriate dues for the current year, shall constitute formalization of membership, and no nomination or election to membership shall be necessary. The annual and life dues shall be fixed by the By-Laws.

Section 5. Any member may become a Life Member upon the payment, at one time, of such sum as shall be fixed by the By-Laws, and shall be exempt from further assessment. He shall receive during his life a subscription to the *Journal* and the *News of the Lepidopterists' Society*. Life Membership fees shall be placed in a permanent Publication Fund.

Section 6. Individuals who have made important contributions to the science of lepidopterology may be elected Honorary Life Members of the Society. There shall not be more than ten living Honorary Life Members.

Section 7. Members one year in arrears in the payment of dues shall be dropped from the rolls by the Secretary.

Section 8. The Executive Council may expel any member of the Society for such cause as it may deem sufficient for expulsion. This action may be taken only after unanimous approval by the members of the Council. Petition for expulsion shall be presented to the Secretary for presentation to the Council. On expulsion, the departing member shall be refunded all dues paid for the current year. An expelled member may be reinstated by unanimous affirmative vote of the Council.

ARTICLE IV. OFFICERS

Section 1. The officers of the Society shall consist of a President, President-elect, three Vice-Presidents (not more than one Vice-President shall reside in one country), a Secretary, a Treasurer, a Secretary-elect and/or a Treasurer-elect, in these two offices, in years when there are incumbents.

Section 2. The business and affairs of the Society, not otherwise provided for, shall be controlled by an Executive Council, consisting of the President, Presi-

dent-elect, the most recent available Past President, three Vice-Presidents, the Secretary, the Secretary-elect, the Treasurer, the Treasurer-elect, and nine other members of the Society. Action on all amendments to the By-Laws and all appointments and elections by the Executive Council shall be obtained by a canvass by the Secretary of all members of the Council.

Section 3. The Executive Council may appoint one or more Assistant Secretaries or Assistant Treasurers to serve during the pleasure of the Council. The offices of Assistant Secretary and Assistant Treasurer may be filled by the same person.

Section 4. The Executive Council shall have power to make and adopt By-Laws for the conduct of the business and affairs of the Society and for the regulations of its procedure not inconsistent with the terms and provisions of the Constitution.

ARTICLE V. ELECTIONS

Section 1. The President shall before the first of July appoint a Nominating Committee who shall nominate no more than two candidates for each elective office to be filled for the ensuing year. Additional candidates may be nominated by submission to the Secretary of written nominations signed by not less than ten members. Ballots containing all nominations shall be mailed in November of each year, setting forth the officers to be elected and the names of those nominated for any office, their names shall be arranged alphabetically on the ballot.

Section 2. Election of Officers. All officers shall be elected by ballot. The President and all Vice-Presidents shall be elected for the term of one year, and shall be eligible to succeed themselves once. The Secretary and Treasurer shall be elected for the term of three years and shall be eligible to succeed themselves twice. The nine other elective members of the Executive Council shall be elected for the term of three years; three of them shall be replaced each year; these members shall not be eligible to succeed themselves. The President, the Secretary and the Treasurer shall be elected a year prior to the time they take office. For each office, the nominee receiving the highest number of ballots shall be elected. The President, Vice-Presidents and newly-elected Members-at-Large of the Executive Council shall take office at the Business Meeting of the

Annual Meeting following their election; in the event that no Annual Meeting is held in a given year, these officers shall assume office on the anniversary of the last Annual Meeting. All other officers shall take office at the beginning of the calendar year for which they are elected.

Section 3. Election of Honorary Life Members. Honorary Life Members shall be nominated by the unanimous vote of the members of the Executive Council. The nominee shall be voted on by mail ballot distributed to all members of the Society and reported in one of the Society's periodicals, and must receive 80% of all ballots cast to be elected. Not more than five Honorary Life Members may be elected at the first Annual Meeting, and not more than two in any one calendar year.

ARTICLE VI. DUTIES OF OFFICERS

Section 1. The President shall preside at all meetings. He shall appoint all committees and be Chairman of the Executive Council and a member *ex officio* of all other committees, except the Editorial Board. He may appoint also delegates to other learned societies, congresses, and conventions.

Section 2. The First Vice-President shall assume the duties of the President in case of his death, resignation, absence, or disability.

Section 3. In case the President and all Vice-Presidents are absent at a meeting, a temporary Chairman may be chosen by a majority vote; he shall be member of the Executive Council unless none is present, in which event another member of the Society may be elected.

Section 4. The Secretary shall keep the minutes of the meetings of the Society and of the Executive Council; shall give notice of the meetings of the Society; shall attend to all general correspondence; shall keep all records and files of the Society; shall prepare and distribute ballots; and shall generally perform all services that may be delegated to him.

Section 5. The Assistant Secretary shall assume the duties of the Secretary in case of the death, resignation, absence, or disability of the Secretary, and shall assist the Secretary as need be.

Section 6. The Treasurer shall receive all monies for the Society and deposit them in the name of the Society in such banking institutions as the Executive Council shall direct. He shall pay therefrom by draft

or check all bills and obligations of the Society; he shall keep an account of all monetary transactions and shall exhibit a statement of them when called for by the President or the Executive Council, and shall make a full report for the preceding calendar year at the annual meeting.

Section 7. The Assistant Treasurer shall assume the duties of the Treasurer in case of the death, resignation, absence, or disability of the Treasurer, and shall assist the Treasurer as need be.

Section 8. At the expiration of his term of office, each officer shall deliver to his successor all books, papers, funds, and vouchers belonging to the Society.

Section 9. The Society shall not and may not make any dividend, gift, division or bonus in money to any of its members.

ARTICLE VII. MEETINGS

Section 1. The annual meeting shall be held in affiliation with the International Congress of Entomology or the annual meeting of the American Association for the Advancement of Science, or at such other time and place as the Executive Council may determine. Notice of said meeting shall be given as provided in the By-Laws.

Section 2. Special meetings of the Society may be called by the Secretary upon the written request of the President or ten active members. Such request shall state the purpose for which the meeting is to be called and the time and place where it is to be held. No other business, except that specified in the call, shall be transacted, except by unanimous consent of the members present.

ARTICLE VIII. PUBLICATIONS

Section 1. The Society shall publish a periodical to be known as the *"Journal of the Lepidopterists' Society,"* a continuation of *The Lepidopterists' News*. The *Journal* shall be devoted to original papers, literature abstracts and other matter of permanent record. Each volume shall be issued for a calendar year, and shall be composed of four numbers. In it shall be published a summary of the proceedings of the annual meetings.

Section 2. The Society shall also issue a periodical to be known as *"News of the Lepidopterists' Society,"* which shall be devoted primarily to notices by members,

lists of new members, announcements of nominations, committee appointments, forthcoming meetings, summaries of the recent field collecting season, and other matter of interest to members but not requiring permanent record. It shall appear at more frequent intervals than the *Journal*. A list of members of the Society shall be issued at least every second year.

Section 3. The Society may issue from time to time serial publications to be known as "*Memoirs of the Lepidopterists' Society*" to contain longer works than are normal for the *Journal* and the *News*. This shall be financed by a special fund, not by the annual dues, and shall be sold separately to members, at a lower price than to non-members.

ARTICLE IX. EDITORIAL BOARD

Section 1. The publications of the Society shall be under the charge of an Editorial Board, consisting of a Chairman and two other at-large members, the Editor of the *Journal*, the Editor of the *News*, the Editor of the *Memoirs*, and the two Associate Editors of the *Journal*. The Chairman may also be one of the above five editors. It shall determine broad publication policies of the Society not otherwise provided for in the Constitution or the By-Laws. It shall consider potential candidates for editorships and then make recommendations to the Executive Council for appointment of the three Editors.

Section 2. The Chairman of the Editorial Board shall be appointed by the Executive Council for the term of three years, and he may be reappointed. The Executive Council shall appoint, on recommendation of the Editorial Board, the three Editors, for terms of three years each; all three Editors may succeed themselves once.

The Associate Editors and other members of the editorial committee of the *Journal* shall be appointed by the *Journal* Editor; their terms shall terminate with his term, but his successor may reappoint any of them.

Editorial committees or staff members of the *News* and the *Memoirs* shall be appointed by the respective Editors, but their terms shall terminate with those of their Editors; they may be reappointed.

ARTICLE X. AUDITING COMMITTEE

Section 1. The President shall appoint an Auditing Committee consisting of three members who shall audit

the accounts of the Treasurer and render their report to the Secretary before December 31st.

ARTICLE XI. LIBRARIAN

Section 1. The Librarian shall be appointed by the Executive Council. The Librarian shall serve for the term of three years, or until his successor shall have been appointed.

Section 2. The Librarian shall have charge of the library of the Society, and of all books, periodicals, reprints, and historical material received by the Society. He may make all necessary rules and regulations for the use of the library, not otherwise provided for in the Constitution or the By-Laws.

ARTICLE XII. AMENDMENTS

Section 1. The Constitution may be altered, amended, or repealed by a two-thirds vote of the members voting by mail ballot. Each proposal for amendment must be signed by not less than five members of the Society and submitted to the Secretary who will promptly transmit it to the Editors of the *Journal* and *News*. Each proposed amendment shall be published in one of the Society's periodicals at least three months before the annual ballot is mailed in November.

Section 2. The By-Laws may be altered, amended, or repealed, by a majority vote of the members voting, at any meeting of the Executive Council or in a mail-canvass of the Council by the Secretary. All changes so validated shall be published in one of the Society's periodicals.

ARTICLE XIII. DISSOLUTION

Section 1. Upon final dissolution or liquidation of the Society, all of its properties and assets remaining after payment of all outstanding liabilities shall be transferred, assigned and paid over as follows: The American Museum of Natural History, Central Park West at 79th Street, New York, New York 10024, and the California Academy of Sciences, Golden Gate Park, San Francisco, California 94118.

If any of the named organizations no longer qualify for exemptions under Section 501 (c) (3) of the Internal

Revenue Code, or are not in existence, or unable or unwilling to accept such assets, such properties and assets shall be distributed to such organization or organizations organized and operated exclusively for charitable or educational purposes as shall at the time qualify as an organization or organizations exempt under Section 501 (c) (3) of the Internal Revenue Code of 1954 (or corresponding provision of any future United States Internal Revenue Laws), as the Executive Council shall determine.

ARTICLE XIV. GENERAL PROHIBITIONS

Section 1. Notwithstanding any provision of the Constitution or By-Laws which might be susceptible to a contrary construction:

(a) the Society shall be organized exclusively for scientific and educational purposes;

(b) the Society shall be operated exclusively for scientific and educational purposes;

(c) no part of the net earnings of the Society shall or may under any circumstances inure to the benefit of any private individual;

(d) no substantial part of the activities of the Society shall consist of carrying on propaganda, or otherwise attempting to influence legislation;

(e) the Society shall not participate in, or intervene in (including the publishing or distributing of statements), any political campaign on behalf of any candidate for public office;

(f) the Society shall not be organized or operated for profit;

(g) the Society shall not:

- (1) lend any part of its income or corpus, without receipt of adequate security and reasonable rate of interest, to;
- (2) pay any compensation, in excess of a reasonable allowance for salaries or other compensation for personal services actually rendered, to;
- (3) make any part of its services available on a preferential basis to;
- (4) make any purchase of securities or any other property, for more than adequate consideration in money or money's worth, from;
- (5) sell any securities or other property for

less than adequate consideration in money or money's worth to; or

(6) engage in any other transactions which result in substantial diversions of its income or corpus to; any officer, member of the Governing Board, or substantial contributor to the Society.

The probitions contained in this subsection (g) do not mean to imply that the Society may make such loans, payments, sales or purchases to anyone else, unless such authority be given or implied by other provisions of the Constitution or By-Laws.

BY-LAWS

ARTICLE I. DUES

Section 1. All dues shall be payable by January 1 of each year, and shall be deemed in arrears on February 15 of that year.

Section 2. Publications of the Society shall not be mailed to any member whose dues are in arrears.

Section 3. The annual dues shall be waived for the Secretary, the Treasurer, and the Editors of the *Journal* and *News*, while they are in office; they shall continue to receive all publications of the Society.

Section 4. Honorary Life Members shall pay no annual dues, but shall receive a subscription to all publications of the Society.

Section 5. Beginning in June 1975, the annual dues for Active Members shall be Thirteen Dollars, U.S.A. ($13.00). Active membership shall include a subscription to the *Journal of the Lepidopterists' Society*, including *Supplements*, and the *News of the Lepidopterists' Society*.

Section 6. Beginning in June 1975, the annual dues for Student Members shall be Ten Dollars U.S.A. ($10.00). Student membership shall include a subscription to the *Journal of the Lepidopterists' Society*, including *Supplements*, and the *News of the Lepidopterists' Society*.

Section 7. Beginning with 1971, the annual dues for Sustaining Members shall be Twenty Dollars, U.S.A. ($20. 00). Sustaining membership shall include a subscription to the *Journal of the Lepidopterists' Society*, including *Supplements*, and the *News of the Lepidopterists' Society*.

Section 8. Beginning in September 1976, Life Members shall pay the sum of Two Hundred Fifty Dollars, U.S.A. ($250.00). Each Life Member shall receive a subscription to the regular Society publications during his life.

Section 9. Beginning in June 1975, institutional subscription fees shall be Eighteen Dollars, U.S.A. ($18.00). This will include the *Journal of the Lepidopterists' Society*, including *Supplements*, and the *News of the Lepidopterists' Society*.

ARTICLE II. GIFTS

Section 1. The Society shall not make gift memberships to individuals or subscriptions to institutions. Society members are encouraged to sponsor gift memberships to deserving lepidopterists in foreign countries whose monetary policy prohibits sending currency out of their country.

Section 2. The Society may exchange issues of the *Journal* for other desirable lepidopterological periodicals upon the approval and recommendation of the Society Librarian. Such acquisitions shall remain in the Society Library, but may be loaned to its members in good standing.

ARTICLE III. REPRINTS

Section 1. Beginning with 1971, gratis separates (tear sheets) to *Journal* authors will be discontinued.

ARTICLE IV. MEETINGS

Section 1. Notice of all meetings of the Society shall be printed in the *News of the Lepidopterists' Society* at least two months in advance thereof.

Section 2. A majority of members present at an annual meeting, or represented by proxy, shall constitute a quorum for the transaction of business, not otherwise provided for.

MEMBERSHIP

A Backward Glance - A Glimpse Ahead

Recently, one of us (HKC) made up a list, year by year from the beginning, of how many members we have had in the Lepidopterists' Society - U.S., non-U.S., and both combined. These data are presented in Table 1 and depicted graphically in Fig. 1.

To anyone with a weakness for numbers, such data provide an almost irresistable temptation to "play". We have yielded to that temptation, and here present some results of that activity. We hope that this effort is not entirely frivolous, for it was motivated in part by curiosity regarding the growth of our Society. Perhaps some light may be shed on how and why we have grown, and on what the future may have in store.

Table 1. The membership of The Lepidopterists' Society, U.S. and foreign, 1947-1970.

	Membership			
Years	U. S.	Foreign	% Foreign	Total
1947	194	26	11.8	220
1948	275	104	27.4	379
1949	290	137	32.1	427
1950	292	149	33.8	441
1951	306	161	34.5	467
1952	340	162	32.3	502
1953	355	165	31.7	520
1954	364	163	30.9	527
1955	370	169	31.4	539
1957	369	173	31.9	542
1958	408	174	29.9	582
1960	436	179	29.1	615
1962	553	223	28.7	776
1964	548	211	27.8	759
1966	628	238	27.5	866
1968	713	258	26.6	971
1970	725	267	26.9	992

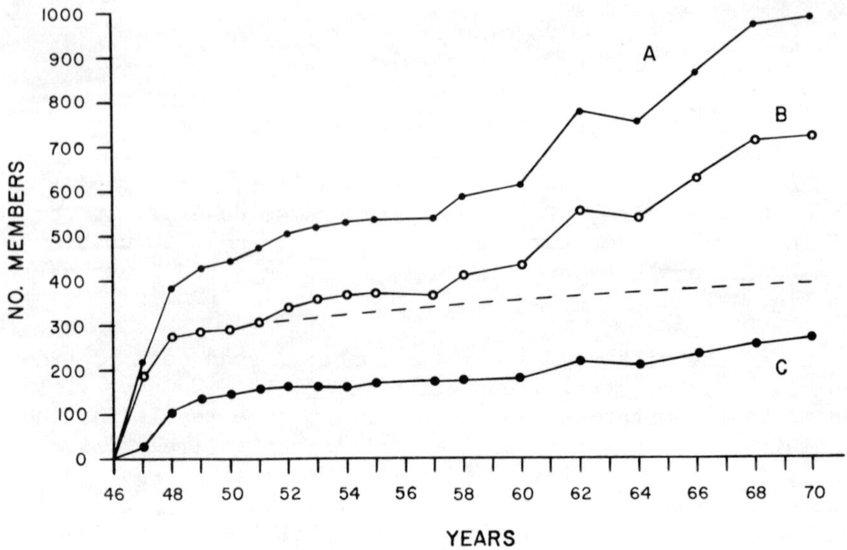

Fig. 1. The total number of members in the Lepidopterists' Society (curve A), with the U.S. (Curve B) and foreign (curve C) contributions to that total, 1947-1970. The dashed line represents the projected course of U.S. membership totals had growth stabilized at the 1950 level (1.99 mpm).

Let us begin with the course of U.S. membership (Fig. 1, curve B). One would expect such a curve to show an initial phase of rapid growth as word spread that the Society existed. Then a stable period should follow, after just about all interested lepidopterists in the country had become members. Thereafter the curve might rise slightly, keeping pace with the rising population of the country.

The actual curve, however, departs to some extent from these expectations. Two small, but significant, upward jumps in the curve are particularly apparent, one in 1952 and another in 1962. In addition, there has been a rather dramatic upward trend in the membership since 1957. What could account for these seeming irregularities?

One important factor in the popularity of any subject is the extent to which writings on that subject, particularly good books, become available. Can it be a coincidence that

Klots's *Field Guide to the Butterflies* appeared in 1951, and the Ehrlichs' *How to Know the Butterflies* appeared ten years later? These two books could well explain the jumps in membership in 1952 and 1962, and thus together would account for perhaps 50 more members than might otherwise have been expected.

The steady rise in members since 1957 is best expressed in terms of another measure, members per million (mpm). This is just what it says, i.e. the number of Society members - U.S. - per million individuals in the entire U.S. population. Before 1951, mpm had stabilized at about 1.95. When Klots's field guide appeared, mpm jumped to 2.24, and then stabilized again. After 1957, however, mpm began to climb, and with negligible exceptions, it has been climbing ever since (Table 2). This trend represents a long-term change in the overall pattern of Society growth, and cannot be correlated with a single publication as were the earlier, smaller increments in the membership.

Table 2. The total U.S. population (in millions), the number of U.S. members in the Lepidopterists' Society, and the resulting Society members per million individuals (mpm) in the U.S. population, 1947-1972. (The U.S. population figures are interpolated between the census years of 1950, 1960 and 1970.)

Years	U.S. Population	Lep. Soc. Members	Members per Million
1947	142.12	194	1.37
1948	144.98	275	1.90
1949	147.84	290	1.96
1950	150.70	292	1.94
1951	153.56	306	1.99
1952	156.42	340	2.17
1953	159.29	355	2.23
1954	162.15	364	2.24
1955	165.01	370	2.24
1956	167.87		
1957	170.73	369	2.16
1958	173.60	408	2.35
1959	176.46		
1960	179.32	436	2.43
1961	181.41		
1962	183.51	553	3.01

Table 2. (continued)

Years	U.S. Population	Lep. Soc. Members	Members per Million
1963	185.60		
1964	187.69	548	2.92
1965	189.78		
1966	191.88	628	3.27
1967	193.97		
1968	196.06	713	3.64
1969	198.16		
1970	200.25	725	3.62
1971	202.34		
1972	204.44	856	4.19

This trend may be related in part, however, to a periodical publication, in this case our own *News/Journal*. It seems quite likely that this publication, as it has expanded and improved over the years, has been attracting an ever-widening audience. Many of these newer members may have primary interests in fields such as genetics, ecology, or behavior, and may have become members of the Society in order to keep abreast of studies involving Lepidoptera in those fields. These persons might not consider themselves lepidopterists, and so they might not have been reached by, or attracted to the Society during its formative years.

In summary then, Society membership in the U.S. seems to have grown under four major influences: normal growth, under which alone we would have had about 400 U.S. members in 1970; a "Klots factor" in 1952 and an "Ehrlich factor" in 1962, which together increased our membership by about 50, so that we would have had approximately 450 U.S. members in 1970; and a long-term influence, perhaps a "Journal factor", which has altered our growth since 1957, so that the actual total in 1970 was a rather remarkable 725 U.S. members.

Thus far we have looked only at the U.S. membership. But the Lepidopterists' Society is really two societies in one. It is both a local organization and an international organization. In the U.S. (and Canada) it serves both functions, as items and articles of local interest may rub shoulers in the *Journal* and *News* with articles of interest to lepidopterists everywhere. Many foreign countries, however, have their own societies and publications that accommodate local interests. Our foreign membership, then, is

mostly made up of (1) individuals in countries that have little in the way of local interest outlets, and (2) individuals whose lepidopterological interests extend worldwide.

Growth in the foreign membership (Fig. 1, curve C) was delayed initially, presumably because less effort was expended in contacting potential foreign members during the first year or two of organization, and because the word would spread more slowly abroad. Thereafter, however, foreign membership grew rapidly, and in 1951 it comprised nearly 35% of the total for the Society. Foreign membership has climbed slowly ever since, though it has not kept pace with the growth in domestic membership.

Little trace of the "Klots and Ehrlich factors" shows in the foreign membership curve, as would be expected (what trace there is probably derives from the Canadian membership). The "Journal factor" may be discernible, though certainly less pronounced than in the U.S. curve.

Predicting where the Society membership will go in the future is, needless to say, an uncertain business. All other factors being equal, it seems fair to surmise that widely useful and popular books on Lepidoptera will, as they have in the past, swell the membership from time to time. It is also likely that the maintenance of a quality *Journal* will do much to ensure our continued viability and growth. Finally, it would appear that the Society may have to direct more attention to its international functions, in order to ensure that the foreign membership will continue to play a substantial and stimulating role in its affairs.

In concluding, we hope that other Society members, twenty-five years hence, may be able to glance backwards and glimpse ahead with the same satisfaction and optimism that we feel now.

Harry K. Clench
and Theodore D. Sargent

Membership Table

These data on number of members, year by year, depict the growth of the Society through 1972. Years shown are those in which membership lists were published, and from which the counts were made. Over the years, individuals from some 78 nations have been members of the Society.

	1947	1948	1949	1951	1952	1953	1954	1955	1957	1958	1960	1962	1964	1967	1968	1970	1972
ARGENTINA	-	2	3	7	6	6	6	6	6	3	3	2	2	1	2	2	2
AUSTRALIA	-	-	2	7	5	4	3	4	4	4	8	11	10	12	14	13	15
AUSTRIA	-	1	2	1	2	2	2	2	2	4	3	3	2	3	3	3	3
BAHREIN	-	-	-	-	-	-	-	-	-	-	1	1	-	-	-	-	-
BELGIAN CONGO	-	-	-	1	1	1	1	-	1	1	1	1	1	-	-	-	-
BELGIUM	1	5	5	5	4	3	2	2	2	2	2	2	3	3	1	1	1
BURMA	-	-	-	-	-	-	-	-	-	-	-	-	1	-	-	-	-
BRAZIL	2	7	9	9	10	10	9	10	11	11	10	10	7	9	9	11	14
BRITISH GUIANA	-	-	-	-	-	-	-	-	1	-	-	-	-	-	-	-	-
B. SOLOMON IDS	-	-	-	-	-	-	-	-	1	-	-	-	-	-	-	-	-
B. WEST INDIES	-	3	3	2	3	3	3	3	2	2	2	2	3	-	-	-	-
CANADA:																	
Alberta	-	-	1	1	2	2	2	1	1	2	3	5	4	8	9	10	8
Br Columbia	1	1	1	2	2	2	2	2	2	1	1	1	1	3	3	6	8
Manitoba	1	2	2	2	2	3	3	3	3	2	3	3	3	3	2	5	5
New Brunswick	-	-	-	-	-	-	-	-	-	-	1	1	1	2	2	1	1
Nova Scotia	2	2	2	2	2	2	2	2	4	5	5	4	1	4	3	3	2
Ontario	4	4	4	9	11	13	16	16	14	13	16	16	19	25	32	40	36
Pr Edward Id	-	1	-	-	-	-	-	-	-	-	-	-	-	-	-	-	-
Quebec	3	3	5	3	3	3	3	5	5	5	6	4	4	3	4	6	10
Saskatchewan	2	2	2	1	1	1	2	1	1	-	2	3	1	-	-	-	1

	1947	1948	1949	1951	1952	1953	1954	1955	1957	1958	1960	1962	1964	1967	1968	1970	1972
CANAL ZONE	-	-	-	-	-	-	-	-	-	-	-	-	1	1	1	1	2
CHILE	-	-	-	1	1	1	1	1	2	2	2	2	1	2	2	1	1
CHINA	-	2	1	2	1	-	-	-	-	-	-	-	-	-	-	-	-
COLOMBIA	-	-	-	-	-	-	-	-	-	2	2	1	1	1	1	1	1
COSTA RICA	-	-	-	-	-	-	-	-	-	-	-	2	1	-	-	-	2
CUBA	-	1	1	1	1	1	1	1	1	1	1	1	1	2	2	1	1
CZECHOSLOVAKIA	-	1	1	5	6	6	6	7	5	5	5	6	7	8	9	9	10
DENMARK	1	2	4	2	2	3	4	4	4	4	4	3	3	4	4	5	5
DOMINICAN REP	-	-	-	-	-	-	-	-	-	-	-	-	-	2	2	2	1
EAST AFRICA	-	-	-	-	-	-	-	-	-	-	-	3	4	2	-	-	-
ECUADOR	-	-	1	1	-	-	-	-	-	-	-	-	-	-	-	-	1
EL SALVADOR	-	-	-	-	-	-	-	-	-	1	1	1	2	2	3	3	3
ENGLAND	4	13	15	16	17	19	16	16	22	22	30	36	34	34	32	28	34
EIRE	-	2	2	-	-	-	-	-	-	-	-	-	-	-	-	-	-
ETHIOPIA	-	-	-	-	-	-	-	-	-	1	1	1	1	1	1	-	-
FINLAND	-	1	2	4	5	5	4	4	4	4	4	4	4	5	5	5	6
FORMOSA	-	-	-	-	-	1	-	-	1	3	3	2	2	2	2	2	2
FRANCE	2	12	13	13	14	12	11	9	11	12	12	14	14	15	16	16	23
GERMANY	-	2	10	14	13	12	11	10	9	9	9	8	11	10	9	10	7
GOLD COAST	-	-	-	-	1	1	1	1	-	-	-	-	-	-	-	-	-
HONDURAS	1	-	-	-	-	-	-	-	-	-	-	-	-	-	-	1	1
HUNGARY	-	1	3	3	3	3	4	3	3	3	3	3	3	3	3	3	2
INDIA	-	1	1	1	2	1	1	1	1	-	1	1	-	-	1	-	-
INDONESIA	-	-	3	2	2	2	2	2	1	1	1	1	1	1	1	2	2
IRAN	-	-	-	-	-	-	-	-	-	-	-	-	-	-	-	-	1

	1947	1948	1949	1951	1952	1953	1954	1955	1957	1958	1960	1962	1964	1967	1968	1970	1972
ISRAEL	-	-	-	-	-	-	-	-	-	-	-	-	-	-	1	-	1
ITALY	-	1	1	1	3	4	4	4	4	4	3	3	2	2	2	5	8
JAMAICA	-	-	-	-	-	-	-	-	-	-	-	-	-	1	2	4	4
JAPAN	-	3	6	13	16	16	14	20	19	15	17	23	19	22	24	24	31
JAVA	-	1	-	-	-	-	-	-	-	-	-	-	-	-	-	-	-
KENYA	-	-	-	-	-	-	1	2	2	2	3	2	2	3	4	2	1
KOREA	-	-	1	1	-	-	-	-	-	-	1	1	-	-	-	-	-
LIBERIA	-	-	-	-	-	-	1	1	-	-	-	-	-	-	-	-	-
LIBYA	-	-	-	-	-	-	-	-	-	-	-	-	-	-	-	1	1
MADAGASCAR	-	-	-	-	-	-	-	-	-	-	-	-	-	1	1	1	2
MALAYSIA	-	-	-	-	-	-	-	-	-	-	-	-	-	2	2	2	3
MALTA	-	-	-	1	1	1	1	1	1	1	1	1	-	-	-	-	-
MEXICO	-	-	1	-	2	1	2	4	3	4	5	6	7	7	8	8	10
NEPAL	-	-	-	-	-	-	-	-	1	1	-	-	-	-	-	-	-
NETHERLANDS	-	2	2	4	4	4	5	5	5	5	4	4	3	2	3	3	4
NEW GUINEA	-	-	-	-	-	-	-	-	-	-	-	1	-	1	5	1	4
NEW ZEALAND	-	-	-	-	1	1	-	-	-	-	-	-	-	1	2	3	4
NICARAGUA	-	-	-	-	-	-	-	-	1	1	1	1	-	-	-	-	-
NIGERIA	-	-	-	-	-	-	-	-	-	-	-	-	-	-	1	-	-
NORWAY	-	-	-	-	-	-	-	-	-	-	-	-	-	-	-	-	1
PERU	-	-	-	-	-	-	-	-	-	-	-	1	1	1	1	1	1
PHILIPPINES	-	1	2	2	2	3	2	2	2	2	1	1	1	1	1	2	3
POLAND	-	-	-	-	-	-	-	-	-	-	-	1	1	-	-	1	1
PORTUGAL	-	1	1	2	1	1	1	1	1	1	1	1	-	-	-	-	-

	1947	1948	1949	1951	1952	1953	1954	1955	1957	1958	1960	1962	1964	1967	1968	1970	1972
PUERTO RICO	-	1	1	1	-	-	-	-	-	1	-	-	-	-	-	-	-
SCOTLAND	-	-	-	-	-	1	1	2	2	2	3	1	1	1	1	1	2
SINGAPORE	-	-	-	-	-	-	-	-	-	-	-	-	-	-	-	-	2
SOUTH AFRICA	-	-	-	1	1	2	2	2	2	2	3	3	2	3	3	3	3
SPAIN	1	4	6	8	4	-	3	2	1	2	2	4	5	6	6	6	6
SUDAN	-	-	-	-	-	-	-	-	-	-	-	-	-	1	-	-	-
SWEDEN	-	2	2	2	2	2	2	2	2	1	2	3	3	3	3	3	4
SWITZERLAND	-	1	3	4	3	2	3	2	3	3	2	3	4	4	3	5	5
THAILAND	-	-	-	-	-	-	-	-	-	-	-	-	-	1	1	-	-
TRINIDAD	-	-	-	-	-	-	-	-	-	-	-	-	-	1	1	1	1
UGANDA	-	-	-	1	1	1	1	-	-	-	-	-	-	-	2	-	2
U.S.S.R.	-	-	-	-	-	-	-	-	-	-	1	2	3	2	2	3	4
VENEZUELA	-	1	1	1	-	-	1	1	1	1	4	1	1	-	-	-	-
WALES	-	-	1	1	-	-	-	-	-	-	-	-	-	-	-	-	-
YUGOSLAVIA	-	-	-	-	1	1	1	1	1	1	1	1	1	1	1	1	1
ZAMBIA	-	-	-	-	-	-	-	-	-	-	-	-	-	-	-	1	1
U.S.A.:																	
Alabama	1	1	1	2	2	2	1	2	2	2	3	1	2	1	2	2	3
Alaska	-	-	-	-	-	-	-	-	-	-	-	-	-	2	2	1	2
Arizona	2	1	1	1	-	1	1	1	2	1	4	8	6	12	15	13	14
Arkansas	-	-	-	-	-	-	-	-	-	1	1	3	2	2	3	4	4
California	33	49	50	46	51	52	60	62	62	69	76	90	100	121	140	142	172
Colorado	3	7	7	6	7	8	8	7	6	9	13	20	13	16	17	20	23
Connecticut	6	6	9	8	8	11	12	12	15	15	17	21	20	23	26	16	15
Delaware	-	-	1	1	1	1	1	1	1	1	1	1	1	3	5	3	2

	1947	1948	1949	1951	1952	1953	1954	1955	1957	1958	1960	1962	1964	1967	1968	1970	1972
Dist Columbia	4	4	4	4	2	2	1	1	2	2	4	6	5	7	9	9	11
Florida	3	4	6	8	7	8	7	10	11	11	12	14	12	12	14	19	29
Georgia	5	6	5	7	7	5	4	3	6	5	8	8	4	8	6	9	11
Hawaii	1	2	2	2	2	2	2	1	-	-	-	-	2	2	-	2	2
Idaho	-	1	3	2	1	2	2	1	2	2	2	2	3	3	2	2	2
Illinois	12	19	24	28	31	32	32	34	35	30	34	36	27	35	37	33	38
Indiana	1	-	2	3	4	6	4	4	3	4	3	4	3	3	7	6	10
Iowa	-	-	1	1	1	1	1	1	2	6	7	9	9	8	10	8	10
Kansas	2	1	4	4	4	4	5	8	9	11	9	11	11	12	9	8	7
Kentucky	2	3	3	4	4	4	6	5	6	5	4	4	5	7	7	6	12
Louisiana	-	-	1	1	1	1	1	1	-	-	1	3	3	1	1	3	8
Maine	2	2	2	2	2	2	2	2	2	2	2	2	3	2	6	3	3
Maryland	1	4	4	6	7	6	7	9	10	10	11	11	10	13	13	17	21
Mass	19	20	19	16	19	20	22	23	18	15	13	16	19	18	19	26	33
Michigan	9	10	12	12	15	13	13	12	13	15	16	20	20	16	17	21	26
Minnesota	1	2	-	1	1	1	3	3	2	3	2	6	7	17	19	14	8
Mississippi	1	1	1	1	2	2	2	2	1	2	2	2	2	3	3	4	5
Missouri	5	5	5	5	5	5	8	9	9	9	10	9	7	7	8	6	4
Montana	-	-	-	1	-	-	-	1	1	2	3	3	2	2	1	1	2
Nebraska	-	-	1	1	2	2	2	2	3	3	3	3	4	2	3	3	3
Nevada	-	-	-	-	-	1	1	1	-	-	-	1	2	2	2	2	3
N Hampshire	2	2	2	2	2	4	4	3	3	3	5	3	1	2	2	2	2
New Jersey	4	8	11	17	21	15	15	11	12	11	15	20	18	20	24	21	30
New Mexico	-	1	2	2	2	2	2	2	2	4	3	4	4	5	4	5	8
New York	24	34	38	39	38	43	39	35	43	53	58	57	54	61	72	80	83

	1947	1948	1949	1951	1952	1953	1954	1955	1957	1958	1960	1962	1964	1967	1968	1970	1972
N Carolina	1	1	-	1	2	5	7	5	6	5	6	7	6	8	8	7	10
N Dakota	-	1	-	1	2	1	2	1	-	2	3	3	2	2	2	1	1
Ohio	9	13	14	12	13	14	11	9	13	13	17	22	21	24	22	33	35
Oklahoma	-	-	1	1	-	-	-	1	1	2	2	1	3	2	2	2	3
Oregon	5	5	3	2	4	5	5	6	6	7	8	13	15	19	21	14	19
Pennsylvania	12	12	13	15	17	21	22	23	25	27	23	36	39	39	48	40	56
Rhode Island	1	1	-	1	1	2	3	3	1	1	1	2	3	2	1	1	1
S Carolina	-	-	-	-	-	-	-	2	1	-	-	1	1	1	2	6	8
S Dakota	-	1	2	1	1	1	1	1	1	-	-	-	-	1	2	2	-
Tennessee	1	-	-	-	2	3	3	3	4	5	5	4	7	4	3	3	3
Texas	8	9	7	9	10	9	8	12	11	13	14	17	19	22	26	29	36
Utah	1	2	3	3	3	3	3	2	2	2	3	6	5	7	5	7	7
Vermont	-	-	-	1	1	1	-	-	-	-	1	1	-	1	2	2	2
Virginia	3	4	1	2	3	5	5	4	4	3	7	7	4	7	9	12	12
Washington	5	6	9	7	10	8	8	9	11	9	12	14	15	13	22	23	22
W Virginia	-	-	-	1	-	-	-	-	1	1	1	1	1	2	1	1	1
Wisconsin	7	9	10	12	13	16	12	14	16	13	21	21	21	24	28	30	32
Wyoming	1	2	2	1	2	2	2	2	2	2	1	1	3	3	4	4	2
TOTALS:																	
U.S.A.	197	258	286	303	333	354	360	366	388	411	467	555	546	629	713	728	856
Canada	13	25	17	20	23	26	30	30	30	28	37	37	34	48	55	71	71
All Others	12	75	109	140	141	136	133	138	144	146	164	185	177	191	204	203	250
TOTAL MEMBERSHIP	222	358	412	463	497	516	523	534	562	585	668	777	757	868	972	1002	1177

FINANCIAL GROWTH

The following tabulation portrays the financial status of the Society over its first quarter century of operations. Assets include cash on hand (checking, savings/Memoirs accounts) and capitalized stocks of back issues (Memoirs, Journal, News). No other properties are owned by the Society at this time except its library. Accounting and auditing procedures for Society funds are contained in Article VI, Section 6 and Article X, Section 1 of the Constitution. The Society has operated as a non-profit organization from the very beginning (letter, 24 Mar 1947). No attempt was made, however, to formalize this status with the Internal Revenue Service until 1968 when Roy O. Kendall, then Treasurer, filed an application for such status with the U. S. Treasury Department. Early in 1973 the application was approved and the Society was so informed.

Table 1. The financial status of the Society year by year, 1947-1972.

Year	Reference	Assets	Income	Expense	P/Loss	Reg.	Sust.	Life
1947	Ed Report	$	$	$	$+ 46	$1	$	$
1948	" "				- 98	1.5	3	
1949	" "		1,086	1,059	+ 27	2	4	
1950	6:121				+ 379	2	4	
1951	6:121		1,508	1,247	+ 261	2	5	
1952	7:132		1,958	1,258	+ 699	3	5	50
1953-54						3	5	50
1955-57						4	10	75
1958	3:1/59	2,882	3,053	3,727	- 674	4	10	75
1959	2:3/60	2,928	2,907	3,297	- 386	4	10	75
1960	2:2/61	2,416	3,555	4,087	- 532	5	10	75
1961	1:2/62	2,615	5,096	5,336	- 241	6	15	75
1962	2:2/63	4,714	5,359	4,631	+ 729	6	15	75
1963	2:3/64	4,860	5,430	5,446	- 17	6	15	75
1964	2:3/65	4,549	6,028	6,881	- 853	6	15	125
1965	2:3/66	5,314	5,941	4,707	+1234	6	15	125
1966	2:2/67	7,556	8,834	6,709	+2125	6	15	125
1967	2:2/68	9,281	7,924	6,308	+1615	6	15	125
1968	1:5/69	6,319	4,885	8,124	-3239	6	15	125
1969						8	15	125
1970	2:2/71	5,651	11,441	12,753	-1312	8	15	125
1971	3:2/72	14,481	12,926	12,471	+ 454	8	15	125
1972	3:6/73	17,663	15,333	15,011	+ 322	10	20	150

PUBLICATIONS

The Society's first publication, which began in 1947, was known as *The Lepidopterists' News*. It evolved from an initial mimeographed (8½ x 11 inch) format (volume 1) to a photolithographed format of the same size (volumes 2-5), and then, beginning in 1952, to a letterpress (6 x 9 inch) format (volume 6 onward). In the early years descriptions of new taxa and proposals of nomenclatorial changes were considered inappropriate. With volume 8 (1954) this policy was changed because the journal had by then reached a stature entirely suitable for such material. Since then many new taxa have been described in its pages.

With volume 13 (1959) the name of the publication was changed to the *Journal of the Lepidopterists' Society*, but numbering of the volumes was kept continuous. At the same time the Society initiated a second publication, the *News of the Lepidopterists' Society*, designed to include matter of more transient interest. Special features included free notices by members, the annual field *Season Summary*, a biennial membership list, announcements of meetings, appointments, candidates for office, and so on. Originally eight numbers of the *News* appeared each year. Beginning in 1971, however, only six issues per year have been printed, a change for the sake of economy. Approximate publication dates are the 15th of January, March, May, July, September and November.

In 1964 the Society initiated a new series, *Memoirs of the Lepidopterists' Society*. These were designed for works much longer than could be accommodated in the *Journal*. Memoirs are financed from a special fund and, unlike the Society's other publications, are not distributed free to members. So far, only one *Memoir* has been published, *A Synonymic List of the Nearctic Rhopalocera* by Cyril F. dos Passos.

Jerry A. Powell, then editor, initiated in 1969 a series of *Supplements* to the *Journal*. These were designed to permit the relatively inexpensive (photo-offset) publication of papers that were too long for the *Journal*, yet shorter than appropriate for the *Memoirs*. Three were issued in 1969, as supplements to *Journal* volume 23, and another three (origi-

nally numbered 1-3, but later re-numbered 4-6) in 1971 as supplements to volume 25. None has been published since 1971. Supplements issued are:

1. Systematic review of the Megathymidae. H. A. Freeman. 1969. 59 pp., 2 pl.

2. Records, new species, and a new genus of Hesperiidae from Mexico. H. A. Freeman. 1969. 62 pp., 15 pl.

3. On the frequency of butterflies in eastern Brazil, with a list of the butterfly fauna of Pocos de Caldas, Minas Gerais. Heinz Ebert. 1969. 48 pp.

4. A new genus of hairstreak from Central and South America (Lycaenidae, Theclinae). S. S. Nicolay. 1971. 39 pp., 4 pl.

5. A biosystematic study of two species of *Callophrys* (*Callophrys*) in California (Lycaenidae). Glenn A. Gorelick. 1971. 41 pp., 21 figs.

6. Biological studies on moths of the genus *Ethmia* in California (Gelechioidea). Jerry A. Powell. 1971. 67 pp., 29 figs.

Although sequential re-numbering of the *Supplements* and the method of citing them, as used in this volume, was explained in the Introduction, the method is repeated here for the reader's convenience. The letter "S" preceding a number indicates one of the *Supplements*. For example: S5:23 refers to the above numbered *Supplement* 5, page 23. Absence of the prefix "S" refers to the *Lepidopterists' News* (vols. 1-12) and/or the *Journal* (vol. 13 forward), e.g. "2:6" or "24:118." Citations referring to the *News of the Lepidopterists' Society* from 1959 forward always give the issue number, page, and year, e.g. "4:3/59" or "1:2/62."

ANNUAL MEETINGS

Article VIII of the Constitution provides for the convening of annual and special meetings as determined by the Executive Council. From the time the Society held its first meeting in 1950, only two annual meetings have been canceled, those of 1959 and 1965. The cancellations have caused some incorrect numbering of meetings as announced in the *News* and in printed programs of annual meetings. These are corrected here.

Arrangement of data for each meeting below is in the following order: Category and number (regular meetings indicated by roman numerals), date, location and local host (chairman), program chairman, program features, total attendance (members & guests) and number of states represented (e.g., "51/8"), number of foreign countries and persons present (e.g., "1-1"), and reference citation.

I. 29-30 December 1950. American Museum of Natural History New York City, N. Y., A. B. Klots. C. L. Remington. Ratification of the Constitution and By-Laws. Symposium: Geographic Subspeciation in Lepidoptera. 51/8, 1-3. (4:37, 5:1)

II. 28-29 December 1951. Chicago Natural History Museum, Illinois, A. K. Wyatt & Henry S. Dybas. ? C. L. Remington. Nomination of first 5 Honorary Life Members. Round-table: Techniques and Objectives of Rearing Lepidoptera. 32/7, 1-1. (6:121)

Special. 21 August 1951. Little Auditorium, Congress Building, Amsterdam, Netherlands. Walter Forster (5:10). Jointly with IX International Congress of Entomology. Symposium: Phylogeny and Classification of Lepidoptera. 41/4, 13-37. (6:28)

III. 2-4 July 1952. Entomology Research Institute, Ottawa, Ontario, Canada, B. P. Beirne. T. N. Freeman. Symposium: Collecting and Field Study of Lepidoptera. 36/6, 2-26. (7:1)

IV. 2-4 July 1953. Los Angeles County Museum of Natural History, California, (?). Fred T. Thorne (6:91). Sympo-

sium: Collecting in Western North America. 30/2. (8:5)

Pacific Slope, 1. 4-6 September 1954. California Academy of Sciences, San Francisco, (?). Fred T. Thorne (8:7). Submitted papers. 20/1. (9:18)

V. 28-30 December 1954. Carnegie Museum, Pittsburgh, Pennsylvania, E. G. Wallace. Harry K. Clench. Submitted papers. 24/5, 2-1. (8:140, 9:17)

Pacific Slope, 2. 3-5 September 1955. San Diego Society of Natural History Museum, California, Robert C. Miller. William Hovanitz (9:20). Submitted papers. 23/3. (9:147)

VI. 28-30 December 1955. Archbold Biological Station, Lake Placid, Florida, Richard Archbold (9:184). ? C. L. Remington. Submitted papers. Round-table: Techniques of Hand-pairing; Collecting of Sphingidae, Heliothid Moths, and *Megathymus* Skippers; UV Light Traps; Preserving Immature Lepidoptera. 26/5. (10:74)

Pacific Slope, 3. 4-5 August 1956. Santa Barbara Museum of Natural History, California, Nelson W. Baker. Fred T. Thorne (9:149). Symposium: Usefulness of Genetics, Ecology and Morphology in the Taxonomy of Lepidoptera (10:80). 34+/3. (10:172)

VII. 27 December 1956. American Museum of Natural History, New York, Alice L. Hopf. Sidney A. Hessel (10:136). Submitted papers. Round-table: Hand-pairing and Rearing Methods; Infra-specific Nomenclature. 20/6, 2-2. (11:158)

Pacific Slope, 4. 3-4 August 1957. Santa Barbara Museum of Natural History, California, A. S. Coggeshall. Nelson W. Baker (10:174). Symposium: Origins and Distribution Patterns of Western Lepidoptera. 35/3. (11:230)

VIII. 27-28 December 1957. U. S. National Museum, Washington, D. C., J. F. G. Clarke. John Fales (12:52). Submitted papers. 53/12, 1-1. (12:52)

Pacific Slope, 5. 23-24 August 1958. Santa Barbara Museum of Natural History, California, Harold Chase. R. H. T. Mattoni (11:231). Symposium: Ecological Studies of Butterflies and Moths. 26/3. (12:209)

IX. 29-30 December 1958. Department of Entomology, Uni-

versity of Kansas, Lawrence, Kent H. Wilson. Paul R. Ehrlich (12:80). Symposium: The Systematics of the Skipper Genus *Megathymus* and Allies. Round-table: Modern Techniques in Systematics. 31/6, 1-1. (13:31)

Pacific Slope, 6. 29-30 August 1959. Santa Barbara Museum of Natural History, California, Nelson W. Baker. Robert L. Langston (12:211). Symposium: Nonvariable Species and Hesperioidea. 26/2. (13:238)

Pacific Slope, 7. 27 August 1960. Morrison Auditorium, California Academy of Sciences, San Francisco, C. Don Mc Neill. J. W. Tilden (13:240). Presidential address delivered by Dr. Walter Forster of Germany. Submitted papers. 24/3, 2-2. (15:200)

X. 28 December 1960. Department of Entomology, American Museum of Natural History, New York, Frederick H. Rindge. ? C. L. Remington. Submitted papers. 40/8, 1-2. (15:63)

XI. 7-10 August 1961. Rocky Mt. Biological Laboratory, Gothic, Gunnison Co., Colorado, C. L. & Jeanne Remington. C. L. Remington. Submitted papers. Round-table: Methodology, including Preservatives, Light-trapping, Photography, and Rearing Techniques. 57+/13. (15:249)

Pacific Slope, 8. 26-27 August 1961. Santa Barbara Museum of Natural History, California, Nelson W. Baker. Lloyd M. Martin (15:202). Symposium: Migration. Open Forum: Techniques for Rearing Lepidoptera, and Preparation and Study of Lepidoptera Genitalia. 36/? (15:251)

XII. Combined with Pacific Slope, 9. 25-26 August 1962. Santa Barbara Museum of Natural History, California, Nelson W. Baker. Fred T. Thorne (15:252). Submitted and invitational papers. ?/? (20:49)

XIII. 1-2 July 1963. Town Hall, Wonderland Shopping City, San Antonio, Texas, Connie Kendall & Ellen Quillin. Roy O. Kendall (15:250). Submitted and invitational papers. 41+/7. (18:117)

Pacific Slope, 10. 24-25 August 1963. Santa Barbara Museum of Natural History, California, Carl Kirkwood. William Hovanitz (20:49). Submitted papers. ?/?. (20:50)

XIV. Combined with Pacific Slope, 11. 18-20 July 1964.

Nevada State Museum, Carson City, Peter J. Herlan. J. W. Tilden (20:50). Symposium: Collecting in Unusual Places. 25+/?. Report not published.

Pacific Slope, 12. 19-21 June 1965. San Diego Natural History Museum, California, Charles Harbison. Fred T. Thorne (20:52,55). Symposium: The Giant Skipper Megathymidae. 60/10. (20:53)

XV. 29 May - 1 June 1966. Entomology Research Institute, Ottawa, Ontario, Canada, Dr. & Mrs. Hardwick. D. F. Hardwick (20:55). Symposium: Lepidoptera Around the World. ?/?, ?-?. Report not published.

Pacific Slope, 13. 11-12 June 1966. Santa Barbara Museum of Natural History, California, Nelson W. Baker. C. L. Hogue (20:54). Symposia: Diurnal Moths of America; Distributional Patterns and Geographic Variability in Butterflies. 38/5, 1-1. (21:210)

XVI. Combined with Pacific Slope, 14. 24-25 June 1967. Oregon State University, Corvallis, Dr. & Mrs. Ernst J. Dornfeld. E. J. Newcomer (21:210). Symposia: Pacific Northwest Lepidoptera; Conservation of Lepidoptera. 40/?. Report not published.

XVII. 15-18 June 1968. Smithsonian Institution, Museum of Natural History, Washington, D. C., Dr. & Mrs. Donald W. Duckworth. Don R. Davis (5:5/67). Symposium: Polymorphism in Lepidoptera. 75/?, 2-3. Report not published.

Pacific Slope, 15. 6-8 September 1968. University of California, Berkeley, Paul A. Opler. Jerry A. Powell (23:279). Symposium: Population Dynamics of Butterflies, and submitted papers. 60/10. (23:278)

XVIII. 14-17 June 1969. Michigan State University, East Lansing, F. W. Stehr. Julian P. Donahue (3:4/69). Symposia: Microlepidoptera, Current Status; Sensory Mechanisms and Communication in Lepidoptera. 66/?. Report not published.

Pacific Slope, 16. 5-7 September 1969. Los Angeles County Museum of Natural History, California, C. L. Hogue. Fred T. Thorne (2:2/69). Submitted papers. 56/6 (7:1/69,24:73)

XIX. Combined with Pacific Slope, 17. 19-20 June 1970.

Nevada State Senate Chambers, Carson City, Peter J. Herlan. J. W. Tilden (5:5/70). Symposia: Lepidoptera of the Great Basin; Conservation and Endangered Species. 53/13. (5:2/70)

XX. 17-20 June 1971. University of Louisville, Kentucky, Dr. & Mrs. C. V. Covell, Jr. Lee D. Miller & John C. Downey (1:3/71). Symposia: Biology of the Genus *Colias*; Collecting Lepidoptera in the Southeastern U.S. 54+/20, 3-6. (4:1/71)

Pacific Slope, 18. 28-29 August 1971. Santa Barbara Museum of Natural History, California, Nelson W. Baker. Robert L. Langston (4:6/71). Submitted papers. 37/4. (6:2/71)

XXI. 22-25 June 1972. Trinity University, San Antonio, Texas, Roy & Connie Kendall. C. L. Remington. Symposia: Recent Studies of Lepidoptera Behavior; Endangered and Extinct Lepidoptera. (Twenty-fifth anniversary meeting.) 100+/23, 2-2. (5:1/72)

Pacific Slope, 19. 1-3 September 1972. Lecture Hall, Oakland Museum, California, ? C. Don MacNeill. C. Don Mac Neill (6:2/71). Symposium: Lepidoptera Conservation. 50+/5. (3:5/72, 1:2/73)

XXII. 21-24 June 1973. Allyn Museum of Entomology, Sarasota, Florida, Mr. & Mrs. A. C. Allyn. Lee D. Miller. Symposia: Introduction to the Neotropics (Mexico and Middle America, South America, the West Indies); Biology of *Heliconius* Butterflies. 104/22, 3-4. (4:1/73)

Pacific Slope, 20. 25-26 August 1973. Santa Barbara Museum of Natural History, Santa Barbara, California, Nelson W. Baker. Charles A. Sekerman. Submitted papers. 49/4. (3:6/73, 5:1/73)

XXIII. Combined with Pacific Slope 21. 22-25 August 1974. Natural History Museum, Los Angeles, California, Julian & Kathy Donahue. Julian P. Donahue. Symposia: Arctic Lepidoptera; Lepidoptera of Baja California, Mexico. 144/22, 1-1. (5:4/74)

XXIV. 21-24 August 1975. University of Massachusetts, Amherst, Dr. & Mrs. Theodore D. Sargent. Theodore D. Sargent. Symposia: Lepidoptera Ecology; Biology of the *Catocala*. 58/18, 2-2. (4:5/75, 1:3/76)

Pacific Slope, 22. 20-22 June 1975. Oregon State University, Corvalis, Dr. & Mrs. Ernst J. Dornfeld, et al. David V. McCorkle. Symposium: Lepidoptera Conservation in Western North America. 22/8. (6:4/75)

XXV. 25-28 June 1976. Wisconsin Center, University of Wisconsin, Madison, Walter L. Gojmerac. William E. Sieker. Submitted papers. 47/14, 1-1. (6:2/76)

Pacific Slope, 23. 2-4 July 1976. Nevada State Museum, Carson City, Gloria Harjes & Peter J. Herlan. Jerry A. Powell. Environmental Physiology of Lepidoptera; Nevada Lepidoptera; and submitted papers. 43/6. (1:6/76, 2:1/76, 6:5/76)

ANNUAL FIELD SUMMARIES

History

The first summary was begun by C. L. Remington and Harry K. Clench, founders of our Society. The summary was to be an attempt to present significant information for all Lepidoptera of America north of Mexico, with each geographic area receiving approximately equal treatment. North America was divided into eight zones and by the time of the 1948 summary each had a corresponding "Zone Coordinator." The succession of Zone Coordinators through the history of the Summary is presented below (Table 1). Coordinators who have devoted eleven or more years of service to the Summary are: J. Donald Eff (Zone III) - 19 years; H. A. Freeman (Zone IV) - 16 years; E. C. Welling (Zone IX), Bryant Mather (Zone VI) and L. P. Grey (Zone VIII) - all 14 years; and E. J. Newcomer (Zone II) - 11 years.

Table 1. The zone coordinators and editors, and their years of service, through the history of the annual field summaries.

Zones	Years & Coordinators
I Southwest	1947-48 L. M. Martin; 1949 L. M. Martin and B. H. Weber; 1950-52 L. M. Martin; 1960-68 F. T. Thorne; 1969-72 R. L. Langston
II Northwest	1947-52, 59-60 J. C. Hopfinger; 1961-71 E. J. Newcomer; 1972 Mike Van Buskirk
III Rocky Mountains	1947-52, 59-69 J. D. Eff; 1970 C. D. Ferris; 1971-72 J. D. Eff
IV Plains	1947-50 D. B. Stallings; 1951-52, 59-72 H. A. Freeman
V Central	1948-52 P. S. Remington, Jr.; 1959-61 E. G. Voss; 1962-70 J. R. Heitzman; 1971-72 M. C. Nielsen

Zones	Years & Coordinators
VI Southeast	1948-52 R. L. Chermock; 1959-72 B. Mather
VII Northeast	1947-50 E. G. Munroe; 1951-52 S. A. Hessel; 1959-72 L. P. Grey
VIII Far North	1948-52 T. N. Freeman; 1960-68 L. D. Miller; 1969-72 K. W. Philip
IX Neotropics	1959-72 E. C. Welling
X South America	1971-72 T. C. Emmel
EDITORS	1957-58 C. L. Remington; 1959-64 J. W. Tilden; 1965-69 E. J. Newcomer; 1970- P. A. Opler

From the first, individual members were encouraged to contribute to the Summary and to keep careful field notes during each collecting season. The number of contributors to the compilation of each zone's report for each Summary are indicated in Table 2. From 1947 through 1952, except 1950, the Summary showed an annual increase in the number of contributors from 59 to 85. From 1953 through 1958 the Summary was discontinued due to increased publication costs. When the Summary reappeared in 1959 the number of contributing members was 56. A steady increase in the number was shown from 1960 through 1965 when the number reached 100. Each succeeding year had more than 100 contributors, peaking at 157 in 1970.

Coverage

In all years, the Summary has presented field information on several aspects of lepidopteran biology, including:

Distribution. (1) Notable locality records. These include new localities as well as those previously known. Most of the records cited have been of species considered "rare" or "choice" by collectors. (2) New county and state records. These have appeared with some frequency. (3) Range extensions. These have been documented for a number of species, among which *Thymelicus lineola* is perhaps the most notable. Others include *Paratrytone melane* and *Dione vanillae* in

Table 2. The number of contributors each year to the annual season summary.

Years	Zones										Total
	Z 1 S.West	Z II N.West	Z III Rocky Mts	Z IV Plains	Z V Central	Z VI S.East	Z VII N.East	Z VIII Far North	Z IX Neotr.	Z X S.Am.	
1947	10	4	9	4	9	6	5	2	-	-	49
1948	7	6	7	3	10	7	12	2	-	-	54
1949	6	7	13	8	17	8	16	2	-	-	77
1950	9	4	12	3	18	7	19	1	-	-	73
1951	9	4	10	8	24	6	24	1	-	-	86
1952	9	7	6	8	23	10	21	1	-	-	85
1959	8	10	(6	)	10	15	6	-	1	-	56
1960	10	11	6	7	11	7	9	2	3	-	66
1961	9	13	7	5	9	9	9	1	4	-	66
1962	17	11	9	6	17	14	16	2	6	-	98
1963	14	15	13	10	16	6	14	-	3	-	91
1964	23	14	11	8	-	16	13	5	7	-	97
1965	18	13	16	7	23	8	9	1	5	-	100
1966	22	11	18	14	29	14	18	5	3	-	134
1967	25	6	18	13	52	17	18	3	3	-	155
1968	31	6	17	13	21	18	19	15	3	-	143
1969	24	3	18	14	18	11	13	18	6	-	125
1970	22	7	15	8	31	13	14	40	7	-	157
1971	15	7	14	11	27	12	17	3	8	2	116
1972	27	10	18	13	22	21	17	1+	3	1	133

California.

Seasonality. (1) Dates of capture. Most records in the Summary include the capture date, and as such, each gives added knowledge about the seasonal occurrence of a species. (2) Effects of weather. Quite frequently the occurrence of unusual weather (such as extreme cold, heat, drought, hurricanes, snow) has been stated as causing a delay or hastening of the emergence of a species. (3) Early or late records. Collecting records representing unusually early or late dates for the presence of a species have appeared with or without citation of possible correlative causes. (4) Extra broods. Occurrences of extra broods for normally univoltine populations have been reported on several occasions. Most notable of these have been the reports of fall broods for desert species in the Anza-Borrego area of San Diego County, California.

Ecology. (1) Population changes. Perhaps more space has been devoted to attempts at recording population changes than any other aspect of lepidopterology. These attempts are usually couched in verbal rather than numeric terms, such as "extremely abundant" and "flew by the thousands," or "did not appear" and "absent this year." Also appearing are attempts to place species into "up" or "good," and "down" or "bad" year categories. Some of these changes have been attributed to the weather, and others to pesticides or habitat destruction. Most have been based on the experience of the individual contributor and almost none on modern population sampling methods.

Biology. From time to time foodplant records or habitat descriptions have been included.

Migration. Numerous reports of migratory species have appeared in past summaries, most of them for *Danaus plexippus*. Movements of emigratory species such as *Vanessa cardui*, *Nymphalis californica*, *Dione vanillae*, *Libytheana bachmanii*, *Phoebis sennae*, and *Ascia monuste* have been reported on numerous occasions. Data on *Danaus plexippus* and *Vanessa cardui*, taken in part from the Summary, were included by C. B. Williams in his book *Insect Migration* (1958).

Strong Points

In assaying the strong points of past summaries, I have taken into consideration not only the number of reports, but how well they seem to have fulfilled their intended goals. This evaluation is personal and subjective and, as such, may not agree with the opinions of others.

Distribution. The main goal of the Summary in the opinion of many members is that it should serve as a source of information regarding new localities for rare or choice species. It is often utilized by collectors to plan field trips for the next season. In this regard the summary has frequently fulfilled its unstated purpose at least in some western states, to judge from the number of collectors one encounters in "well-reported' localities.

Seasonality. The range of dates of occurrence for the common species in an area, especially when compared to the range of dates in previous years, provides a valuable information store which may be relied upon by future students and collectors.

Migration. The observations of migratory and emigratory Lepidoptera is one of the most important features of the Summary when taken on a year to year basis.

Weak Points

Overall. The three most obvious and most frequently cited weaknesses of the Summary have been (1) emphasis on butterflies and skippers with a corresponding underemphasis on the remainder of the order, (2) emphasis on rare species, and (3) uneven geographical coverage. While the last weakness is unavoidable due to the uneven geographic distribution of the Society membership, some attempts could be made to improve with respect to the first two defects. Any improvement lies with the source, i.e. the total sum of field data from which contributors select what they deem most worthy of reporting. Then the regional coordinators select what they deem most worthy of including in their zone reports. Since most of the coordinators are primarily interested in butterflies, and since the fairest way to sort through the

sometime voluminous information submitted by contributors is to select the reports for what they know are "good" species, we end up in the kettle we are in.

Distribution. A visit to one of the continent's major collections, such as the American Museum of Natural History, the U. S. National Museum, the Carnegie Museum, or the Canadian National Collection, will demonstrate the futility of reporting distribution records for North American butterflies and macro-moths. The task is essentially complete, and most (not all) efforts directed at increasing our knowledge of the distribution of these insects in our Summary have been redundant. Any professional lepidopterist, when investigating the distribution of insects, places his supreme confidence in museum material, and treats with justifiable suspicion all literature reports, which are not supported by material deposited in some major institutional collection.

Seasonality. The attempts in the Summary at correlating weather phenomena with the appearance of Lepidoptera in the field are of dubious value. These correlations are usually anecdotal observations, and are not based upon a knowledge of the insect's physiological characteristics, nor presented in a quantitative manner.

Since some individuals of a population must appear each year, mate, and deposit eggs to ensure the future survival of the population (by definition), statements such as "*Xus albus* did not fly this year" or "was absent in Iowa" only indicate that the contributor was not successful in locating the individuals that did fly or were present.

Ecology. The correlation of population levels with habitat change, pesticide usage, or weather cycles, while usually well-intended, have been largely of no value. Such changes should be backed by yearly quantitative figures and directly associated with the inferred phenomenon by the use of proper controls.

Future

In recent years our society has become justifiably concerned with the preservation and maintenance of our

precious environmental heritage. At several of the annual meetings we have discussed the problem of how the Society might make a meaningful contribution to the conservation movement and, further, how we might relate our knowledge of changes in population levels of Lepidoptera to changes in environmental quality.

As one possible solution to this problem, we might be able to make our Summary a more accurate indicator of changes in populations of Lepidoptera as they occur through the years. Most important to the accumulation of this sort of knowledge is the careful monitoring of species which occur in the immediate vicinity of our own "home base." If we can demonstrate correlations between changes in Lepidoptera populations and changes in the environment, either positive or negative, then we will have used our own special skills to contribute to the knowledge of the effects of man's alterations of the natural environment.

Paul A. Opler

LIBRARY

The library of The Lepidopterists' Society is housed in Room 242 of the Science Building, California State University-Northridge, Northridge, California 91324 USA. Its location here is something of a historical accident. Originally, publications received in exchange for the *Lepidopterists' News* were kept by Dr. Remington, the first editor and one of the Society's founders, at Yale University. As a graduate student there, I began writing abstracts for the Recent Literature section of the *News*, and eventually became Associate Editor in charge of this section, continuing preparation and assembly of abstracts in Jamaica and at Yale until 1958. When I came to San Fernando Valley State College (now CSU-Northridge) in that year, it seemed reasonable to have the library go with me, in order to facilitate my continuing preparation of the Recent Literature section. I do not now recall what formalities led to my being appointed librarian (12:160).

Unfortunately, the flood of scientific literature which has appeared since World War II led eventually to the preparation of far more abstracts than could possibly be published in the *Journal*. Even so, our aim of providing complete coverage of the world literature on Lepidoptera was far from being realized, in spite of the invaluable assistance of Dr. Josef Moucha (Czechoslovakia), Dr. Eugene G. Munroe (Canada), Dr. A. Diakonoff (Netherlands), and other devoted contributors. (I should like to call the attention of members who have missed the Recent Literature section to the new Journal, *Entomology Abstracts*, which in the last few years has been providing the most complete, prompt, and convenient coverage of entomological literature available anywhere.)

The original reason for the location of the library here has therefore disappeared. In addition, the space available is finite, and the lack of money for binding periodicals to prevent their deterioration is a matter for concern. Neither of these problems will become critical for some years, but the Society might well give some thought to the future of this collection.

Meanwhile, the library is, I regret to say, hardly used by members. This may be due in part to the presence of

excellent libraries locally (for instance, at the University of California-Berkeley, and the Los Angeles County Museum). Lack of publicity may also be responsible, though efforts have been made at meetings and in the *News* to call members' attention to its availability. Judging from the membership list, there should be many people in this area who would find the library useful (though the only amateurs I seem to meet are beetle men). Let us make a further attempt to arouse interest.

The library contains complete, or nearly complete, sets of some 40 periodicals devoted in whole or in part to entomology, for the last 10-25 years (depending on the journal), and incomplete sets of as many more. In addition there is a large collection of reprints and books donated by authors and members. [It is hoped that all *Journal* authors will send two copies of their reprints to the Society library.] Any of this material is available on loan to members; those in the area are welcome to visit and use it, and others may borrow anything we have for the cost of postage.

LIST OF SOME SERIALS IN THE SOCIETY LIBRARY

Acta Entomologica Bohemoslovaca - Czechoslovakia, vols. 44-61, formerly *Casopis Ceskoslovenska Spolecnosti Entomologicke*

Acta Zoologica Lilloana - Argentina, 1963 +

Alexanor - France, 1961 +

Amateur Entomologists' Society Bulletin - England, 1960 +

Annalen des naturhistorischen Museums in Wien - Austria, 1951 +

Annales Zoologici Fennici - Finland, 1 (1964) +

Annales Zoologici - Poland, 1947 +, formerly *Annales Musei Zoologici Polonici*

Beiträge zur Entomologie - Germany, 1951 +

Butterflies and Moths - Japan, vols. 1-21, 1970, (continued as *Tyo to Ga*, and *Lepidopterological Society of Japan, Transactions*)

Coleopterists' Bulletin - USA, 1947-57

Deutsche entomologische Zeitschrift - Germany, 1954 +

Entomologia Sinica - China, 1947-51

Entomological News - USA, 1957 +

Entomologicheskoe Obozreni (Revue russe d'entomologie) - USSR, 1947 +

Entomologische Berichten - Netherlands, 1950 +

Entomologische Nachrichten - Germany, 1947 +

Entomologisk Tidskrift - Sweden, 1957 +

Entomologiske Meddelelser - Denmark, 1957 +

Entomologist - England, 1959, 1961 part only

Entomologists' Gazette - England, 1950 +

Entomologist's Record - England, 1962 +

Entomophaga - France, 1956-70

Fragmenta Faunistica - Poland, 1946 +

Insecta Matsumurana - Japan, 1952 +

Keys for the Identification of Polish Insects - Poland, 1954 +

Kontyu - Japan, 1970 +

Lambillionea - Belgium, 1947 +

Lepidoptera - Denmark, 1946-50, and new series vol. 1+ (1965+)

Lepidopterological Society of Japan, Bulletin; Journal - 1946 +

Mitteilungen der Münchner Entomologischen Gesellshaft - Germany, 1943 +

Mushi - Japan, 1950 +

Nachrichtenblatt der bayerischen Entomologen - West Germany, 1952 +

Notulae Entomologicae - Finland, 1962 +

Opuscula Entomologica - Sweden, 1959-69

Polskie Pismo Entomologiczne - Poland, 1967-69

Revue française de Lepidopterologie - France, 1947-57

Royal Entomological Society of London, Proceedings - England, Ser. B, 1961 +

Shikoku Entomological Society, Transactions - Japan, 1950+

Societas Entomologica Cechosloveniae, Casopis (Acta) - Czechoslovakia, 1947 +

South London Entomological and Natural History Society, Proceedings and Transactions - England, 1952 +

Vsesoyuznoe Entomologicheskoe Obshchestvo Trudy - USSR, 1963 +

Wasmann Journal of Biology (was *Wasmann Collector*) - USA, 1948 +

Wiener entomologische Gesellschaft, Zeitschrift - Austria, 1945 + (vol. 56 +)

Zeitschrift der Arbeitsgemeinschaft oesterreichischer Entomologen - Austria, vols. 12-18, 1966, formerly *Entomologisches Nachrichtenblatt* - Switzerland, vols. 1-11

Peter F. Bellinger

DISTINGUISHED SERVICE

The Society has two principal methods of recognizing outstanding and distinguished service in the field of lepidopterology: (1) electing individuals to Honorary Life Memberships, and (2) awarding individuals the Karl Jordan Medal.

HONORARY LIFE MEMBERSHIPS. Persons who have made important contributions to the science of lepidopterology may be elected Honorary Life Members. Such members are nominated by the unanimous vote of the Executive Council, and then must receive at least 80% of all mail ballots cast by the membership. Not more than two may be elected in any one year (Art. V, Sec. 3, Constitution), and the number of living Honorary Life Members is restricted to ten (Art. III, Sec. 6). These members are exempt from paying dues and they receive, without charge, copies of all Society publications (Art. I, Sec. 4, By-Laws).

The following is an alphabetical list of all honorary life members through the 1972 nominations/elections. Following each name is the lifespan if deceased, the year of election to Honorary Life Membership (with reference citation), and the country in which the individual resides. Also, if deceased, known obituary citations are given.

BRAUN, Annette F., 1960 (6:1/60) USA
BROWN, F. Martin, 1973 (27:86) USA
CARPENTER, G. D. Hale (1883-1953), 1952 (6:121) ENGLAND; obituary, portrait, bibliography: 8:31; *Brit. Med. J.* 1:406
COMSTOCK, John A. (1883-1970), 1969 (2:1/69) USA; obituary, portrait 25:215
DOS PASSOS, Cyril F., 1973 (27:86) USA
EVANS, William H. (1876-1956), 1952 (6:121) ENGLAND; obituary, portrait, bibliography: 10:193.
FORBES, William T. M. (1885-1968), 1952 (6:121) USA; obituary: *Entomol. News* 8(4)
FORD, Edmund B., 1960 (6:1/60) ENGLAND
FORSTER, Ernst W., 1969 (2:1/69) GERMANY
GEROULD, John H. (1867-1961), 1959 (3:1/59) USA; obituary: 5&6:1/61; 12:190
HERING, Erich M. (1893-1967), 1956 (10:80) GERMANY; obituary, portrait: 22:123

THE LEPIDOPTERISTS' SOCIETY
EST. 1947

This certifies that

has been duly elected

HONORARY LIFE MEMBER

of the Lepidopterists' Society by vote of the membership of the Society in the year   and is entitled to all of the rights and privileges appertaining thereto.

President

Secretary

Fig. 1. The certificate awarded to Honorary Life Members of the Society.

JANSE, Antonie J. T. (1877-1970), 1970 (6:1/70) SOUTH AFRICA; obituary, portrait: 25:211

JORDAN, Heinrich Ernst Karl (1862-1959), 1952 (6:121) ENGLAND; obituary: 12:198; obituary, portrait, bibliography: *Biograph. Mem. Fellows Royal Soc.* 6

KLOTS, Alexander B., 1972 (2:16/72) USA

MC DUNNOUGH, James H. (1877-1962), 1952 (6:121) CANADA; obituary, portrait, bibliography: 16:209; *Entomol. News* 41:179
MUNROE, Eugene G., 1972 (2:16/72) CANADA
REMINGTON, Charles L., 1966 (3:1/66) USA
WILLIAMS, C. B., 1959 (3:1/59) SCOTLAND

KARL JORDAN MEDAL. This medal is awarded in recognition of original research, and is not intended to be a career award as in the case of Honorary Life Memberships. The prize consists of an engraved silver medal, a $1000 cash award and travel expenses for the recipient(s) to accept the award. It was proposed by Mr. A. C. Allyn in recognition of the Society's 25th Anniversary meetings in San Antonio, Texas. It will be funded, on a year-to-year basis for the present, through the Allyn Museum of Entomology. The procedures and criteria for awarding the Jordan Medal are (26:207):

1. An award committee will be established consisting of:
   A. One representative to be designated by the Allyn Museum of Entomology;
   B. The President of the Lepidopterists' Society, unless he is affiliated with the Allyn Museum of Entomology, in which case the First Vice President of the Society will be part of the award committee; and
   C. An elected member of the Lepidopterists' Society to be known as "the Jordan Medal Representative" who will serve a one-year term and will not be affiliated with the Allyn Museum of Entomology.
2. The medal will be awarded only by the *unanimous* vote of the committee. This, of course, will result in the medal not being awarded in some years, but it is felt that the requirement for unanimity of the committee will be a safeguard in assuring that only work of the highest quality is awarded.
3. The award will be for original research, not for the compilation of already known facts. The judgment of where the line must be drawn between research and compilation will be left to the discretion of the committee and may be expected to change from year to year as different people are involved.
4. The award may be based upon a single piece of research or on a series of interrelated works, and the

work must be at least three, but not more than twenty-five years old.

Fig. 2. The Karl Jordan Medal.

5. Jointly awarded prizes will be allowed only in instances of coauthorship, and the cash prize for that year will be divided equally between the workers or their heirs. In instances of jointly awarded prizes, duplicate medals will be presented, and travel expenses for more than one worker will be allowed.
6. No member of the Jordan Medal award committee will be eligible to receive the prize during his tenure, but will not be ineligible in succeeding years.
7. The recipient of the Jordan Medal need not be a member of the Lepidopterists' Society. It is hoped, however, that if the recipient is not a Society member that he may elect to become one.

8. Nominations for recipients of the Jordan Medal will be accepted from any member of the Lepidopterists' Society and should be sent to: Jordan Medal Awards Committee, Allyn Museum of Entomology, 3701 Bay Shore Road, Sarasota, Florida 33580 USA.

The first Karl Jordan Medal was awarded 23 June 1973 to M. Henri Stempffer of Paris, France, for various works on the Lycaenidae, especially of Africa, and in particular his "Genera of the African Lycaenidae (Lepidoptera: Rhopalocera)," Bull. British Museum (Nat. Hist.) 10 (Suppl.). 322 pp., illustrated (1967). The Awards Committee consisted of: Lee D. Miller, Allyn Museum representative; J. F. Gates Clarke, President of the Lepidopterists' Society; and Frederick H. Rindge, Jordan Medal Representative of the Society.

The second Karl Jordan Medal was awarded 24 August 1974 to Frederick W. Stehr, Michigan State University, East Lansing, for his "A Revision of the Genus *Malacosoma* Hübner in North America (Lepidoptera: Lasiocampidae): Systematics, Biology, Immatures, and Parasites," Bull. U.S. Nat. Mus., No. 276, 321 pp., illustrated (1968). The Awards Committee consisted of: Lee D. Miller, Allyn Museum representative; Harry K. Clench, President of the Lepidopterists' Society; and W. Donald Duckworth, Jordan Medal Representative of the Society.

No medal was awarded in 1975 or 1976. Those serving on the Awards Committee for these years were: Lee D. Miller, Allyn Museum Representative; Harry K. Clench and Ronald W. Hodges, Society Presidents; Jerry A. Powell and Cyril F. dos Passos, elected Representatives of the Society.

PERSONNEL

Officers and Editors

PRESIDENTS

Listed here in chronological order through 1976 are those who have served as President of the Society. Biographical data for these and others who have served the Society in an official capacity may be found under the title Biographical Sketches.

1951 James H. McDunnough, *Canada*
1952 H. E. Karl Jordan, *England*
1953 William T. M. Forbes, *USA*
1954 A. Diakonoff, *Netherlands*
1955 Lauro P. Travassos, *Brasil*
1956 Thomas N. Freeman, *Canada*
1957 Alexander B. Klots, *USA*
1958 Norman D. Riley, *England*
1959 Eugene G. Munroe, *Canada*
1960 E. Walter Forster, *Germany*
1961 Takashi Shirôzu, *Japan*
1962 John A. Comstock, *USA*
1963 H. B. D. Kettlewell, *England*
1964 John G. Franclemont, *USA*
1965 Frederick H. Rindge, *USA*
1966-67 Don B. Stallings, *USA*
1968 F. Martin Brown, *USA*
1969 David F. Hardwick, *Canada*
1970 Edmund B. Ford, *England*
1971 Charles L. Remington, *USA*
1972 Lloyd M. Martin, *USA*
1973 J. F. Gates Clarke, *USA*
1974 Harry K. Clench, *USA*
1975 André Blanchard, *USA*
1975/76 Ronald W. Hodges, *USA*
1976/77 Stanley S. Nicolay, *USA*

In addition to the above, 58 others have served as Vice-President during the same period. Their names are given in alphabetical sequence together with the year they served

and the country which they represented: Shigeru A. Ae (1972) *Japan,* Ramon Agenjo (1965) *Spain,* Burchard Alberti (1967) *Germany,* Hans G. Amsel (1964) *Germany,* Carlos R. Beutelspacher (1976) *Mexico,* Ceslau M. Biezanko (1963) *Brasil,* Annette F. Braun (1954) *USA,* Don Alberto Breyer (1955) *Argentina,* Lincoln P. Brower (1970) *USA,* Keith S. Brown (1972) *Brasil,* Hahn W. Capps (1957) *USA,* Georg Christensen (1958) *Denmark,* Austin H. Clark (1950-51) *USA,* Cyril A. Clarke (1966) *England,* Ian F. B. Common (1957,65) *Australia,* Charles F. Cowan (1973) *England,* Romualdo F. d'Almeida (1952) *Brasil,* Hubert DeLesse (1961) *France,* Donald R. Davis (1975) *USA,* Tarsicio Escalante (1962) *Mexico,* Maria Etcheverry (1967) *Chile,* Hugh A. Freeman (1971-72) *USA,* Lancelot A. Gozmany (1956) *Hungary,* Lionel P. Grey (1952) *USA,* Kenneth J. Hayward (1951) *Argentina,* Claude Herbulot (1960) *France,* Erich M. Hering (1953) *Germany,* Sidney A. Hessel (1969) *USA,* Howard E. Hinton (1965) *England,* William Hovanitz (1975) *USA,* Hiroshi Inoue (1962) *Japan,* Frank M. Jones (1954) *USA,* Julian Jumalon (1971) *Philippines,* Sergius G. Kiriakoff (1963) *Belgium,* Josef W. Klimesch (1968) *Austria,* Zdravko Lorkovic (1959,66) *Yugoslavia,* Olaf H. H. Mielke (1975) *Brasil,* Jose Oiticia (1950) *Brasil,* Yoshio Okada (1953) *Japan,* Kenelm W. Philip (1971) *USA,* Dalibor Povolny (1957) *Czechoslovakia,* Jules C. E. Riotte (1974) *Canada,* Miriam Rothschild (1976) *England,* Kazuo Saitoh (1964) *Japan,* Theodore D. Sargent (1976) *USA,* Demetrius G. Sevastopulo (1967) *Kenya,* Henri Stempffer (1968) *France,* Esko Suomalainen (1966) *Finland,* W. H. T. Tams (1960) *England,* Fred T. Thorne (1958) *USA,* James W. Tilden (1970) *USA,* G. van Son (1961) *South Africa,* L. Vari (1974) *South Africa,* Leonila Vazquez G. (1958,69) *Mexico,* Pierre E. L. Viette (1955,64) *France,* Edward G. Voss (1961) *USA,* Georg H. G. Warnecke (1956) *Germany,* Carrington B. Williams (1969) *Scotland.*

SECRETARIES

1947-50 Charles L. Remington
1951-54 Frederick H. Rindge
1955-57 John G. Franclemont
1958-63 Paul R. Ehrlich
1964-71 John C. Downey
1972-76 Lee D. Miller
1977- Julian P. Donahue

TREASURERS

1947-50	Charles L. Remington
1951-52	J. Benjamin Ziegler
1953-58	Sidney A. Hessel
1959-65	George Ehle
1966-68	Roy O. Kendall
1969	John S. Buckett
1969-74	Stanley S. Nicolay
1975-	John M. Snider

EDITORS OF NEWS/JOURNAL

1947	Charles L. Remington, Harry K. Clench (co-editors)
1948-63	Charles L. Remington
1964-69	Jerry A. Powell
1970-71	David F. Hardwick
1972-74	Theodore D. Sargent
1975-	George L. Godfrey

EDITORS OF NEWS

1959-64	James W. Tilden
1964-70	Erval J. Newcomer
1970-73	Charles V. Covell, Jr.
1973-76	Ronald Leuschner
1977-	Jo Brewer (Mrs. William D. Winter, Jr.)

EDITORS OF SUPPLEMENTS

1969-70	Jerry A. Powell
1971-	Lee D. Miller

MANAGERS OF MEMOIRS

1962-74	Sidney A. Hessel
1975-	Charles V. Covell, Jr.

Charter Members

A Charter Member is anyone who joined the Society in 1947 (1:12). In the following list, deceased members are noted by the abbreviation (d.), and the year of death is given, if known. Unless otherwise indicated, the individual is assumed to be alive as of 31 December 1976. Those who were still shown in the 30 September 1976 Membership List are marked (*).

Ackermann, Otto (d.)	*Pennsylvania*
*Adams, Joseph W.	*Pennsylvania*
Adelphe, David P. (d.)	*(Ontario) CANADA*
Ahlen, Charles	*New York*
*Albright, Ray	*Oregon*
Alexander, Charles P.	*Massachusetts*
Allison, A. Crawford G.(Mrs.)	*Pennsylvania*
Anderson, Andrew (d.)	*Washington*
Araujo, R. L.	*BRASIL*
Arnhold, Francis R. (d. 1959)	*Wisconsin*
Avinoff, Andrey (d. 1949)	*New York*
Bailey, Norman S.	*Massachusetts*
Baker, Clement W. (d. 1950?)	*Ohio*
*Baker, James H.	*Oregon*
*Bauer, David L.	*California*
Bauer, William R.	*California*
Beal, John L.	*North Carolina*
Beebe, Ralph (d.)	*Michigan*
Bell, Ernest L. (d. 1964)	*New York*
Berry, Dean F. (d. 1949)	*Florida*
Bishop, John A.	*Kentucky*
Boswell, John S. C.	*Tennessee*
*Braun, Annette F.	*Ohio*
Bridwell, Louis H. (d. 1951)	*Texas*
Bristol, Maurice L. (d. 1975)	*Illinois*
*Brower, Auburn E.	*Maine*
*Brown, F. Martin	*Colorado*
Bruggemann, Paul F.	*(Saskatchewan) CANADA*
Brumbaugh, John M.	*Texas*
Buchholz, Otto (d. 1958)	*New Jersey*
*Calkins, Virgil F.	*Washington*
Carpenter, A. J. (d.)	*Massachusetts*
Carpenter, Frank M.	*Massachusetts*
Carpenter, S. C.	*Connecticut*

Carr, Robert	*Ohio*
Cary, Margaret M. (d. 1969)	*Pennsylvania*
Casselberry, Raymond C. (d. 1954)	*New York*
*Chermock, Ralph L.	*Alabama*
*Christensen, Georg	*GREECE*
Clark, Austin H. (d. 1954)	*District of Columbia*
Clay, Charles I.	*California*
Clement, Ian D.	*Massachusetts*
*Clench, Harry K.	*Pennsylvania*
Comstock, John A. (d. 1970)	*California*
Coher, Edward I.	*Massachusetts*
*Cook, Carl	*Kentucky*
Cook, William C. (d. 1967)	*Washington*
Crawford, John W.	*Massachusetts*
Creelman, James L.	*California*
Crickmer, Noel	*California*
*Davies, Thomas W.	*California*
Davis, John E.	*Oregon*
Dickenson, Ronald	*California*
*dos Passos, Cyril F.	*New Jersey*
*Downey, John C.	*Iowa*
Dreisbach, Robert R.	*Michigan*
*Eff, J. Donald	*Colorado*
*Ehrlich, Paul R.	*California*
Eustis, Henry W. (d. 1951)	*Georgia*
*Evans, William H.	*District of Columbia*
Fattig, Perry W. (d. 1953)	*Georgia*
Fauteux, Jacques M.	*Massachusetts*
*Ferguson, Douglas C.	*District of Columbia*
Ferguson, Elias A.	*Ohio*
*Field, William D.	*District of Columbia*
Field, W. L. W. (d.)	*Massachusetts*
Fitch, Richard J. (d. 1961)	*(Saskatchewan) CANADA*
Forbes, William T. M. (d. 1968)	*New York*
*Ford, Edmund B.	*ENGLAND*
*Ford, Robert J.	*California*
*Franclemont, John G.	*New York*
*Frechin, Donald P.	*Washington*
*Frederick, Albert C.	*Texas*
Frederick, Edward J.	*California*
*Freeman, Hugh A.	*Texas*
Freeman, Thomas N. (d. 1975)	*(Ontario) CANADA*
Friday, F. W. (d.)	*California*
Fuller, Stanley F. (d. 1966)	*Virginia*
Gerhard, William J. (d. 1958)	*Illinois*
Gerould, John H. (d. 1961)	*New Hampshire*
Glasgow, Clyde L. (d. 1951)	*Wyoming*

*Glenn, Murry O. — *Illinois*
*Glick, Perry A. — *Texas*
*Gottschalk, Carl W. — *North Carolina*
Gray, Percy H. H. (d. 1971) — *(Quebec) CANADA*
*Grey, Lionel P. — *Maine*
Griewisch, Louis W. — *Wisconsin*
Grimshawe, Florence M. — *Florida*
Guedet, Edward F. (d. 1967) — *California*
*Guppy, Richard — *(British Columbia) CANADA*
Hall, Arthur D. (d. 1949) — *(Nova Scotia) CANADA*
*Hammer, William A. — *California*
*Hardwick, David F. — *(Ontario) CANADA*
*Harris, Lucien,Jr. — *Georgia*
Heid, Graham — *California*
*Heineman, Bernard — *New York*
Hessel, Sidney A. (d. 1974) — *Connecticut*
Holley, F. E. — *Illinois*
Hopfinger, John C. (d. 1961) — *Washington*
Hovanitz, William — *California*
Hulbirt, Lowell H. — *California*
[=Edwin Ray Hulbirt] (d. 1965) — *California*
Hynes, Vonta P. — *Michigan*
*Irwin, Roderick R. — *Illinois*
Jay, William — *Pennsylvania*
Johnston, Edward C. (d. 1951) — *Washington*
Johnston, William M. — *Massachusetts*
Jordan, H. E. Karl (d. 1959) — *ENGLAND*
Kaiser, Justus — *New York*
Keith, Edward D. (d. 1949) — *Rhode Island*
*Keji, Joseph A. — *New York*
Kellner, John J. — *New York*
Key, Edward S. — *Ohio*
*Kimball, Charles P. — *Massachusetts*
*King, Harold L. — *Florida*
Kiriakoff, Sergius G. — *BELGIUM*
Kirkwood, Carl W. (d. 1972) — *California*
*Klots, Alexander B. — *Connecticut*
Lauck, Albert G. (d.) — *Illinois*
Learned, Elmer T. — *Massachusetts*
Lennox, Donald J. — *California*
Lindsey, Arthur W. (d. 1963) — *Ohio*
Mack, Myrtle A. — *California*
*Macy, Ralph W. — *Oregon*
*Martin, Lloyd M. — *Arizona*
*Mather, Bryant — *Mississippi*
*Mattoni, Rudolf H. T. — *California*
McAlpine, Wilbur S. — *Michigan*

McDunnough, James H. (d. 1962)	*New York*
McElhose, Arthur L. (d. 1954)	*Illinois*
*McElvare, Rowland R.	*North Carolina*
*McHenry, Paddy B.	*California*
Meiners, Edwin P. (d. 1960)	*Missouri*
*Merker, Charles G.	*Pennsylvania*
Meyer, William T.	*California*
Michener, C. D.	*New York*
Moeck, Arthur H. (d. 1971)	*Wisconsin*
Moyer, Howard C.	*Pennsylvania*
Mueller, Paul H.	*New York*
*Munroe, Eugene G.	*(Ontario) CANADA*
Murphy, Gardner (d.)	*Massachusetts*
Nabokov, Vladimir	*New York*
*Napier, Arthur H.	*Pennsylvania*
Nelson, Carl J. E.	*Colorado*
Newman, John H.	*Michigan*
*Nicolay, Stanley S.	*Virginia*
*Nielsen, Mogens C.	*Michigan*
O'Byrne, Harold I. (d. 1951)	*Missouri*
Orchard, Charles D.	*Texas*
Panske, Leonard G.	*Illinois*
Park, Francis E.	*Massachusetts*
Parker, Dwight E.	*Arizona*
*Peters, George	*Pennsylvania*
Pickel, Benjamin H.	*Missouri*
Price, Homer F.	*Ohio*
*Quelch, C. S.	*(Manitoba) CANADA*
*Rawson, George W.	*Maryland*
Reese, Jean	*Maryland*
Reeves, Robert T.	*California*
Reid, Robert H.	*California*
*Remington, Charles L.	*Connecticut*
*Remington, Jeanne E.	*Connecticut*
Remington, P. Sheldon (d. 1975)	*Connecticut*
Renk, John J.	*Wisconsin*
*Riley, Norman D.	*ENGLAND*
*Rindge, Frederick H.	*New York*
*Roberds, Joseph	*California*
*Robert, John H.	*SPAIN*
Robinson, Paul F.	*Massachusetts*
Rogers, W. Prescott (d. 1953)	*Massachusetts*
*Rupert, Laurence R.	*New York*
Salazar, M. (Sr.)	*HONDURAS*
Schoenherr, William H.	*Illinois*
Schroeter, Otto H. (d. 1957?)	*Connecticut*

*Shappirio, David G.	*Michigan*
*Sheppard, Arthur C.	*(Quebec) CANADA*
*Shoumatoff, Nicholas	*New York*
*Sieker, William E.	*Wisconsin*
Smalley, Stephen B.	*Ohio*
Smith, Claude I. (d. 1949)	*California*
Smith, Marion E.	*Massachusetts*
Smith, M. Eugene	*Georgia*
*Spelman, Morris	*Florida*
Sperry, John L. (d. 1954)	*California*
Stauffacher, Emil W. (Mrs.) (d. 1962?)	*Wisconsin*
*Stallings, Don B.	*Kansas*
*Stempffer, Henri	*FRANCE*
Sternitzky, Robert F.	*Arizona*
Stevens, Herbert (d.)	*ENGLAND*
*Struck, B.	*Texas*
Sweadner, Walter R. (d. 1951)	*Pennsylvania*
Sweetman, Harry F.	*Minnesota*
Thomas, George W.	*Missouri*
Thomson, Samuel C. (d. 1973)	*Texas*
*Thorne, Fred T.	*California*
Tietz, Harrison M. (d. 1964)	*Texas*
*Tilden, James W.	*California*
*Towers, Abner A.	*Georgia*
Travassos, Lauro P. (d. 1970)	*BRASIL*
Travis, Dorothy	*Massachusetts*
Troyer, Sprague	*(Ontario) CANADA*
Varnier, M.	*FRANCE*
Vogel, Harold A.	*Michigan*
*Voss, Edward G.	*Michigan*
Weber, Bernie H.	*California*
Werner, Floyd G.	*Massachusetts*
*Whittaker, Robert H.	*New York*
Wilhelm, Herman P. (d. 1970)	*Connecticut*
Williams, Joseph L. (d.)	*District of Columbia*
*Wilson, Kent H.	*Oklahoma*
Wind, Robert G. (d. 1975)	*(Chiapas) MEXICO*
Wood, Carroll E., Jr.	*Virginia*
Woodcock, Harold E. (d.)	*Illinois*
Woodcock, Joseph M.	*New York*
Wyatt, Alex K. (d. 1971)	*Illinois*
Young, Frank N.	*Indiana*
*Zappalorti, Michael A.	*New York*
*Ziemer, Sylvester E.	*Wisconsin*

INSTITUTIONAL SUBSCRIBERS IN 1947

American Entomological Society	*Pennsylvania*
Bio-metal Associates	*California*
Butterfly Store	*New York*
Oregon State College Library	*Oregon*
University of Kansas Library	*Kansas*

Obituaries/Biographies/Bibliographies

Listed here are all lepidopterists for whom an obituary notice has appeared in the Society's periodicals. In some instances only the announcement of death was given; in others, a complete biography, portrait and bibliography. Available data, with appropriate reference citations, are shown following each name.

ABBOT, John (1751-183?), portrait 2:28, 43, 108; 4:25.
ARNHOLD, Francis R. (1904-1959), 15:124.
AVINOFF, Andrey (1884-1949), 3:11, 77, 114, portrait 4:7.
BARNES, William (1860-1930), portrait 3:53, 17:117.
BARON, Oscar T. (1847-1926), including travels 19:35.
BAYER, Lutz (d. 1973), 3:6/73.
BELL, Ernest L. (1876-1964), 19:190.
BERRY, Dean F. (d. 1949), 4:31, 36.
BEURET, Henry (1901-1961?), 5&6:2/61.
BÖRNER, Carl J. B. (1880-1953), 2:61, 3:11, 8:30.
BOSCH, C. (d. 1948), 2:49.
BOWER, Harold M. (1888-1963), 18:192.
BRIDWELL, Louis H. (d. 1951), 5:126.
BRISTOL, Maurice L. (1890-1975), 30:152.
BROOKS, Cecil J. (1875-1953), 9:22.
BROOKS, George S. (1872-1947), 2:1.
BRYK, Felix (1882-1957), 11:55.
BUCHHOLZ, Otto (1874-1958), 13:27.
BURDICK, William N. (1887-1960), 5&6:2/61, 16:246.
BURK, Jakob (1899-1962?), 1:1/62.
CARPENTER, G. D. Hale (1882-1953), 7:64, portrait, bibliography 8:31.
CARR, F. M. B. (d. 1967?), 3:3/67.
CARY, Margaret (1882-1969), 2:6/69.
CASSELBERRY, Raymond C. (1900-1954), 8:30.
CHERMOCK, Franklin S. (d. 1967?), 1:1/68.
CLARK, Austin H. (1880-1954), 8:168, portrait, bibliography 9:151.
CLARK, Gowan C. (1888-1964), portrait 18:195.
COCKERELL, Theodore D. A. (1866-1948), portrait 2:14.
COLBERT, A. Steven (d. 1948) 2:79.
COMSTOCK, John A. (1883-1970), portrait 25:215.
COMSTOCK, William P. (1880-1956), 13:30.
COOK, John H. (1878-1958), 12:207.
COOK, William C. (1895-1967), portrait 21:279.
COWPER, Denis (d. 1974), 6:1/75, 1:1/76.
CRETSCHMAR, Max (1901-1961), 1:1/62.

D'ALMEIDA, Romualdo F. (1891-1969), 1:4/71, portrait 28:293; bibliography 29:40.
DANILEVSKI, Alexandr S. (1911-1969), portrait 24:70.
DANNENBERG, K. (d. 1948), 2:61.
DAVIDSON, William M. (1887-1961), 1:1/62, 16:250, 17:117.
DE JOANNIS, Joseph R. P. (1854-1932), portrait 3:77.
DE LA TORRE BUENO, Jose R. (1872-1948), 2:42.
DE LATTIN, Gustaf (d. 1968), 2:6/69.
DE LESSE, Hubert (1914-1972), 5:3/72, bibliography 26:268.
DERENNE, F. (d. 1948), 2:80.
DJAKONOV, Aleksandr M. (1888-1956), 11:55.
DOETS, Cornelis (1894-1952), 7:24.
DOMINICK, Richard B. (1919-1976), 3:4/75.
DORANT, Georges (d. 1948), 2:80.
DUERCK, H. (d. 1948), 2:49.
DUPONT, F. (d. 1942), 2:79.
DYAR, Harrison G. (1866-1929), 1:30.
EDWARDS, Henry (1830-1891), portrait 2:7.
EDWARDS, William H. (1822-1909), 1:8, portrait 2:54, home 3:61.
EGGERS, F. (d. 1948), 2:61.
EUSTIS, Henry W. (1877-1951), 6:77.
EVANS, William H. (1876-1956), 10:101, portrait, bibliography 10:193.
FARQUHAR, Donald W. (d. 1963?), 5:2/63.
FATTIG, Perry W. (1881-1953), 8:30.
FEDERLEY, Harry (1879-1951), bibliography 5:126, 6:57.
FORSYTH, Marguerite S. (1889-1952), 5:126, 6:76.
FOX, Richard M. (1911-1968), bibliography 22:192.
FRENCH, George H. (1841-1935), portrait 2:40.
FUKUSHIMA, I. (d. 1948?), 2:95.
GAEDE, Max (d. 1948), 2:61.
GEROULD, John H. (1867-1961), 5&6:1/61, 12:190.
GLASGOW, Clyde L. (1870-1951), 4:110.
GLOOR, Eugene E. (d. 1949), 3:64, 72.
GLOVER, Townend (1833-1883), portrait 1:1/71.
GOTLIEB, Stanley R. (1937-1971?), 1:5/72.
GRAY, Percy H. B. (1891-1971), bibliography 6:2/71, 26:66.
GROTE, Augustus R. (1841-1903), portrait 2:17.
GUNDER, Jeane D. (1888-1948), 2:105.
HAIG-THOMAS, Peter (d. 1959?), 2:1/59.
HAYWARD, Kenneth J. (1891-1972), 5:3/72, portrait 28:79.
HALL, Arthur D. (d. 1949), 3:114.
HEMMING, Arthur F. (1893-1964), 18:237.
HERING, Erich M. (1893-1967), portrait 22:123.
HESSEL, Sidney A. (1907-1974), 5:9/74.

HEWES, Lawrence I. (d. 1950), 4:48.
HOLLAND, William J. (1848-1932), 1:98.
HOPFINGER, John C. (1888-1961), portrait, bibliography 5&6:2/61, 16:147.
HOWARD, Leland O. (1858-1950), 4:69.
HÜBNER, Jacob (1761-1826), portrait, bibliography 3:107.
HULBIRT, Edwin Ray (1886-1965), 19:243, 29:130.
HULST, George D. (1846-1900), portrait 2:66, 8:46.
IMMS, Augustus D. (d. 1949), 3:42.
INGHAM, Charles H. (1904-1957), portrait 11:169.
IWASE, Taro (d. 1970?), 3:1/70.
JACKSON, Thomas H. E. (1903-1968), bibliography 23:131.
JACOBSON, Edward R. J. (d. 1945), 2:79.
JANSE, Antonie J. T. (1877-1970), portrait 25:211.
JELINEK, Anton (d. 1957?), 11:96.
JOHNSON, Frank (1878-1949), 4:30.
JOHNSTON, Edward C. (1888-1951), 5:16, 66.
JONES, Frank Morton (1868-1962?), 7:2/62, 12:190.
JORDAN, H. E. Karl (1862-1959), 12:198, 2:1/59.
KALIS, J. P. A. (1899-1949), 4:48.
KAUTZ, Hans (1870-1954), 9:157.
KAY, S. W. (d.1963?), 5:2/63.
KEITH, Edward D. (d. 1949), 3:42, 60.
KILMAN, Leroy N. (1874-1954), 8:30.
KIRKWOOD, Carl W. (d. 1972), 2:16/72, 29:130.
KOEHLER, Wilhelm (d. 1948), 2:49.
KOVACS, L. (d. 1972?), 3:3/72.
KOZHANCHIKOFF, N. B. (d. 1958), 12:160.
KREMKY, Jerzy (d. 1948), 2:49, 3:4.
KUSNEZOV, Nicholas J. (1873-1948), 2:110, portrait, bibliography 3:29.
LAMEERE, Auguste (1864-1948?), 2:80.
LE CERF, F. (d. 1944), 2:80.
LECLERC, ? (General) (d. 1948), 2:80.
LEDERER, Gustav (1892-1962), 4:1/62.
LE MAGNAN, ? ? (d. 1948), 2:80.
LE MARCHAND, S. (1876-1953), 7:166.
LEMMON, Helen Lee ((1916-1967), 22:196.
LENGYEL, Julius F. (d. 1972?), 3:3/72.
LHOMME, Léon (d. 1949), 3:38, 60, 4:9.
LINDSEY, Arthur W. (1894-1963), portrait, bibliography 6:1/59, 17:181.
LOELIGER, Robert (1886-1952), 6:78, 7:23.
MAC DONALD, J. J. (d. 1972), 5:3/72.
MASLOWSKI, Ledwig (d. 1948?), 3:4.
MASLOWSKI, Marjane (d. 1948?), 3:4.
MC DUNNOUGH, James H. (1877-1962), portrait, biblio-

graphy, 16:209, 17:99, 117, 7:1/62.
MC ELHOSE, Arthur L. (1887-1954), 8:30.
MEDER, O. (d. 1948), 2:61.
MEINERS, Edwin P. (1893-1960), portrait, bibliography 16:71.
MELANDER, Axel L. (1878-1962), 7:1/62.
MELL, Rudolf (d. 1970?), 7:2/70.
MINOT, George R. (1886-1950), 4:9, 24.
MOECK, Arthur H. (d. 1972?), 3:3/72.
MOSHER, Edna (1879-1972?), 3:3/72.
MOUCHA, Josef (d. 1973), 3:6/73.
MÜLLER-STÜSSI, Hans (1916-1962?), 1:1/62
MURTFELDT, Mary Esther (1839-1913), portrait 2:83.
MUSPRATT, Vera M. (d. 1963?), 5:2/63.
OBRAZTSOV, Nicholas S. (1906-1966), portrait, bibliography 20:255.
O'BYRNE, Harold I. (1898-1951), 4:69, 110, portrait, bibliography 5:11.
OLTHOF, J. (d. 1945), 2:79.
OVERLAET, F. G. (d. 1958), 12:160.
PACKARD, Alpheus S., Jr. (1839-1905), 1:39, portrait 2:54.
PEET, Max Minor (d. 1949), 3:35.
PENDLEBURY, H. M. (d. 1945), 2:79.
PETERSON, Alvah (1888-1972), 5:3/72, portrait 27:87.
PHILLIPPS, ? ? (d. 1948), 2:49.
PHILLIPS, Leonard S. (1908-1968), bibliography 22:209.
PLATT, Ernest E. (1874-1966), portrait 23:135.
POITRAS, David (d. 1971), 2:4/71.
PRAVIEL, G. (d. 1944), 2:80.
PRONIN, Georg (1898-1962), bibliography 21:74.
PRZEGENDZA, ? ? (d. 1948), 2:49.
QUILLIN, Ellen S. (1888-1970), 5:6/70.
RAU, Phil (1885-1948), 2:62, 76.
REAKIRT, Tryon (1844- ?), 18:211.
REMINGTON, P. Sheldon (d. 1975), 4:6/75.
RILEY, Charles V. (1843-1895), 1:56, portrait 2:54.
RIMSKY-KORSAKOFF, M. N. (d. 1951), 5:40.
RÖBER, J. (d. 1948), 2:49.
ROEPKE, Walter K. J. (1882-1961), portrait 14:241.
ROGERS, William P. (1887-1953), 7:64, portrait 8:44.
ROMANISZYN, Jan (d. 1948?), 3:4.
ROMIEUX, Jean (d. 1951), 5:52.
ROMINE, Ray (1910-1954), 9:22.
ROSIER, J. P. (d. 1943), 2:79.
SCHAWERDA, ? ? (d. 1948), 2:49.
SCHROETER, Otto H. (d. 1957?), 11:96.
SCHRYVER, Clark D. (1883-1959), 7:1/59, 1:1/60.

SCHULTZE, Willy (d. 1948), 2:49.
SCHWANWITSCH, Boris N. (d. 1958), 12:160.
SCOTT, Arthur H. (d. 1957?), 11:96.
SCOTT, R. S. (d. 1972), 5:3/72.
SCUDDER, Samuel H. (1837-1911), 1:17, portrait 2:54.
SETTE, Oscar E. (d. 1972), 5:3/72, 29:130.
SHIBUYA, J. (d. 1948?), 2:95.
SILVESTRI, Filippo (d. 1949), 3:77.
SKINNER, Henry (1861-1926), 1:69.
SMITH, Claude I. (d. 1949), 3:114.
SMITH, John B. (1858-1912), 1:81, 8:46.
SNODGRASS, Robert E. (1875-1962), 7:1/62.
SPERRY, Grace H. (d. 1950), 4:69.
SPERRY, John L. (1894-1954), 8:30.
STAUFFACHER, Emil W. (Mrs.) (d. 1962?), 5:3/62.
STEINHOFF, Gordon R. (d. 1963?), 5:2/63.
STINSON, Walter C. (1898-1958), 6:1/59, portrait 15:121.
STRANAHAN, L. H. (d. 1963?), 2:5/64.
STRECKER, Ferdinand H. H. (1836-1901), portrait 2:98.
STUGLIK, Zdzislaw (d. 1948?), 3:4.
SÜFFERT, ? ? (d. 1948), 2:61.
SWEADNER, Walter R. (1903-1951), 4:69, 110, 5:40, portrait, bibliography 5:57.
SYMMES, John C. (d. 1973), 3:6/73.
TALBOT, George (1882-1952), 7:24.
THIRKELL, H. G. (d. 1963?), 5:2/63.
THOMANN, H. (1873-1959?), 6:1/59.
TOLL, Sergiusz (1893-1961), portrait 17:47.
TOXOPEUS, Lambertus J. (1894-1951), 5:16, 36.
TURNER, Alfred J. (1861-1947), portrait 2:48, 82.
TURNER, Henry J. (1856-1950), 5:36.
VERITY, Roger R. (d. 1959?), 4:1/59.
VOGT, Charles J. (1885-1964), 18:239.
VON BARTHA, ? ? (Colonel) (d. 1948), 2:49.
VON DER GOLTZ, H. (d. 1948), 2:49.
VON ROSEN, Kurt (d. 1948), 2:49, 61.
VON STERNECK, ? ? (d. 1948), 2:49.
VON STUBENRAUCH, ? ? (d. 1948), 2:49.
WALTON, Munroe L. (1890-1974), 29:130.
WARNECKE, Georg (1883-1962), 1:1/63.
WATARI, Masami (1897-1953), 7:166.
WHEELER, George (1858-1947), 2:67.
WILLIAMS, Evelyn G. (d. 1952?), 5:126, 6:78.
WILLIAMS, Roswell C. (1869-1946), 1:7.
WIND, Robert G. (1912-1975), portrait, bibliography 30:239.

WYATT, Alex K. (1878-1971), 5:2/71, portrait, bibliography 28:368.
WYATT, Colin (1909-1975), 1:1/76.
ZAWERCIE, ? ? (d. 1948?), 3:4.
ZERNY, H. (d. 1948), 2:49.
ZIKÁN, W. (d. 1949?), 4:24.

Disposition of Collections

The following is a list of the names of lepidopterists whose collections, large or small, have been deposited in a major museum and the fact recorded in one of the Society's periodicals. In each instance the name of the recipient is given, if known. Private or institutional collections of Lepidoptera which have been destroyed or badly damaged for some reason are also given, when this has been recorded in the Society's periodicals.

ARNHOLD, Francis R. - to Peabody Mus., Yale Univ., 5550, including 1450 *Catocala* (15:124).

AVINOFF, Andrey - to Mus. Acad. Sci. Leningrad, 80,000 Palearctic butterflies, including some 20 types and paratypes; Carnegie Mus., 14,000 Jamaican Lepidoptera (4:7).

BARNES, William - to U. S. National Mus., about 465,000 (3:53).

BELL, Ernest L. - to American Mus. N. H., many hesperiids (19:190).

BERRY, Dean F. - held for sale (2:11).

BIRD, Henry - to American Mus. N. H., 1118 including 32 holotypes, 10 allotypes, 63 paratypes; 400 reared parasites (14:82).

BOCK, Theodore - held for sale (4:73).

BOWER, Harold M. - to Los Angeles Co. Mus., about 40,000 (18:192).

BROOKS, Cecil J. - to British Mus. (NH) and Castle Mus., Norwice (9:22).

BROOKS, George S. - to Manitoba Mus. (2:1).

BUCHHOLZ, Otto - to American Mus. N. H., 124,985 including 6 holotypes, 3 allotypes, 551 paratypes (12:208, 13:27).

BURDICK, William N. - to Univ. Colorado Mus., about 20,000 (16:246).

CARR, F. M. B. - put-up for sale (3:3/67).

CHNEOUR, Alexander - to American Mus. N. H., 806 Tunisian macrolepidoptera (21:180).

COOK, John H. - to A. C. Frederick, New York (12:207).

DE JOANNIS, Joseph (R. P.) - to Mus. National, Paris (3:77).

DE LESSE, Hubert - to Entomol. Lab., Mus. National, Paris (26:270).

DOETS, Cornelis - to Zool. Mus., Amsterdam, macrolepidoptera; Leiden Mus., microlepidoptera (7:24).
DOWNEY, John C. - to Allyn Mus. Entomol., Sarasota, Florida, immatures (5:3/71).
DUPONT, F. - to Agricultural Coll., Wageningen, Sphingidae; Buitenzorg Mus., other Lepidoptera (2:79).
EDWARDS, Henry - to American Mus. N. H., including types (2:7).
EDWARDS, William H. - to Carnegie Mus., Rhopalocera; British Mus. (NH), Heterocera (1:17, 24).
EHRLICH, Paul R. - to American Mus. N. H., 2193 including 122 paratypes, 132 genitalic slides of NA *Erebia* (14:82).
FERGUSON, Douglas C. - to U. S. National Mus., more than 51,000 (25:168).
FREDERICK, Albert C. - to American Mus. N. H., 6459 including more than 700 *ex* John H. Cook collection (21:180).
FREEMAN, Hugh A. - to American Mus. N. H., 2353 Megathymidae, including 9 holotypes, 14 allotypes, 1 neallotype, and 638 paratypes (24:208).
GAGARIN, ? ? - purchased by James R. Neidhoefer for Milwaukee Public Mus., Wisconsin, 16,000 Brazilian (6:1/69).
GEROULD, John H. - to Peabody Mus., Yale Univ. (8:173).
GALACZ, Andreas - offered for sale, about 1500 Indian butterflies (7:7/68).
GILLHAM, Nicholas W. - to Peabody Mus., Yale Univ., mainly Hesperiidae (6:2/63).
GROTE, Augustus R. - to American Mus. N. H., most of Grote & Robinson types; personal collection to British Mus. (NH) (2:17).
GUNDER, Jeane D. - to American Mus. N. H., more than 28,000 (2:105).
HAIG-THOMAS, Peter - to British Mus. (NH) (2:1/59).
HARVARD Mus. Comp. Zool. - to Peabody Mus., Yale Univ., about 800 vials of immatures, including about 60 vials by Edwards (8:104).
HERING, Erich M. - to Mus. Zool. Berlin, 2900 reared, including 425 types (22:123).
HOBERT, Kurt - held for sale (4:73).
HOPFINGER, John C. - to U. S. National Mus., about 18,000 (6:1/64).
HOWE, William H. - to American Mus. N. H., 7374 (21:180).
HULBIRT, Edwin Ray - to Citrus Junior Coll., Azusa, California, 18,000 (3:1/63, 5:2/64), 5000 unmounted (19:243); loaned to Los Angeles Co. Mus. (29:130).

HULST, George D. & John B. Smith - to American Mus. N. H. 32,022 including 2200 types, about 1200 holotypes (8:46).

IBARRA, Mariana - to American Mus. N. H., 12,546 (21:180).

INGHAM, Charles H. - to Peabody Mus., Yale Univ. (12:94).

IRWIN, Roderick R. - to Illinois Nat. Hist. Survey, Urbana, 3800, plus 63 *ex* William Henry Edwards collection (25:83, 26:198).

JACKSON, Thomas H. E. - to British Mus. (NH), about 65,000; National Mus. of Kenya, about same number (23:131).

JACOBSON, Edward R. J. - to Leiden Mus. (2:79).

JOHNSON, Frank - to American Mus. N. H. (4:30).

JOHNSTON, Edward C. - to Donald P. Frechin (5:66).

JONES, Frank M. - to Peabody Mus., Yale Univ., major portion, 9200; to U. S. National Mus., Psychidae (8:47).

KALIS, J. P. A. - to Lord Rothschild and British Mus. (NH) (2:79, 4:48).

KIRKWOOD, Carl W. - to Peabody Mus., Yale Univ., part; remainder to Los Angeles Co. Mus. (12:94, 29:130).

KLAGES, ? - to Cornell Univ., including paratypes (6:36).

KLOTS, Alexander B. - to American Mus. N. H., 3468 Pieridae (21:180).

LAUCK, Albert G. - held for sale, about 11,200 (6:55).

LE MARCHAND, S. - Paris Mus., large series French microlepidoptera (7:166).

LE MOULT, Eugene - sold at auction in Paris, many types; part to Allyn Mus. Entomol. (8:4/67, 3:4/68).

LUCHT, H. - destroyed by looters (2:79).

MC ALPINE, Wilbur S. - to U. S. National Mus., 12,000 including 19 types; American Mus. N. H., 5 types; British Mus. (NH) 3 types; Carnegie Mus., 6 types; Los Angeles Co. Mus., 2 types (27:240).

MC DUNNOUGH, James H. - to Canadian Nat. Coll. and Nova Scotia Mus. (16:209).

MC ELHOSE, Arthur L. - to Chicago Nat. Hist. Mus. (8:30).

MC FARLAND, Noel - to Los Angeles Co. Mus., 8420, including 3315 preserved immatures (26:108).

MEINERS, Edwin P. - to Univ. Missouri, 30-40,000 (16:71).

MOECK, Arthur H. - to Milwaukee Public Mus. (5:6/68).

MURTFELDT, Mary E. - to Cornell Univ., microlepidoptera (2:83).

MUSEUM COLLECTIONS DESTROYED OR BADLY DAMAGED:

Abbeville, France, destroyed WW-II (2:80).

Augusburg, Germany, destroyed WW-II (2:49).

College of Agriculture, Laguna, Phillippines, destroyed WW-II (2:61).

Dresden, Germany, partially lost WW-II (2:49).
Hamburg, Germany, destroyed WW-II (2:49).
Instituto de La Salle, Bogota, Colombia, destroyed by fire (riotous mob) 10 April 1948 (2:61).
Societa Entomologica, Genova, Italy, destroyed WW-II (2:95).
Société Entomologique du Nord de la France, destroyed WW-II (2:80).
Stettin, Germany, destroyed WW-II (2:49).
University of Caen, France, destroyed WW-II (2:80).
Warsaw Zoological (Lepid.), Poland, destroyed WW-II (3:4).
Santa Barbara, California, large coll. of butterflies destroyed by fire (5:1/62).

NEWCOMER, Erval J. - to U. S. National Mus., 3000; remainder to Washington State Univ. (27:240).

O'BYRNE, Harold I. - to Univ. Missouri, about 4000 (4:110, 5:40).

OLTHOF, J. - to Buitenzorg Mus. (2:79).

ORCHARD, Charles D. - to U. S. Fourth Army Medical Lab., Ft. Sam Houston, Texas, about 60,000 (6:3/65). [Including specimens *ex* Samuel C. Thomson and B. Struck collections, personal communication 29 August 1971.]

PEET, Max M. - to Univ. Michigan Mus. (3:35).

PHILLIPS, Leonard S. - to Buena Vista Coll., Storm Lake, Iowa, 5600 (22:209).

PLATT, Ernest E. - to Grahamstown Mus., reared adults; Transvaal Mus., preserved immatures, notebooks, negatives, photos (23:135).

POWELL, Jerry A. - to Univ. California, Berkeley, about 33,200 (15:203).

PRIVATE COLLECTIONS DESTROYED OR BADLY DAMAGED:
Dervaux of Riom, France, lost or heavily damaged WW-II (2:80).
Draudt of Darmstadt, Germany, destroyed WW-II (2:49).
Fremont of Bordeaux, France, lost or heavily damaged WW-II (2:80).
Hajkowicz, M. of Warsaw, Poland, destroyed WW-II (3:4).
Klener, J. of Warsaw, Poland, destroyed WW-II (3:4).
Morault of Nantes, France, lost or heavily damaged WW-II (2:80).
Nieniewski, C. of Warsaw, Poland, destroyed WW-II (3:4).
Pfaff of Frankfurt-am-Main, Germany, destroyed WW-II (2:61).
Phillips of Cologne, Germany, destroyed WW-II (2:49).
Prack of Frankfurt, Germany, badly damaged WW-II (2:49).
Teplicki, S. of Warsaw, Poland, destroyed WW-II (3:4).

Till of Frankfurt-am-Main, Germany, destroyed WW-II (2:61).
Urbahn of Stettin, Germany, destroyed WW-II (2:61).
Von Frohrich of Aachen, Germany, destroyed WW-II (2:61).

REAKIRT, Tryon - to Chicago Mus., incorporated with Strecker's (18:211).

RILEY, Charles V. - to U. S. National Mus. (1:56).

RINDGE, Frederick H. - to American Mus. N. H., 97,199 (21:180).

ROGERS, William P. - to Fall River Pub. Library, Massachusetts (8:44).

ROSIER, J. P. - to Buitenzorg Mus. (2:79).

ROUSSEAU-DECELLE, ? ? - sold at auction in Paris, large number of types and paratypes (4:3/66).

SEITZ-MARTEN EXPEDITION - sold, 30,000 (6:87).

SETTE, Oscar E. - to Los Angeles Co. Mus. (5:3/72, 29:130)

SHAPIRO, Arthur M. - to Cornell Univ., 13,000 Rhopalocera (25:82); Univ. California, Davis, 22,000 (6:4/74).

SMITH, John B. & George D. Hulst - to American Mus. N. H. 32,022 including 2200 types, about 1200 holotypes (8:46).

SPERRY, John L. - to American Mus. N. H. (8:30).

STINSON, Walter C. - to Michigan State Univ., East Lansing, 7000 (6:1/59, 15:121).

STRANAHAN, L. H. - sold, about 300 (3:3/64).

STRECKER, F. H. Herman - to Chicago Mus. (2:98).

STRUCK, B. - (see ORCHARD, Charles D., above).

THOMSON, Samuel C. - (see ORCHARD, Charles D., above).

TOLL, Serguisz - to Rostov Municipal Mus., 40,000; Warsaw Mus., 100,000 (17:47).

TURNER, Alfred J. - to Mus., Div. Economic Entomol., Canberra, Australia (2:82).

VALLINS, F. T. - to U. S. National Mus., 22,000 Palaearctic Lycaenidae (27:240).

VAN SOMEREN, V. G. L. - to American Mus. N. H., 22,931 East African & Ethiopian; British Mus. (NH), types and about 300,000 (2:3/59, 2:2/60, 25:143).

WALTON, Munroe L. - to Los Angeles Co. Mus. (29:130).

WIELGUS, Ronald S. - to Arizona State Univ., Tempe (8:5/67).

WILLIAMS, Roswell C. - to Philadelphia Acad. Nat. Sci. (1:7).

BIOGRAPHICAL SKETCHES

Brief biographical data are given for all individuals who served the Society in an official capacity during its first quarter century, to the extent that such data were provided the compiler. The format used in preparing the biographical sketches follows: Name (in alphabetical sequence); title; brief address (complete address may be found in the Society's current Membership List); major lepidopterological interests; date and place born; education (dates and institutions from which academic degrees received); marital status (name of spouse and date married); names of children; service to the Lepidopterists' Society (offices held and dates); other memberships and affiliations with other scientific organizations (positions held, experience, etc.); honors or awards received; patronyms; and publications (books, monographs, periodicals to which contributed).

Abbreviations shown in BioSciences Information Service, List of Serials, 1971, have been used when citing periodicals in which papers have been published. All serials in which a given biographee has published are not listed in each case; but an attempt has been made to cite those containing the greatest number or the most significant papers. Complete lists of publications furnished by participating biographees will be filed for future reference. Authors and biographees are encouraged to provide the Society Librarian with periodical up-dates of their publications, including copies of reprints.

Some additional specific abbreviations used in these biographical sketches are:

aptd = appointed
assoc = associate or association
asst = assistant
auth = author
biblio = bibliography
biog = biography
biol = biology
ch = child or children
chm = chairman
com = committee

contrib	=	contributor
corr	=	corresponding
d	=	deceased
dt	=	daughter
ed	=	editor or editorial
educ	=	education
ex	=	executive
exp	=	experience
exped	=	expedition(s)
hon	=	honors or honorary
inv	=	invitational
ling	=	linguist
M	=	married
M-a-l	=	member at large
mem	=	member
mil svc	=	military service
monog	=	monograph(s)
obit	=	obituary
orgin	=	organization
pres	=	president
proc	=	proceedings
res	=	research
ret	=	retired
s	=	son
sci	=	science or scientist
sec	=	secretary
sust	=	sustaining
treas	=	treasurer
V Pres	=	vice presidnet
zool	=	zoology (zoological)

AE, Shigeru Albert (Ph.D.), Prof. Biology, Nansan University, Showaku, Nagoya, JAPAN. Genetics, ecology, evolution. 20 February 1927. EDUC.: Graduate, Zoology Inst., Univ. Kyoto, JAPAN, 1952; Ph.D. (biol.) Univ. Notre Dame, USA, 1957; D.Sc., Univ. Kyoto, 1961. LEPID. SOC.: M-a-l Ex. Council, 1963-65.

AGENJO, Ramón (Director) Instituto Español Entomología, J. Gutiérrez Abascal 2, Madrid 6, SPAIN. Lepidoptera of Spain. 21 January 1908, Santander, SPAIN. EDUC.: Licentiate in Laws, Univ. Madrid, 1931. M., one s. LEPID. SOC.: V. Pres., 1965. OTHER: Real Sociedad Española de Hist. Nat., gen. sec., 1939. EOS, Redaction Council, 1941-; GRAELLSIA, co-founder, 1943; EOS and GRAELLSIA, Director of reviews, 1967-. HON.: Alonso de Herrera Award, 1950; Academician, Seven-hundredth Acad. for Agriculture, Turin, 1970. Hon. Mem., Finish Entomol. Soc. AUTH.: Fáunla Lepidopterológica Almeriense, 1952; Catálogo Ordenador de los Lepidópteros de España, 1946-66. MONOG.: La Familia Lymantriidae, 1957. CONTRIB.: About 200 papers and 100 reviews appearing in Bol R Soc Esp Hist Nat Secc Biol; EOS Rev Esp Entomol; Zeit Wien Entomol Ver; Graellsia; Arbor Kornickie; Broteria Ser Cienc Nat; Bol Serv Plagas Forest; Proc (VI, VIII, IX, XI, & XII) Int Congr Zool; and others.

ALBERTI, Burchard (Dr.), 34 Gottingen, GERMANY. Phylogenetics. 21 January 1898, Berlin, GERMANY. EDUC.: Degree in Chem., Berlin, 1919; Dr., ibid., 1924. LEPID. SOC.: V. Pres., 1967. OTHER: Several European sci. soc. and organizations; asst. to Prof. Erich M. Hering, Zool. Mus., Humboldt Univ., Berlin, 1952-65. CONTRIB.: About 125 papers papers published in Bull Soc Entomol Mulhouse; Dtsch Entomol Z; Entomol Z; Rev Zool Bot Afr; Opusc Zool (Budap); Entomophaga; Beitr Entomol; Redia; and others.

AMSEL, Hans Georg (Ph.D.), 75 Karlsruhe-Durlach, Kastell St., 13, GERMANY. Systematics (palaearctic microlepidoptera). 29 March 1905. EDUC.: Ph.D., Berlin, 1933. LEPID. SOC.: V. Pres., 1964. OTHER: Finish Entomol. Soc., Correspond. Mem., 1968. EXPED.: Palestine, Afghanistan, Sardinia, Dalmatia, Iran. AUTH.: Microlepidoptera Palaearctica, 4 vols., first vol. 1965. CONTRIB.: About 120 papers, predominantly on palaearctic microlepidoptera.

BELLINGER, Peter F. (Ph.D.), Biology Dept., California State Univ.-Northridge, Northridge, Calif., USA. Collembola. 15 June 1921, New Haven, Conn. EDUC.: B.A., Yale Univ., 1942; Ph.D., ibid., 1952. M. Carolyn Jones, 11 April 1953. CH.: Christina; Frederick. LEPID. SOC.: Asst. Ed. *News* and *Journal*, 1951-69; Librarian, 1958- . OTHER: Amer. Assoc. Adv. Sci.; Entomol. Soc. Amer. CONTRIB.: Bull Conn Agr Expt Station; Bull Soc Zool Fr; Entomol News; J NY Entomol Soc; Lepid News; Nat Hist Soc Jamaica; Psyche; and others.

BIEZANKO, Ceslau Mário (Prof. Dr.), Pelotas, Rio Grande do Sul, BRAZIL. Lepidoptera of Brazil. 22 September 1895, Kielce, POLAND. EDUC.: Agr. Eng., University Poznañ, POLAND; ibid., four more years specializing in Natural Sci. and Chemistry. M. Joan M. Lúcy Mantovani, 1928, Beaulieu, France. EXP.: Prof. Chem. & Biology, Army Sch. Officers, Bydgoszcz, Poland. Taught Chem. & Nat. Hist., Lemos High Sch., Rio Grande, R. S., Brazil; also, taught at Pelotense Coll. and Santa Margarida High Sch., Pelotas, Brazil. Prof. (Entomol.) Eliseu Maciel Sch. Agr., and Sch. Pharmacy (Botany), 1936- . LING.: Portuguese, French, Spanish, Italian, German, Polish, Russian, and well versed in Latin and Greek. LEPID. SOC.: V. Pres., 1963. OTHER: Affiliated with entomol. soc. Poland, Spain, Brazil, Argentina, and France. PATRONYMS: *Biezankoia* Strand, 1936, *Plusia (Euchalcia) biezankoi* Alberti, 1965, *Zygaena cynarae biezankoi* Alberti, 1968, *Polyortha biezankoi* V.O. Becker, 1970, *Euphyes subferruginea biezankoi* Mielke, 1972 (Lepidoptera); *Apanteles biezankoi* Blanchard, 1960, *Ephuta*

biezankoi Casal, 1965 (Hymenoptera); *Willistonia biezankoi* Blanchard, 1960, *Archytas biezankoi* Guimarães, 1961, *Lastauropsis biezankoi* Carrera & Papavero, 1962 (Diptera); *Theridion biezankoi* Levi, 1963 (Araneae). CONTRIB.: Many papers (about 300) published in Agros (Pelotas); Arq Entomol Ser A & B; Escola de Agronomia "Eliseu Maciel" Bol; Rev Fac Agr; and others.

BLANCHARD, André, V. Pres. (ret.) Res. & Development, Schlumberger Well Services, Houston, Texas, USA. Taxonomy, distribution, Heterocera of Texas. 11 June 1896, Marennes, FRANCE. EDUC.: Grad. Ecole Navale (French Naval Acad.), 1917; Engineer, Ecole Supérieure de Radiotélégraphie, 1922. M. Marguerite Normand, 22 April 1921 (deceased, 1958). CH.: Simone; Michelle. 2nd M. May Elise Moraud, 9 May 1959. HON.: Officer de la Legion d'Honneur; Croix de Guerre, 1917. LEPID. SOC.: Sustaining Mem.; M-a-l Ex. Council, 1971-73; Pres., 1975. OTHER: Lepid. Found.; Entomol. Soc. Washington. PATRONYMS: *Basilodes blanchardi* Hogue, 1965; *Grotella blanchardi* McElvare, 1966; *Stenoporpia blanchardi* Rindge, 1968; *Opsigalea blanchardi* Todd, 1966; *Scoparia blanchardi* Munroe, 1972; *Sphingicampa blanchardi* Ferguson, 1972. CONTRIB.: J Lepid Soc; Proc Entomol Soc Washington.

BRAUN, Annette Frances (Ph.D.), 5956 Salem Road, Mt. Washington, Cincinnati, Ohio, USA. Microlepidoptera. 28 August 1884, Cincinnati, Ohio. EDUC.: B.A., Univ. Cincinnati, 1906; M.A., ibid., 1908; Ph.D., ibid., 1911. LEPID. SOC.: Charter Mem.; V. Pres., 1954; Hon. Life, 1960. OTHER: Amer. Assoc. Adv. Sci., Fellow; Cincinnati Mus. Nat. Hist., Life Trustee; Entomol. Soc. Amer., Past Charter Mem.; Ohio Acad. Sci., Fellow; Entomol. Soc. Canada; Sigma Xi. CONTRIB.: About 70 papers, mostly on microlepidoptera, published in Ann Entomol Soc Amer; Can Entomol; Cincinnati Soc Nat Hist; Entomol News; Ohio J Sci; Trans Amer Entomol Soc; Proc US Nat Mus; Pan-Pac Entomol; J Acad Nat Sci Phila; Proc Calif Acad Sci; Amer Entomol Soc Memoirs 13, 18 & 28; Lepid News.

BREYER, Don Alberto (1890-1963), (obit. and portrait, *Rev. Soc. Entomol. Arg.* 26:1-4; biblio., *Argentine Entomol. Biblio.*). Lepidoptera of Argentina (geographic variation and distribution). 24 March 1890, Buenos Aires, d. 12 February 1963, ibid. EDUC.: Elem. and second. studies, Colegio Goethe. LEPID. SOC.: V. Pres., 1955. OTHER: Entomol. Soc. Argentina, Treas., Mem. Advisory Bd., Pres., and Hon. Pres.; Roy. Entomol. Soc., Fellow; Argentine Soc. Geographic Studies; Ornithological Soc. La Plata; Assn. Friends Nat. Parks; Linnean Soc. Lyon; Entomol. Soc. Brazil, Chile, Uruguay, and Munich. CONTRIB.: About 40 papers, cited in biblio noted above.

BROWER, Auburn Edmond (Ph.D.), Augusta, Maine, USA.

Taxonomy, life histories (especially *Catocala* and Aegeriidae). 22 May 1898, Springfield, Missouri. EDUC.: B.S., Southwest Missouri State Coll., 1927; A.B., ibid., 1927; Ph.D., Cornell Univ. 1932. M. Lurana Cole VanDoren, 22 August 1932. CH.: Edmond Dehm; John Harold; Lurana Katherine. LEPID. SOC.: Charter Mem.; M-a-l Ex. Council, 1954-55, 1969-71. OTHER: Entomol. Soc. Amer.; Entomol. Soc. Can.; Amer. Ornith. Union. PATRONYMS: *Graphium marcellus* ab."broweri" Gunder, 1927; *Crambus browerellus* Klots, 1942; *Nymphula broweri* Heinrich, 1940; *Manhatta broweri* Heinrich, 1956; *Catocala connubialis* f. "broweri" Muller, 1960. CONTRIB.: Many papers on Lepid. and other insects published in Entomol News; Bull Brooklyn Entomol Soc; Lepid News; J NY Entomol Soc; J Lepid Soc; Trans Amer Entomol Soc.

BROWN, F. Martin (Sc.D.), Colorado Springs, Colorado, USA. LEPID. SOC.: Charter Mem.; Asst. Ed., 1948-59; M-a-l Ex. Council, 1960-62; Pres., 1968; Obit. Ed., 1972- ; Hon. Life, 1973.

BROWN, Keith S., Jr. (Ph.D.), Prof. Post-Graduate Courses, Biochemistry, Centro de Pesquisas de Produtos Naturais, Universidade Federal do Rio de Janeiro, BRAZIL. Insect chemistry, variation, distribution and behavior studies. 1 October 1938, Chicago, Illinois. EDUC.: B.S. (Chem.) California Inst. Tech., 1959; Ph.D., (Organic Chem./Biochem./ Pharmaceutical Chem.) Univ. Wisconsin, 1962. M. Kay Fowell, 23 June 1962. CH.: Eric Nelson, Julia McLennan,

George Gardner, Elizabeth Fessenden. LEPID. SOC.: Ed. Com. (*Journal*), 1972- ; V. Pres., 1972.

OTHER: Amer. Chem. Soc.; Chem. Soc. (London); Amer. Assoc. Adv. Sci.; Soc. Tropical Biol.; Soc. Systematic Zool.; Soc. Brasileira Entomol.; Sigma Xi; Tau Beta Pi. Pesquisador-Conferencista, Conselho Nacional de Pesquisas, Brazil. Res. Assoc., Stanford Univ., Calif. PATRONYMS: *Sarmientoia browni* Mielke. CONTRIB.: More than 40 papers, including 60 photographs and legends *in* Brazilian Insects and their Surroundings, Koyo Shoin Co., Ltd., Tokyo, 1971. Other papers published in Arch Biochem Biophys; J Chromatogr; J Amer Chem Soc; Tetrahedron Letters; J Lepid Soc; Syst Zool; An Acad Bras Cienc; Rev Bras Biol; Entomol News; Zoologica; and others.

BURNS, John M. (Ph.D.), Assoc. Curator Lepidoptera, Dept.

of Entomology, United States National Museum, Smithsonian Institution, Washington, D. C., USA. Evolutionary biology (diurnal Lepid.). 6 June 1932, Rochester, New York. EDUC.: A.B. (Biol.), Johns Hopkins Univ., 1954; M. A. (Zool.), Univ. Calif., Berkeley, 1957; Ph.D., ibid., 1961. M. Sarah F. Nance, 1954. CH.: Laura; Karen; Colin. LEPID. SOC.: Nom. Com., 1966 (chm.); M-a-l Ex. Council, 1972-74; Ed. Com. (*Journal*), 1972- . OTHER: Amer. Assoc. Adv. Sci.; Amer. Soc. Naturalists; Cambridge Entomol. Club; Genetics Soc. Amer.; Lepid. Foundation; Phi Beta Kappa; Sigma Xi; Soc. Study Evol.; Soc. Systematic Zool. CONTRIB.: About 25 papers appearing in Can Entomol; Entomol News; Isozyme Bull; J Lepid Soc; Proc Nat Acad Sci; Psyche J Entomol; Quarterly Rev Biol; Science; Sci Monthly; Evolution; Syst Zool; Univ Calif Publ Entomol.

CLARKE, Cyril Astley (Res. Fellow), Nuffield Unit of Medical Genetics, Univ. Liverpool, and President of the Royal College of Physicians of London, ENGLAND. Hybridization (*Papilio*). 22 August 1907. EDUC.: M.D., Univ. Cambridge, 1937; Sc.D., ibid., 1963. HON.: Fellow, Roy. Coll. Phys., 1949; President, Roy. Coll. Phys., 1972; Cmdr., Order Brit. Empire, 1969; Fellow, Roy. Sci. Soc., 1970;

Fellow, Roy. Coll. Obst. and Gynec., 1970; D.Sc., Edinburg Univ., 1971, Leicester Univ. 1971, and Univ. East Anglia, 1973. M. Frieda (Féo) Margaret Mary Hart, 1935. CH.: three s. LEPID. SOC.: M-a-l Ex. Council, 1959-61; V. Pres., 1966. CONTRIB.: Many papers, about 80 on Lepidoptera, published in Ecology; Entomologist; Entomol Rec & J Var; Entomol Z; Evolution; Heredity; J Genet; J Lepid Soc; Lepid News; Nature (Lond); Proc Int Cong Genet (Montreal); Proc Roy Entomol Soc Lond; Proc Roy Soc Med; Philos Trans Roy Soc Lond; and others.

CLENCH, Harry Kendon, Section of Insects & Spiders, Carnegie Museum of Natural History, Pittsburgh, Pennsylvania, USA. Taxonomy (Lycaenidae). 12 August 1925, Ann Arbor, Michigan. EDUC.: B.S. (Zool.) and M.S. (Zool.) Univ. Michigan 1950, 1951. M., one dt. LEPID. SOC.: Co-founder; Co-ed., 1947; Orgn. Com. 1950; Ed. Com (*Journal*) 1955-63; M-a-l Ex. Council, 1968-70; Pres., 1974. OTHER: Amer. Assoc. Adv. Sci.; Lepid. Res. Foundation; Soc. Syst. Zool.; Ecol. Soc. Amer.; Amer. Soc. Naturalists. AUTHOR: Theclini and Lycaenini *in* How to Know the Butterflies, W.C. Brown, Dubuque, Iowa, 1961. Lepidoptera *in* The Encyclopedia of the Biological Sciences (Ed.: Peter Gray) Reinhold, New York, 1961. Lycaenoidea *in* Butterflies of Liberia, Entomol. Soc. Amer., Memoir 19, 1965. CONTRIB.: Over 100 papers, appearing in Ann Carnegie Mus; Ann Mag Nat Hist; Bull Brooklyn Entomol Soc; Bull Mus Comp Zool; Can Entomol; Ecology; Entomol News; J Lepid Soc; J NY Entomol Soc; J Res Lepid; Lepid News; Mem Soc Cuban Hist Soc; Mitt Munchen Entomol Ges; Pan-Pac Entomol; Neotropica; Psyche J Entomol; Rev francaise d' Entomol; Rev Zool Bot Afr; Tijd v Entomol; Torreia; Veröff Zool Staatssaml München; Vidensk Medd Dansk Nat Foren; and others.

COMMON, Ian Francis Bell (Dr.), Division of Entomology, CSIRO, Canberra City, AUSTRALIA. Taxonomy; biology. 23 June 1917, Alford, Lincolnshire, ENGLAND. EDUC.: B.A. (1st Hons.) Univ. Queensland, 1941; B. Agr. Sc., ibid., 1945; M.A., ibid., 1946; B. Agr. Sc. (1st Hons.), ibid.,

1947; M. Agr. Sc., ibid., 1953; D. Agr. Sc., ibid., 1969. M. 1946. CH.: Frances Margaret (Mrs. V. Fifield); Jennifer Helen. LEPID. SOC.: V. Pres., 1957; 1st V. Pres., 1965. OTHER: Entomol. Soc. Queensland, Sec., 1939-40; Australian Entomol. Soc., Regional Councilor 1965-69, V. Pres. and Chm. Executive Com., 1969-72; Ecol. Soc. Australia, Councilor; Australian Inst. Agr. Sci.; Linnean Soc. New South Wales; Royal Entomol. Soc. Lond. PATRONYMS: *Heliocoverpa armigera commoni* Hardwick, 1965; *Epermenia (Epermeniola) commonella* Gaedike, 1968; *Commonia* Kiriakoff, 1968. AUTH.: Australian Moths, 1963; Australian Butterflies, 1964; Lepidoptera *in* The Insects of Australia, 1970. CO-AUTH.: Butterflies of Australia, Angus & Robertson (Sydney), 1972. MONOG.: Yponomeutidae (Lepid.) Heard Island, 1970. CONTRIB.: More then 30 papers published in Aust J Biol Sci; Aust J Zool; Aust Nat Hist; Emp Cotton Grow Rev; J Aust Entomol Soc; J Entomol Soc; J Lepid Soc; Nature (Lond); Pac Insects; Proc Linn Soc NSW: Proc Xth Int Congr Entomol; Queensl Agric J; Queensl J Agric Sci; Trans Roy Entomol Soc Lond.

COMSTOCK, John Adams (D.Sc.) (1883-1970). 30 January 1883, Evanston, Illinois, d. 26 December 1970, Del Mar, California. Life histories. EDUC.: M.A. (Hon.), Occidental College, 1924; D.O., Coll. Osteopathic Physicians & Surgeons, 1945; D.Sc. (Hon.) ibid., 1945; M.D., California College Medicine, 1962; D. Sc., ibid, 1962. M. Caroline Townsend, 1905. CH.: John Sterling; Jean; Bette. 2nd M. Ruth Mary Gard, 1928. LEPID. SOC.: Charter Mem.; 1st V. Pres., 1955-56; Pres., 1962; Hon. Life, 1969; Nom. Com., 1957. OTHER: Chicago Entomol. Soc., Sec., 1898-1903; Chicago Acad. Sci., Entomol. Recorder, 1903-04; Lorquin Entomol. Club (now Society), Founding Mem., first Pres., 1917; Royal Entomol. Soc. Lond., Fellow 1919. Southwest Mus. Los Angeles, Asst. Director, 1919; ibid., Director, 1921-26; Southern California Acad. Sci., Sec. 1921; ibid., Ed., *Bulletin*, 1921-62; ibid., Pres., 1926-27; ibid., Fellow, 1929; Los Angeles Co. Mus., Acting & Asst. Director, 1928-38; ibid., Director Sci., 1939-42; ibid., Chief Curator, Sci. Div., 1942-48; ibid., Curator Emeritus, 1948-1970; ibid., Research Associate (Entomol.) 1958-70. HON. MEM.: Pi Gamma Mu; Phi Sigma; Sigma Xi; San Diego Soc. Nat. Hist., President Fellows, 1958; San Diego Mus., Board of Directors,

1957-70; ibid., Hon. Curator (Entomol.), 1969; ibid., mem. Emeritus, 1970. PATRONYMS: *Plebejus shasta comstocki* Fox, 1924; *Polites sabuleti comstocki* Gunder, 1925; *Philotes sonorensis* f. "comstocki" Gunder, 1925; *Limenitis lorquini* ab. "comstocki" (Gunder), 1925; *Speyeria callippe comstocki* (Gunder), 1925; *Agraulis vanillae incarnata* ab. "comstocki" (Gunder), 1925; *Papilio bairdii rudkini* f. "comstocki" Chermock and Chermock, 1937; *Callophrys comstocki* Henne, 1940. AUTH.: Butterflies of California, 1927. CONTRIB.: 220 papers, mostly on life histories of Lepidoptera, published in Ann Inst Biol Mexico; Bull So Calif Acad Sci; Can Entomol; Entomol News; J Lepid Soc; J Res Lepid; Lepid News; Pan-Pac Entomol; Southwest Sci Bull; The Lepidopterist; Trans San Diego Soc Nat Hist.

COVELL, Charles VanOrden, Jr. (Ph.D.), Dept. Biology, University Louisville, Louisville, Kentucky, USA. Taxonomy, life histories (Geometridae). 10 December 1935, Washington, D. C., EDUC.: B.A. (English), Univ. North Carolina, 1958; M. S. (Entomol.), Virginia Polytechnic Inst., 1962; Ph.D. (Entomol.), ibid., 1965. M. Elizabeth Gore Barnes, 23 August 1958. CH.: Charles VanOrden III; Robert Barnes. LEPID. SOC.: Ed. *News*, 1970-73; Ed. Board, 1970-73. OTHER: Alpha Epsilon Delta; Amer. Entomol. Soc.; Amer. Mosq. Control Assn.; Can. Entomol. Soc.; Entomol. Soc. Ontario; Entomol. Soc. Washington; Kentucky Acad. Sci., Ed. Board, 1970-; Kentucky Soc. Nat. Hist., Zool. V. Pres., 1968-70; National Audubon Soc., Louisville Chapter, Pres. 1972-73; Nat. Geographic Soc.; Phi Sigma, Pres. Virginia Polytechnic Inst. chap., 1963-64; Royal Entomol. Soc. Lond., Fellow; Sigma Xi. AWARDS: Sigma Xi, research, 1963; National Science Foundation, systematics, 1969; Danforth Found. Assoc. Program, 1970- . CONTRIB.: About 25 papers published in Ann Entomol Soc Amer; Entomol News; J Lepid Soc; Lepid News; Ky Naturalist; Mosquito News; Trans Amer Entomol Soc; Trans Wisc Acad Sci Arts & Letters; Va J Sci.

DE LA TORRE y CALLEJAS, Salvador Luis (Prof.), Sciences Faculty, Universidad de La Habana, CUBA. Lepidoptera of Cuba; *Trichogramma*. 18 June 1910, Habana, CUBA. EDUC.: Dr. (Nat. Sci.), Univ. Habana. POST GRAD.: Smithsonian Inst. (Lepid. & Mollusks), 1956; Acad. Nat. Sci. Phila., and Amer. Mus. Nat. Hist. (Lepid.), 1957; Prague, Bratislava and

Bruno, Czechoslovakia (Electron Microscopy), 1964; Moscow, Leningrad, Kiev and Cherkasy, Ukraine, USSR (*Trichogramma*); Eberswalde and Berlin, Germany, and Bratislava, Czechoslovakia, 1968-69. AWARDS: First National in Natural History, 1959, for paper "Reconsideracion taxonomica de las especies cubanas del genera *Kricogonia* Reakirt." LEPID. SOC.: M-a-l Ex. Council, 1962-64. CONTRIB.: More than 60 papers published in J Lepid Soc; J NY Entomol Soc; Lepid News; Mem Soc Cuban Hist Nat "Felipe Poey"; Proc Entomol Soc Wash; Rev Inst Matanzas; Rev Univ Oriente; and others.

DE LESSE, Hubert (Dr.), (1914-1972), (obit. portrait and bibliography, *J Lepid. Soc* 26:268-274); Rhopaloceran chromosomes. 7 April 1914, Paris, FRANCE; d. March 1972. LEPID. SOC.: V. Pres., 1961.

DIAKONOFF, Alexey Nikolaevich (Dr.), Rijksmuseum van Natuurlijke Historie, Leiden, NETHERLANDS. Biology; morphology (microlepidoptera). 1 March 1907, Leningrad (St. Petersburg), USSR. EDUC.: D.Sc., Univ. Amsterdam, 1939. M. Christina Aaltina Tonnis, 20 April 1911 (d. 13 July 1946). CH.: Nikolai; Irina. LEPID. SOC.: Adviser to founders, 1947; organized special Lepid. Soc. meeting with IX Int. Congr. Entomol., Amsterdam, 1951; V. Pres., 1952; Pres., 1954, Nom. Com., 1957. OTHER: Netherlands Entomol. Soc., Ed.-in-Chief, Mem. Merit, Life. PATRONYMS: *Bactra diakonoffi* Amsel, 1958; *Eboda diakonoffi* Razowski, 1966. AUTH.: The genera of Indo-australian and Papuan Tortricidae 1939; Tortricidae from Madagascar, 1960. MONOG.: The Microlepidoptera of New Guinea, 1-5, 1952-55; Microlepidoptera of the Philippine Islands, 1968; De Nederlandse Tortriciden, 1968; Revision So. Asiatic Olethreutini (in press). CONTRIB.: More than 190 titles in Ann Soc Entomol Fr; Bull Br Mus (Nat Hist) Entomol; Entomol Ber (Amst); J Lepid Soc; Lepid News; Proc Entomol Soc Wash; Rec Indian Mus; Temminckia; Tijdschr Entomol; Treubia; US Nat Mus Bull; Verhand Zool Mus Leiden; and others.

DOMINICK, Richard B. (M.D.) (1919-1976), Res. Assoc., The Charleston Mus., Charleston, South Carolina, USA. Ecology, taxonomy, life histories. 17 September 1919, New York, New York; d. 4 May 1976, The Wedge Plantation, McClellanville, South Carolina. EDUC.: B.A., Yale Univ., 1942;

M.D., Columbia Univ., 1951. M. Tatiana Djeneeff Waldron, 1966. CH.: Christopher and Elizabeth Blackwell, Oliver and Julia Dominick, Stephanie and Victoria Waldron. LEPID. SOC.: Sustaining Mem.; M-a-l Ex. Council, 1971-73. OTHER: Ed. and Photographer, Moths of America North of Mexico; Entomol. Soc. Amer.; Can. Entomol. Soc.; The Explorers Club, Fellow, V. Pres., 1966-1971. PATRONYMS: *Scoparia dominicki* Munroe, 1972; *Dasychira dominickaria* Ferguson (in press); *Synanthedon dominicki* Duckworth & Eichlin, 1973. CONTRIB.: J Lepid Soc, News Lepid Soc.

DONAHUE, Julian P., Asst. Curator Entomol. (Lepid.), Los Angeles Co. Mus. Nat. Hist., California, USA. Life histories; larval taxonomy. 23 April 1940, College Station, Texas. EDUC.: B.S. (w/hon), Michigan State Univ., 1963; M.S. (Entomol.), ibid., 1965. FELLOWSHIPS: National Sci. Foundation (cooperative grad.), 1963-64; ibid., (grad. teach.), 1965; ibid. (grad.), 1965-68. GRANTS: Sigma Xi Resa Grant-in-Aid of research on Indian Lepid. M., no ch. LEPID. SOC.: M-a-l Ex. Council, 1971-73; Ed. Com. (*Journal*) 1972- ; Nom. Com., 1973 (chm.). OTHER: Bombay Nat. Hist. Soc., Life Mem.; Bengal Nat. Hist. Soc.; Royal Entomol. Soc. Lond., Fellow; Michigan Entomol. Soc., Ex. Sec., 1964-68; Ed. Newsletter, 1964-68; Ed. Bd., 1966- ; Gov. Body, 1964-68, 1970-71; Pres., 1970-71; Entomol. Soc. Amer.; Soc. Systematic Zool.; Amer. Assoc. Adv. Sci.; Sigma Xi. PATRONYMS: *Calephelis perditalis donahuei* McAlpine, 1971. CONTRIB.: J Bengal Nat Hist Soc; J Bombay Nat Hist Soc; J Lepid Soc; Mich Entomol; and others.

DOWNEY, John C. (Ph.D.), Prof. & Head, Dept. Biology, University Northern Iowa, USA. Rhopalocera (Lycaenidae). 12 April 1926, Eureka, Utah. EDUC.: B.S., Univ. Utah, 1949; M.S., ibid., 1950; Ph.D., Univ. Calif. (Davis), 1957. M. Norine Margaret Simpson, 25 June 1959. CH.: John, Jr.; Mike; Mary Ann; Dennis; Patrick. LEPID. SOC.: Charter Mem.; Sec. 1964-71; Nom. Com., 1972 (chm.). OTHER: Entomol. Soc. Amer.; Ill. Acad. Sci.; Ill. Mosquito Control Assn.; Iowa Acad. Sci.; Lepid. Res. Foundation, Charter

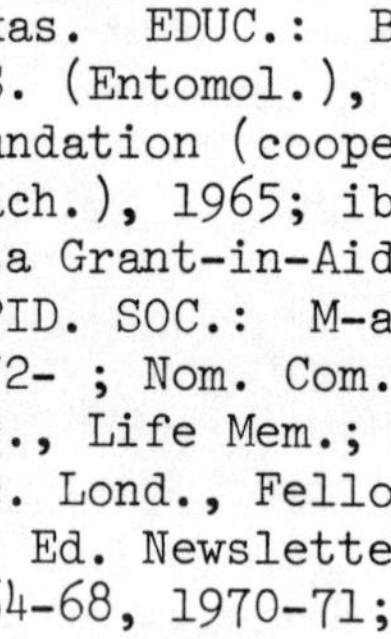

Mem.; Midwest Benthological Soc.; Pac. Coast Entomol. Soc.; Soc. Study Evolution; Soc. Syst. Zool., Charter Mem.; So. Calif. Acad. Sci. HON.: Phi Kappa Phi; Phi Eta Sigma; Phi Sigma; Sigma Xi (So. Ill. Univ. Chap.; V. Pres., 1967-68; Pres., 1969). RES. GRANTS: Amer. Philosophical Soc., 1959; National Sci. Foundation, 1959-62, 62-64, 64-66, 66-68. CONTRIB.: More than 45 papers, mostly on Lycaenidae, published in Ann Entomol Soc Amer; Bioscience; Bull Entomol Soc Amer; Bull S Calif Acad Sci; Ecology; Encycl Brittanica; Entomol Ber (Amst); Entomol News; J Lepid Soc; J Res Lepid; News Lepid Soc; Pan-Pac Entomol; Proc Ill Mosq Contr Assoc; Proc (No. Centr. Br.) Entomol Soc Amer; Syst Zool; Trans Amer Entomol Soc; Trans Ill Acad Sci; Yearb Amer Philos Soc.

EHLE, George E., Lancaster, Pennsylvania, USA. Life histories (Rhopalocera). 27 July 1913, Scranton, Pennsylvania. EDUC.: B.S. (Chem.), Penn. State Univ., 1936. M. Betty Ruth Cannon, 25 December 1938. CH.: Robert; David; Richard; Marguerite. LEPID. SOC.: Treas., 1959-65; Nom. Com., 1967 (chm.). OTHER: Muhlenberg Bot. Club, Treas., 1948-60. CONTRIB.: Lepid. News.

EHRLICH, Paul R. (Ph.D.), Director, Grad. Study, Dept. Biological Sciences, Stanford Univ., California, USA. Evolution and morphotaxonomy (Rhopalocera). 29 May 1932, Philadelphia, Pennsylvania. EDUC.: A.B., Univ. Penn., 1953; M.A., Univ. Kans., 1955; Ph.D., ibid., 1957; Dr. Humane Arts (Hon), Univ. Pacific. M. Anne Howland, 1954. CH.: one dt. LEPID. SOC.: Charter Mem.; Sec., 1958-63; M-a-l Ex. Council, 1966-68. OTHER: Soc. Study Evolution, V. Pres., 1970; Amer. Soc. Naturalists; Ecol. Soc. Amer.; Fed. Amer. Scientists; Soc. Syst. Zool., Ed. Board, 1964-67; Calif. Acad. Sci., Fellow; Amer. Mus. Nat. Hist., Hon. Life Mem.; Int. J. Environmental Sci., Ed. Board, 1969- ; Int. Studies Assn., Com. Mem., 1970- ; Int. Assn. Ecology, Mem. & Act. Committeeman, 1969-70; Rocky Mt. Biol. Lab., Trustee, 1971; Co-Chm., Res. Com., 1973-. HONORS & GRANTS: Univ. Kansas, Fellow 1954-55; Sigma Xi, Res. Grant, 1955; National Science Foundation, Predoct. Fellow, 1955-57; National Science Foundation, Postdoct. Fellow, 1965-66. CO-AUTH.: How to Know the Butterflies, Wm. C. Brown Co., 1961; The Process of Evolution, McGraw-Hill, 1963. CONTRIB.: About 100 papers, published in numerous scientific journals including Ecology;

Evolution; J Lepid Soc; Nature (Lond); Oecology (Berlin); Science.

ETCHEVERRY, Maria (Dra.), Irarrázaval 1628, Dep. 94, Santiago 11, CHILE. Life histories (macrolepidoptera). 20 April 1928, Chuquicamata, Antofagasta, CHILE. EDUC.: Prof. (Biol. & Chem.) Univ. de Chile, 1949; M.S., Univ. Wisconsin, 1959; Ph. D. (Biol.), Univ. de Chile, 1962. LEPID. SOC.: 1st V. Pres., 1967. OTHER: Prof. Biol & Chem., Liceo de Ninas Nº 1 & 8, Santiago, 1950-58; Univ. Chile: Asst. ad-honorem, 1947-49; Laboratorist; Prof. Laboratorist; Prof. of Seminar; Custodian Entomol. Coll.; and Researcher, 1950-63; Prof. Entomol., 1964- ; Chm., Centro de Estudios Entomológicos, 1964- . CONTRIB.: About 40 papers mostly on Lepidoptera, published in Ann Univ Chile; Farmacia Chilena; Proc Int Congr Zool; Publ Cent Estud Entomol Univ Chile; Rev Chil Entomol; Rev Universitaria; and others.

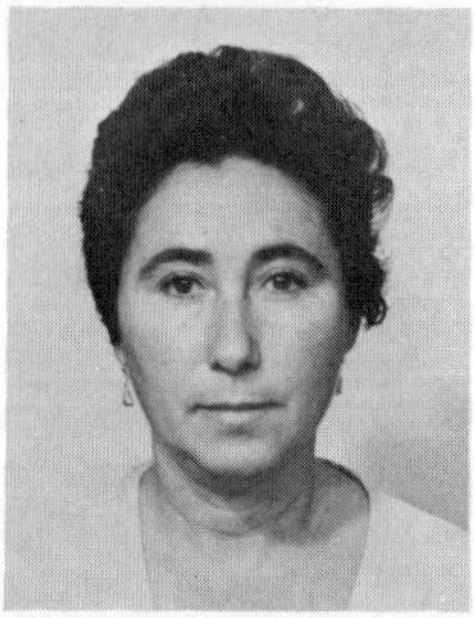

FEDERLEY, Harry (Dr.) (1879-1951). (Obit., biblio. and portrait, *Lepid. News* 6:57-60.) Genetics and experimental Lepidopterology. 22 March 1879, Viipuri, former SE FINLAND; d. 13 November 1951, Helsinki. EDUC.: Dr., Univ. Helsinki, 1907; Licendia docendi (Zool.) 1909, and (Genetics), ibid., 1915. Aptd. Prof. Genetics, 1923. LEPID. SOC.: M-a-l Ex. Council, 1952 (elected but died before taking office). CONTRIB.: Thirty-two papers on Lepidoptera (Lepid. News 6: 59-60).

FERGUSON, Douglas Campbell (Ph.D.), Systematic Entomol. Lab., USDA, Washington, D. C., USA. Life history, distribution, taxonomy. 17 February 1926, Halifax, N. S., CANADA. EDUC.: B.Sc., Dalhousie Univ. Halifax, 1950; M.S., Cornell Univ., 1956; Ph.D., ibid., 1967. M. Charlotte Hawthorn Studd, 11 May 1953. CH.: Stephanie; Caroline. LEPID. SOC.: Charter Mem., M-a-l Ex. Council, 1956-58, 1973-75; Nom. Com., 1963, 1964 (chm), 1968 (chm). OTHER: Entomol. Soc. Amer.; Entomol. Soc. Canada; Entomol. Soc. Wash.; Mich. Entomol. Soc.; Nova Scotian Inst. Sci.; Sigma Xi; Soc. Syst. Zool. AUTH.: Fasc. 20.2, Saturniidae, *in* Moths of America

North of Mexico. CONTRIB.: About 30 papers, published in Bull Brooklyn Entomol Soc; Can Entomol; Discovery (New Haven); J Lepid Soc; Lepid News; Peabody Mus Nat Hist Yale Univ Bull; Postilla; Proc Nova Scotian Inst Sci.

FORBES, William Trowbridge Merrifield (Ph.D.) (1885-1968) (brief biog., *Entomol. News* 80 (4), April 1969). Former Prof., Dept. Entomol., Cornell Univ., New York, USA. EDUC.: B.A., Amherst College, 1906; Ph.D., Clark Univ., 1910. LEPID. SOC.: Charter Mem.; Honorary Life Mem., 1952; Pres., 1953. AUTH.: The Lepidoptera of New York and Neighboring States. Vols. I-IV, Cornell Agric. Exp. Station Mems. CONTRIB.: About 150 papers, mostly on morphology and classification of Lepidoptera.

FORD, Edmund B. (D.Sc.), Former Fellow & Dean, All Souls College, Oxford, ENGLAND. Genetics. 23 April 1901. EDUC.: M.A., Oxford; D.Sc., ibid.; Hon. D.Sc., Liverpool; HONS.: Fellow Roy. Soc; Darwin Medallist Roy. Soc.; Weldon Medallist and Univ. Prizeman, Oxford; Foreign Mem., Finish Acad., Medallist of Helsinki Univ.; Prof. Emeritus, Ecological Genetics, Univ. Oxford; established the science of Ecological Genetics, 1971, Blackwell Scientific Publ., Oxford.

LEPID. SOC.: Charter Mem.; M-a-l Ex. Council, 1955-1957; Honorary Life Mem., 1960; 1st V. Pres., 1968; Pres., 1970. AUTH.: Mimicry, with G. D. H. Carpenter, 1933; The Study of Heredity, 1950, 2nd ed.; Genetic Polymorphism, 1965; Mendelism and Evolution, 1965, 8th ed.; Genetics for Medical Students, 1973, 7th ed.; Butterflies, 1967, 3rd reprint, 3rd ed.; Moths, 1972, 3rd ed.; Ecological Genetics, 1975, 4th ed. (Polish and French translations). CONTRIB.: A large number of scientific research articles, including some on archaeology.

FORSTER, Ernst Walter (Ph.D.), Director, Zoological Coll., State of Bavaria, Maria-Ward-Strasse 1b, Munich 19, GERMANY. Zoogeography. 12 July 1910, Augsburg, GERMANY. EDUC.: Pr., Univ. Munich and Königsberg, 1930-36; Ph.D., Univ. Munich, 1936. M. Elisabeth Mayer, 1936. CH.: Ernst; Gerhard; Gertraud. LEPID. SOC.: V. Pres., 1951; Pres., 1960; Hon. Life Mem., 1969. OTHER: Soc. de Ciencias Nat. La Salle Caracas, Corr. Mem. (Hon.), Librarian and Secretary, 1938-

1961; Munich Entomol. Soc., First Chm., 1962- ; German

Entomol. Soc.; Vienna Entomol. Soc. PATRONYMS: Many, largely European. AUTH.: With Th. A. Wohlfahrt, The Butterflies of Middle Europe (4 vols. todate), Franck'sche Verlagshandlung, Stuttgart. Knauer's Book of Insects, 1968, Droemersche Verlagsanstalt, Munich. CONTRIB.: About 50 scientific papers appearing in Ann Naturhist Mus Wien; Bol Entomol Venez; Entomol Rundsch; Entomol Z; Fragmenta Faun Hung; Freude am Leben; J Lepid Soc; Kosmos (Warsaw); Munich Entomol Ges; Reichenbachia; Zeitschr Wien Entomol Ges; and others.

FRANCLEMONT, John G. (Ph.D.), Prof. Entomol., Cornell

Univ., Ithaca, New York, USA. Taxonomy, life histories (Noctuoidea). 15 April 1912, Buffalo, New York. EDUC.: B.S., Cornell Univ., 1935; Ph.D., ibid. 1953. LEPID. SOC.: Charter/Sustaining Mem.; M-a-l Ex. Council, 1951-53; Sec., 1955-57; Pres., 1964. OTHER: Amer. Entomol. Soc.; Cambridge Entomol. Club; Can. Entomol. Soc.; Entomol. Soc. So. Africa, Fellow; Entomol. Soc. Washington, D. C.; Kansas Entomol. Soc.; Lepid. Res. Foundation; N. Y. Entomol. Soc.; Pac. Coast Entomol. Soc.; Royal Entomol. Soc. London, Fellow; Soc. Syst. Zool.; Soc. Study Evolution; Soc. Bibliography of Nat. Hist.; So. Calif. Acad. Sci., Fellow; Amer. Assoc. Adv. Sci. HONS.: Phi Kappa Phi; Sigma Xi. AFFIL.: Collab., Systematic Entomol. Lab., U. S. Dept. Agric., 1952- ; Res. Assoc. Dept. Entomol., Smithsonian Inst., 1971- . GRANTS: Amer. Philosophical Soc., 1955, 1958, 1960, 1963; U. S. Dept. Agric., 1965-70. AUTH.: Fasc. 20 Pt. 1, Mimallonoidea and Bombycoidea, *in* The Moths of America North of Mexico. CONTRIB.: About 50 papers, appearing in Ann Entomol Soc Amer; Bull Brooklyn Entomol Soc; Bull Mus Comp Zool; Bull Zool Nomenclature; Can Entomol; Entomologist; Entomol News; J Lepid Soc; J NY Entomol Soc; Lepid News; Proc Entomol Soc Wash; Science; and others

FREEMAN, Hugh Avery, Chm., Sci. Dept., Hillcrest H. S., Dallas, Texas, USA. Taxonomy (Hesperiidae of Mexico).

7 October 1912, Conway, Arkansas. EDUC.: A.B., Hendrix Coll., Arkansas, 1936; M.S., So. Methodist Univ., Dallas, 1938. M. Erna Louise Morris, 1940. CH.: Linda Louise; Julia Anne; Avery Gilbert. LEPID. SOC.: Charter Mem.; Coord. Ann. Summary, 1951-52, 59-7-; Ed. Com. (*News*) 1959- ; M-a-l Ex. Council, 1962-64; 1st V. Pres., 1971; V. Pres., 1972. OTHER: Lepid. Res. Foundation; Tri-Beta; Amer. Mus. Nat. Hist., Patron. PATRONYMS: *Euptychia gemma freemani* (Stallings & Turner), 1947; *Agathymus freemani* Stallings & Turner, 1960; *Vinius freemani* Miller, 1970; *Calephelis freemani* McAlpine, 1971. CONTRIB.: About 70 papers, mostly on Hesperiidae, describing many new taxa, published in Amer Mus Novit; Bull Brooklyn Entomol Soc; Can Entomol; Entomol News; J Lepid Soc; Lepid News; Tex J Sci; and others.

FREEMAN, Thomas Nesbitt (Ph.D.) (1883-1975). Formerly, Entomol. Res. Institute, Canada Dept. Agric., Ottawa, CANADA. Rhopalocera (Arctic). 16 October 1911, Saskatoon, Saskatchewan, CANADA; d. May 1975, Ottawa, Ontario, CANADA. EDUC.: B.S.A., Ontario Agr. Coll., 1934; M.Sc., Univ. Colorado, 1936; Ph.D., Univ. Toronto, 1946. M. A. Gertrude Munro, 12 March 1938. CH.: Marianna; Leslie. LEPID. SOC.: Charter Mem.; Coord. Ann. Summary, 1948-52; Orgn. Com., 1950; M-a-l Ex. Council, 1951; 1st V. Pres., 1953; Pres., 1956; Nom. Com., 1968, 1969 (chm). OTHER: Ontario Entomol. Soc.; Can. Entomol. Soc.; Lepid. Foundation, Assoc. Ed., *J. Res. Lepid.*, 1969-75. PATRONYMS: *Diarsia pseudorosaria freemani* Hardwick, 1950; *Erebia epipsodea freemani* Ehrlich, 1955; *Nepytia freemani* Munroe, 1963. CONTRIB.: About 100 papers, published in Ann Report Ontario Entomol Soc; Can Entomol; Entomol News; J Lepid Soc; J Res Lepid; Lepid News; Pan-Pac Entomol; Proc (Xth) Int Cong Zool; Sci Agric; and others.

GOZMÁNY, Lancelot Anthony (Ph.D.), Természettudományi Múzeum Allattára, Budapest VIII, Baross-utca 13, HUNGARY. Taxonomy (microlepidoptera). 9 November 1921, Budapest, HUNGARY. EDUC.: LL.D., P. Pázmány Univ. Sci., Budapest, 1955 Ph.D., Hungarian Acad. Sci., 1963. M. 1944. CH.: Mary; Gabor; Christine. 2nd M. 1964. LEPID. SOC.: V. Pres., 1956. OTHER: Hungarian Entomol Soc., Sec., 1966-69; Hungarian Biol. Soc.; Münchener Entomol. Ges., Münich.

HON.: Frivaldszky Commemorative Plaque (silver), 1969, conferred by Hungarian Entomol. Soc.

LING.: English, German, Italian, French and Latin. Commissioned by Hungarian Acad. Sci. to write a 7 language dictionary (Latin, German, English, French, Hungarian, Russian, Spanish) of the names of European animals. PATRONYMS: *Martyrhilda gozmanyi* Balogh, 1951; *Sterrhopteryx gozmanyi* Kovacs, 1954; *Talaeporia tubulosa gozmanyi* Sieder, 1955; *Catoptria gozmanyi* Bleszynski, 1956; *Bactra gozmanyana* Toll, 1958; *Repetekia gozmanyi* Amsel, 1958; *Amselina gozmanyi* Amsel, 1959; *Nepticula gozmanyi* Szöcs, 1959; *Coleophora gozmanyi* Toll, 1960; *Euchromius gozmanyi* Bleszynski, 1961; *Gozmanyia* Roesler, 1964; *Rhaphidothrix gozmanyi* Mahunka, 1967; *Hyssia gozmanyi* Kovacs, 1968; *Gozmanytinea* Capuse, 1969; *Monopis gozmanyi* Capuse, 1969; *Staudingerian gozmanyella* Roesler, 1970; *Leioptilus gozmanyi* Bigot, 1970; *Geotrupophorus gozmanyi* Mahunka, 1970; *Eupithecia gozmanyi* Vojnits, 1972. AUTH.: Fauna (Lepidoptera) Hungariae (Vol. 16), 1955, Hungarian Acad. Sci. MONOGR.: with L. Vári, The Tineidae of the Ethiopian Region. CONTRIB.: About 100 papers, published in Acta Biol Acad Sci Hung; Acta Zool Acad Sci Hung; Allatt Közl Budapest; Ann Hist/Nat Mus Natl Hung; Ann Mus R Afr Cent Ser Quarte Zool; Arnoldia (Rhod); Bull Mus Hist Nat Belgrade; Dtsch Entomol Z; Entomologist; Entomol Z; Folia Entomol Hung; Fragm Faun Hung; Lepid News; Lorenco Marques; Opuscula Zool (Munich); Rev Entomol Mocambique; Sitzungsber Öesterr Akad Wiss; Zschr Wien Entomol Ges; and others.

GREY, Lionel Paul, Route 1, Box 216, Lincoln, Maine, USA.

Systematics (Argynninae). 23 November 1909, Houlton, Maine. EDUC.: Grad., Farmington State Teacher's College, Maine, 1931. M. Ruth Delano, 17 June 1934. CH.: Linda; Stephen; Prescott; Aldana (d.). LEPID. SOC.: Charter/Sustaining Mem.; 1st V. Pres., 1952; Nom. Com., 1958 (chm). PATRONYMS: *Speyeria atlantis greyi* Moeck, 1950. AUTH.: Genus *Speyeria in* How to Know the Butterflies, Wm. C. Brown, 1961. CONTRIB.: About 20 papers, appearing in Amer Mus Novit; Bull Zool Nomenclature; J Lepid Soc; J Res Lepid; Lepid News; Proc Roy Entomol Soc Lond; Trans Amer Entomol Soc.

HACKMAN, Walter L. V. (Ph.D.), Curator Entomol., Zool. Mus., Helsinki Univ., FINLAND. Taxonomy (Finish microlepidoptera). 7 October 1916, Helsinki, FINLAND. EDUC.: Cand. Phil., Helsinki Univ., 1940; Ph. D., ibid., 1949. LEPID. SOC.: M-a-l Ex. Council, 1952-54. OTHER: Lect., Zool., Helsinki Univ., 1950- ; Curator, Division of Entomol., Zool. Mus., ibid. 1957- . CONTRIB.: Several papers, mainly on the taxonomy of Finish microlepidoptera, 1941-1954. More recent papers largely on Diptera.

HARDWICK, David Francis (Ph.D.), Director, Biosystematics Res. Inst., Dept. Agr., Ottawa, Ontario, CANADA. Taxonomy, life histories. 22 March 1924, Edmundston, New Brunswick, CANADA. EDUC.: B.A., Univ. Saskatchewan, 1946; M.Sc., McGill Univ., 1948; Scholarship, Agr. Inst. Canada, 1948; Ph.D., Univ. Calif., Berkeley, 1955. M. Verna Walker, May 1955. CH.: Koralee Anne; Julia Dawn; Frances Jean. LEPID. SOC.: Charter Mem.; Nom. Com., 1960 (chm); M-a-l Ex. Council, 1965-67; Pres., 1969; Ed. (*Journal*), 1970-71. OTHER: Can. Entomol. Soc.; Michigan Entomol. Soc.; Soc. Syst. Zool. CONTRIB.: About 40 papers, published in Bull Zool Nomenclature; Can Entomol; J Lepid Soc; Mem Entomol Soc Can; and others.

HAYWARD, Kenneth John (1891-1972). (Obit., brief biog. and portrait, *J. Lepid. Soc.* 28:79-80.) Former Prof. Entomol., National Univ., Tucumán, ARGENTINA. Neotropical Rhopalocera. 7 March 1891, Pitney, Somerset, ENGLAND; d. 21 May 1972, Tucumán, ARGENTINA. EDUC.: King's School, Bruton, England, 1899-1910. Worked in Egypt, 1912-23 (except 1914-19) when on active duty with British Army. Became resident of Argentina, 1923; entomologist with Argentine Ministry of Agr., 1933-34; Prof. Entomol., National Univ., Tucumán, Argentina, 1935-72. LEPID. SOC.: V. Pres., 1951. AWARDS: Holmberg prize, 1931; Argentine National prize, 1946-48; Royal Entomol. Soc., Fellow, 1921; Zool. Soc., Fellow, 1924; Royal Geograph. Soc., Fellow, 1928; Degree in Biol., Honoris Causa, National Univ., 1950. Mem. several nat. sci. soc. MONOG.: Argentine Nymphalidae, 1931; Argentine Pyrrhopyginae and Pyrginae (Hesperiidae), 1948.

CONTRIB.: More than 300 papers, published in various scientific journals.

HERRERA GONZÁLEZ, José (Prof. Emeritus), Entomol., Univ. Chile, Santiago, CHILE. Rhopalocera (Genetics). 9 October 1913, Mejillones, Antofagasta, CHILE. ACTIVITIES: Prof. Biol., Chem. & Phys. Educ. in Punta Arenas and Santiago, 1939-1957; Prof. Entomol., Univ. Chile, 1946-1963; Sec., Fac. Philos. & Educ., Univ. Chile, 1957-60; Supt. Educ., Min. Educ., 1960-63; Prof. Emeritus, Entomol., Univ. Chile, 1964. LEPID. SOC.: M-a-l Ex. Council, 1953-55. CONTRIB.: About 25 papers on Lepidoptera, published in An Univ Católica; An Univ Chile; Proc US Nat Mus; Pub Cent Estud Entomol Univ Chile; Rev Chil Entomol; Rev Universitaria; and others.

HESSEL, Sidney Adolphus (1907-1974). (Obit., biog., *Discovery* 10(1):49-50.) Former Res. Afflt., Entomol., Peabody Mus., Yale Univ., New Haven, Connecticut, USA. Macrolepidoptera Northeast US, especially biology of *Papipema* and *Catocala*. 5 January 1907, New York, N.Y.; d. 11 November 1974, Washington, Connecticut. EDUC.: A.B., Harvard Univ., 1928; M.B.A., ibid., 1940. M. Beatrice Hofman, 23 June 1932. CH.: John H.; Mary H. Dodge; Susan J. EXP.: Retired businessman; brokerage, banking, and importing raw materials. Lieut. USNR (1944-46). LEPID. SOC.: Charter/Life Mem.; Treas., 1953-58; Asst. Treas. 1959-68; 1st V. Pres., 1969; *Memoirs* Mgr., 1959-74; Mgr. back issues (*News/Journal*) 1958-74. OTHER: Visiting Com., Mus. Comp. Zool., Harvard Univ., 1960-74; Mem. Board & Com. Chm., Steep Rock Assoc. (natural use, specified land areas), 1957-74; Board Mem., Shepaug Valley Archeol. Soc.; Board Mem. other local civic org.; coll. classic and Lepidoptera stamps; Sigma Xi. EXPED.: Cuba, 1928, 1929; Canada, 1932, 1939, 1941, 1943, 1954; Guatemala, 1940; Bermuda, 1961; Hawaii, 1944-45; Jamaica W. I., 1953, 1959, 1962; Kenya, Tanzania, Uganda, Zaire, Ethiopia, 1960. PATRONYMS: *Callophrys* (*Mitoura*) *hesseli* Rawson and Ziegler, 1949; *Automeris iris hesselorum* Ferguson, 1971. CONTRIB.: Bull Brooklyn Entomol Soc; J Lepid Soc; J NY Entomol Soc; Lepid News.

HINTON, Howard Everest (Sc.D.), Prof. and Head, Dept. Zool., Univ. Bristol, ENGLAND. Taxonomy, physiology, evolution. 24 August 1912, Matehuala, MEXICO. EDUC.: B.Sc., Univ. California (Berkeley), 1934; Ph. D., Univ. Cambridge, 1939; Sc.D., ibid., 1957. LEPID. SOC.: V. Pres., 1965. OTHER: Australian Acad. Sci., Senior Fellow, 1963; British Entomol. & Nat. Hist. Soc., V. Pres.; J Insect Biochem., Ed.; J Insect Physiol., Ed.; Proc. Roy. Entomol. Soc., Ed.; Roy. Entomol. Soc. Lond., Pres. 1969-70; Royal Soc. Lond., Fellow 1961; ibid., Assoc. Ed. 1968-70; Soc. British Entomol., Pres. 1954-55. APTD.: Junior Curator, Univ. Mus. Zool, Cambridge, 1937-39; Asst. Keeper, British Mus. (Nat. Hist.), 1939-49; Lecturer, Univ. Bristol, 1949-51; Reader, ibid., 1951-64; Prof., ibid., 1964-70; Chm., ibid., 1970- . INV. READER & LECTURER: Danish Zool. Soc., Copenhagen, 1960; Chinese Acad. Sci., 1960; Czechoslovak Acad. Sci., 1961 & 1971; State Agric. College, Holland, 1961; International Cong. Entomol., Stockholm, 1948; ibid., Amsterdam, 1952; ibid., Montreal, 1956; ibid., Vienna, 1960; Venezuela, 1971; Iowa, USA, 1971; Australia, 1972. CONTRIB.: Numerous papers in many scientific journals. Developed widely recognized system of setal nomenclature and classification of insect pupae. Work on plastron respiration and diffraction colours.

HOGUE, Charles Leonard (Ph.D.), Senior Curator, Entomol., Los Angeles Co. Mus. Nat. Hist., California, USA. 4 February 1935, Caruthersville, Missouri. EDUC.: B.S., (Agr. Entomol. & Parasit.) Hon., Univ. California (Los Angeles), 1957; Ph.D., ibid., 1962. M. Barbara Mitchell. CH.: James Norman; Brian Thomas; Carolyn Lee. LEPID. SOC.: M-a-l Ex. Council, 1968-69; Nom. Com., 1969. OTHER: Assn. Tropical Biol.; Entomol. Soc. Amer.; Lepid. Foundation; Lorquin Entomol. Soc., Pres. 1959; Soc. Systematic Zool.; So. Calif. Acad. Sci.; XII Int. Congr. Entomol. London, Fellow; Sigma Xi. GRANTS: So. Calif. Acad. Sci./Amer. Assoc. Adv. Sci., 1960; Univ. Calif. (Los Angeles), 1961; Entomol. Soc. Amer., 1964; Amer. Philo. Soc., 1966, 1968; Nat. Geograph. Soc., 1970. CONTRIB.: Over 40 papers, published in Amer Mus Novit; Lepid News; Univ Calif Publ Entomol; J Lepid Soc; An Inst Biol Univ Nac Auton

Mex; Los Ang Cty Mus Contrib Sci; J Res Lepid; Contrib Amer Entomol Inst Ann Arbor; Ann Entomol Soc Amer; Bull So Calif Acad Sci; Nat Hist; and others.

JORDAN, Heinrich Ernst Karl (Dr.) (1861-1959). (Biog., biblio. and portrait, *Biograph. Mem. Fellows Roy. Soc.,* Vol. 6, November 1960.) 7 December 1861, Almstedt near Hildesheim, Hanover, GERMANY; d. 12 January 1959, Tring, GERMANY. EDUC.: Degree (bot. and zool.) Summa cum Laude, Univ. Göttingen, 1886. M. Minna Brünig, 1891. LEPID. SOC.: Charter Mem.; Hon. Life Mem., 1952; Pres., 1952. OTHER: Zool. Mus., Tring, Staff entomol.; Int. Entomol. Cong., Founder and mem., Hon. Life Pres., 1948.

JUMALON, Julian N., Assoc. Prof., Dept. Biol., Univ. San Carlos, Cebu City, PHILIPPINES. Life histories (Philippine Rhopalocera). 18 October 1909, Manila, PHILIPPINES. EDUC.: Cert. Painting, Univ. Philippines, 1936; M.S. (biol.), honoris causa, Univ. San Carlos, 1960. M. Felicitacion N. Alcantara, June 1938; eight ch. LEPID. SOC.: V. Pres., 1971. OTHER: Lepid. Soc. Japan; Philippine Assoc. Entomol.; Soc. Hispano-Luso-Americana de Lepid., Spain, Hon. Mem.; Biol. Teachers' Assoc., Cebu Chap.; Cebu Art Assoc., Pres.; Cebu Hist. Soc., Pres.; Philippine Lepid. Soc., Pres. HON.: Philippine Inst. Architects, Cebu Chap.; Invit. speaker, Lepid. Soc. Japan, 1970 Silver Anniversary; Southeast Asia Treaty Org., Fellow, 1972. CONTRIB.: About 25 papers, in which several new taxa described, published in J Lepid Soc; Leyte-Samar Stud; Philippine Entomol; Philippine Sci; Philippine J Sci. Also about 200 popular articles on exped., nature, conservation and art, in local papers and magazines.

KENDALL, Roy O'Neal, San Antonio, Texas, USA. Ret. Fed. Civil Service (Dept. Defense - Air Force). Life histories, spatial and temporal distribution, especially Texan Lepidoptera. 21 May 1912, Ingalls, Gray County, Kansas. EDUC.: U/Grad., Louisiana State Univ., Baton Rouge; Trinity Univ., San Antonio, Texas. M. Conway B. Alford, 3 May 1947. CH.: (stepsons) G. L. Montgomery, B. G. Montgomery. LEPID. SOC.: Treas.,

1966-68; Nom. Com., 1970, 1971 (chm), 1972; Ed. Com. (*Journal*), 1972- ; M-a-l Ex. Council, 1973-76. OTHER: Lepid. Res. Foundation, Charter Mem.; Florida State Coll. Arthropods, Div. Plant Industry, Florida Dept. Agr. & Consumer Services, Res. Assoc., 1970-. PATRONYMS: *Megathymus coloradensis kendalli* Freeman, 1965; *Zamagiria kendalli* Blanchard, 1970; *Orgyia definita kendalli* Riotte, 1972. CONTRIB.: About 35 papers, dealing largely with larval foodplants of Lepidoptera, published in J Lepid Soc; J Res Lepid; Psyche J Entomol; Trans San Diego Soc Nat Hist.

KETTLEWELL, Henry Bernard Davis (Dr.), Senior Res. Officer, Genetics Lab., Dept. Zool., Oxford, ENGLAND. Polymorphism, industrial melanism. 24 February 1907, Howden, East Yorkshire, ENGLAND. EDUC.: M.A., Oxford; M.B., Cambridge; B.Chir., ibid.; Mem. Roy. Coll. Surgeons, England; Licentiate Roy. Coll. Physicians, London. Med. practice, Cranleigh, Surrey, 1933-49. Cape Town Univ., South Africa, 1949-54. EXPED.: Kalahari, Belgian Congo, and Knysna Forest, among others. M. 25 April 1936. CH.: Dawn Margaret; David Henry Wiltshire. LEPID. SOC.: Pres., 1963. OTHER: British Entomol. & Nat. Hist. Soc.; Roy. Entomol. Soc., Council, 1948-49; Nuffield Res. Fellow., 1956-64. HON.: Darwin Medal, Acad. Sci., Russia, 1959; Mendel Medal, Czechoslovakia, 1968. Co-founder, Rothschild-Cockayne-Kettlewell Coll. Brit. Lepid., now the National Coll. PATRONYMS: Many from Africa, South America and elsewhere. AUTH.: Your Book of Butterflies and Moths, Faber & Faber Ltd., 1963; Charles Darwin and his World (with J. Huxley), Thames & Hudson, 1965; contrib. to Ecological Genetics and Evolution, ed. E. R. Creed, Blackwell, 1971; The Evolution of Melanism, Clarendon, Oxford, 1973; contrib. to Evolution: the Modern Synthesis, 3rd ed., ed. J. Huxley (in press); Mimicry and Camouflage *in* Encyclopaedia Italiana (in press). CONTRIB.: About 75 papers, many on hereditary and adaptive melanism, published in Adv Sci; Ann Rev Entomol; Discovery (Lond); Endeavour (Lond); Entomologist; Entomol Rec & J Var; Heredity; J Lepid Soc; Nature (Lond); New Sci; Proc Roy Soc Lond B, Biol Sci; Science; Sci Amer; Amer Zool; and others.

KLIMESCH, Josef Wilhelm (Ph.D.) (Ret.), Linz a.d. Donau, Donatusgasse 4, AUSTRIA. Life histories, genetics, taxonomy. 5 May 1902, České Budějovice (Budweis), CZECHOSLOVAKIA.

EDUC.: Trieste and Linz (1909-23); Ph.D. (zool. and bot.), Univ. Graz, 1950. M. LEPID. SOC.: V. Pres., 1968. OTHER: Vienna Entomol. Soc., Hon. Mem., 1954. EXPED.: Dalmatia, Macedonia, Hungary, Turkey, Canary Islands, Spain, Mallorca, Switzerland, South France, Corse, Italy, Sicily, and many areas of Austria. HON.: Vienna Mus. Nat. Hist., Corresp. 1950; Government of Austria Superior (O. ö. Landesmuseum, Linz), Sci. Consultant, 1956- ; Univ. Innsbruck, Excellenti in litteris, 1958. AUTH.: Microlepidoptera *in* The Land Animals of the Middle Hohe Tauern Mountains, H. Franz, Memorial Austrian Acad. Sci., Vienna, 1949; Classification of Lepidoptera (Pyralidina, Tortricina, Tineina, Eriocraniina and Micropterygina) *in* The Northeast Alps in the Mirror of their Land-Animal World, Universitätsverlag Wagner, Innsbruck, 1961; The Lepidoptera of Macedonia, IV Microlepidoptera, Prirodonaučen Musej Skopje, 1968. CONTRIB.: More than 70 papers on Microlepidoptera, describing 54 new species.

KLOTS, Alexander Barrett (Prof.), Dept. Entomol., Amer. Mus. Nat. Hist., New York, USA. (Former Prof., City Coll. N.Y.). Taxonomy, correlation Arctic & Alpine faunas. Born 1903, New York, New York. EDUC.: Trinity School; Blair Acad.; Ph.D., Cornell Univ., 1931. M. Elsie Broughton, 1927. CH.: One s., one dt. MIL. SVC.: (USAAF) 1943-46, Major (malaria & mosquito control). LEPID. SOC.: Charter Mem.; Pres., 1957; Hon. Life Mem., 1972. OTHER: Amer. Assoc. Univ. Prof.; Amer. Soc. Naturalists; Entomol. & Nat. Hist. Soc. So. Lond., Fellow 1955- ; Entomol. Soc. Amer., Fellow 1945, Ed. Board, 1950-53, Com. Entomol. Nomenclature, 1964; Explorers Club; Int. Cong. Entomol. (IV, X, XI, XII); Int. Cong. Zool. (XIV, XV, XVI); New York Entomol. Soc., Hon. Life Mem., 1970; Roy. Entomol. Soc. Lond., Fellow, 1956- ; So. Calif. Acad. Sci., 1935-41; Soc. Syst. Zool. HON.: Sigma Xi; Phi Kappa Phi; Hon. Life & Assoc. Patron, Amer. Mus. Nat. Hist., 1947- ; N.Y. Entomol. Soc., Pres., 1937, Trustee, 1955- . PATRONYMS: *Lucaena heteronea klotsi* Field, 1936; *Euphydryas chalcedona klotsi* dos Passos, 1938; *Klotsa* Hemming, 1964?; *Ixala klotsi* Sperry, 1940; *Miraxis*

klotsi Bleszynski, 1962; other patronyms include a mutillid wasp, a pentatomid bug, an *Aedes* mosquito, and a flea. EXPED.: United States, Canada, England, France, Germany, Austria, Denmark, several Caribbean islands, British Guiana, Mexico, and Brazil. AUTH.: A Field Guide to the Butterflies of Eastern North America, 1951, Houghton Mifflin, Boston. The World of Butterflies and Moths, 1957, Harrap, London; McGraw-Hill, New York. Living Insects of the World (with Elsie B. Klots), 1959, Chanticleer Press and Doubleday, New York. 101 Questions Answered About Insects (with Elsie B. Klots), 1961, Dodd Mead, New York. The Community of Living Things in the Desert (with Elsie B. Klots), 1956, Creative Educational Society, Mankato, Minnesota; 2nd printing, 1959; 2nd edition, 1960. Insects of North America (with Elsie B. Klots), 1971, Doubleday, New York. Lepidoptera, *in* Encyclopaedia Britannica, *in* Encyclopaedia Americana, and *in* Tuxen, Taxonomists' Glossary of Genitalia in Insects, 1956, 1970. CONTRIB.: Many papers, describing about 30 North American butterflies and moths, and published in Amer Mus Novit; Bull Amer Mus; Bull Assoc Minn Entomol; Bull Brooklyn Entomol Soc; Biol Rev; Bull So Calif Acad Sci; Bull Zool Nomenclature; Entomol Amer; Entomol News; Evolution; J NY Entomol Soc; Lepid News; Pan-Pac Entomol; Proc Roy Entomol Soc Lond; Proc X, XI, XIII Int Cong Entomol; Proc XIV Int Cong Zool; Psyche J Entomol; Syst Zool; Copeia.

LORKOVIĆ, Zdravko (Ph.D.), Prof. Biol., Zagreb Med. Facility, YUGOSLAVIA. Lepidoptera. 3 January 1900, Zagreb, Croatia, YUGOSLAVIA. EDUC.: Diplome, Faculty Nat. Sci., Univ. Zagreb, 1924; D.Sc., Ljubljana, 1928. M. 30 July 1928. LEPID. SOC.: 1st V. Pres., 1959, 1966. OTHER: Entomol. Soc. Finland; Jugoslavenska Akademija Znanosti i Umjetnosti, Zagreb; Soc. Entomol. Jugosl., Zagreb; Soc. Nat. Sci. Croatica, Zagreb; Zool. Acad., Agra, India. HON.: Rugjer Bŏsković, Zagreb, 1965. PATRONYMS: *Leptidea duponcheli lorkovici* Pfeiffer, 193?; *Hipparchia statilinus lorkovici* Moucha, 1965. CONTRIB.: About 50 papers, publisted in Acta Soc Entomol Yugosl; Acta Soc Sci Nat; Biol Glasnik; Boll Soc Entomol Ital; Bull Acad Sci Yugosl; Bull Soc Zool Fr; Bull Zool Nomenclature; Chromosoma; Cold Spring Harbor Symp Quant Biol; Entomol Tidskr; Glasnik Yugosl Entomol Drustva; Godisnjak Sveucilista; Int Entomol Zeits

Guben; J Lepid Soc; Lambillionea Bruxelles; Mitt Munch Entomol Ges; Mitt Schweiz Entomol Ges; Nachrichtenb Bayerischen Entomol Munch; Physiologia Comparata et Ecologia Leiden; Rev Suisse Zool; Uppsala Univers Arsskrift; Verh Zool-Bot Ges Wien; Verh Int Entomol Kongr Wien; Z Lepid Krefeld; Z Osterr Entomol Vereines Wien; and other European journals.

MAC NEILL, C. Don (Ph.D.), Senior Curator, Natural Sci. Div., Oakland Mus., California, USA. Life histories, systematics, ecology. 1924, San Francisco, California. EDUC.: B.S., Univ. California (Berkeley), 1950; Ph.D., ibid., 1960. M. Grace F. Beckberger, 1945. CH.: 3 dts., Nora and Daren (twins) 1953; Linn, 1956. LEPID. SOC.: M-a-l Ex. Council, 1966-68; Nom. Com. 1970 (chm). OTHER: Soc. Syst. Zool.; Ecol. Soc. Amer.; Biosystematics; Calif. Acad. Sci.; Pac. Coast Entomol. Soc., past Pres.; The Nature Conservancy, Bd. Trustees; Calif. Native Plant Soc.; American Fern Soc. CONTRIB.: Entomol News; J Lepid Soc; Pan-Pac Entomol; Proc Calif Acad Sci; Proc NY Entomol Soc; Univ Calif Publ Entomol; Wasmann J Biol.

MARTIN, Lloyd Milo, Prescott, Arizona, USA. Life histories, taxonomy. 24 May 1912, Bayard, Nebraska. EDUC.: Grad., So. Pasadena H.S., 1930. M. Dorothy Ethel Hitchcock, 16 July 1938. CH.: William Lloyd. LEPID. SOC.: Charter Mem.; Ann. Summary Coord., 1947-51; M-a-l Ex. Council, 1951-52; Nom. Com., 1958, 1960, 1966; Ed. Com. (*News*), 1959- ; V. Pres., 1970; Pres. 1972. OTHER: Lorquin Entomol. Soc., 1927- ; Sec., 1937-48; Pres., 1948-57; Treas., 1959-69; hon. longest active mem., 1968; Hon. Life Mem., 1970. So. Calif. Acad. Sci., 1938- ; Hon. Life Mem., 1963; Asst. Sec., 1949-65; Membership Sec., 1951; Sec., 1953; bd. directors, 1953-65. Soc. Syst. Zool., 1951- . Western Reg. Conf. Amer. Assoc. Mus., Charter Mem. HON.: Man of the Year, 1958, Optimist Club, Los Angeles, Calif. PATRONYMS: *Colias harfordii martini* Gunder, 1931; *Boloria ingens martini* Gunder, 1934; *Megathymus yuccae martini* Stallings & Turner, 1943; *Chlorochlamys martinaria* Sperry, 1948; *Philotes battoides martini* Mattoni, 1954; *Plumipalpia martini* Munroe, 1961; *Hesperia pahaska martini* MacNeill, 1966; *Cisthene martini* Knowlton, 1967; *Papilio indra martini* Emmel & Emmel, 1966. CONTRIB.: Bull S Calif Acad Sci.

MATHER, Bryant, 213 Mt. Salus Drive, Clinton, Mississippi, USA. Lepidoptera of Mississippi. 27 December 1916. EDUC.: A.B., Johns Hopkins Univ., 1936; grad. work, ibid., and at Amer. Univ. M. Katharine Kniskern, 27 March 1940. LEPID. SOC.: Charter/Life Mem.; Ed. Com. (*News*), 1959- ; Nom. Com., 1969; M-a-l Ex. Council, 1970-72. OTHER: Amer. Assoc. Adv. Sci., Fellow; Amer. Inst. Biol. Sci.; Amer. Mus. Nat. Hist., Life Mem., 1969, Assoc. Patron, 1972; Can. Entomol. Soc.; Fla. State Coll. Arthropods, Res. Assoc., 1969; Lepid. Res. Foundation; Miss. Acad. Sci.; Nat. Hist. Soc. Md.; Phi Beta Kappa; Sigma Xi; Gamma Alpha; Amer. Concrete Institute, Pres. 1963-64. PATRONYMS: *Chimoptesis matheri* Powell, 1964; *Neohermes matheri* Flint, 1965; *Microcrambus matheri* Klots, 1968; *Enosis matheri* Freeman, 1969; *Calephelis matheri* McAlpine, 1971. CO-AUTH.: The Butterflies of Mississippi, 1958. CONTRIB.: Lepid News; J Lepid Soc; J Res Lepid; Md Nat; J Miss Acad Sci.

MC DUNNOUGH, James Halliday (Ph.D.), (1877-1962). (Obit., biblio., and portrait, *J. Lepid. Soc.* 16:209-228.) Taxonomy (proposed almost 1500 names). 10 May 1877, Toronto, Ontario, CANADA; d. 23 February 1962. EDUC.: M.A., Queen's Univ., Kingston, Ontario, 1909; Ph.D., Kaiser Wilhelm Inst., Berlin, Germany, 1909. M. Margaret Berteles, 22 October 1909. LEPID. SOC.: Charter Mem.; Pres. pro temp., 1950; Pres., 1951; Hon. Life Mem., 1952. OTHER: Amer. Mus. Nat. Hist., Res. Assoc., 1946; Nova Scotia Mus. Sci., Res. Assoc., 1950; Entomol. Soc. Amer., Fellow; Roy. Soc. Can., Fellow; N.Y. Entomol. Soc., Honorary Mem.; Can. Entomol., Editor 1921-38. PATRONYMS: *Euphydryas colon* ab. "mcdunnoughi" Gunder, 1928; *Hemaris diffinis mcdunnoughi* Clark, 1927; *Euxoa mcdunnoughi* Cook; *Catocala mcdunnoughi* Brower. AUTH.: Checklist of the Lepidoptera of Canada and the United States of America, Part 1, Macrolepidoptera, 1938, Part 2, Microlepidoptera, 1939, Mem. So. Calif. Acad. Sci. 1 (1) and 2 (1). CONTRIB.: 313 papers, of which 213 appeared in Can Entomol; the remainder in some 23 other scientific journals.

MC GUFFIN, William Clayton (Ph.D.), Entomol. Res. Inst., Ottawa, Ontario, CANADA. Biology, ecology, morphology, and systematics. 14 February 1913, London, Ontario. EDUC.: B.A., Univ. Western Ontario, 1936; M.A., ibid., 1937; Ph.D., Univ. Minnesota, 1954. M. 27 May 1939. CH.: William Cecil. LEPID. SOC.: M-a-l Ex. Council, 1969-71; Nom. Com., 1971. OTHER: Entomol. Soc. British Columbia; Can. Entomol. Soc.; Entomol. Soc. Ontario; Professional Inst. Pub. Serv. Can.; Soc. Study Evolution. CONTRIB.: About 30 papers, published in Can Entomol; Mem Entomol Soc Can (8, 50, 86); J Lepid Soc.

MILLER, Lee D. (Ph.D.), Allyn Mus. Entomol., Sarasota, Florida, USA. Life histories, zoogeography, taxonomy, morphology. 1 June 1935, Des Moines, Iowa. EDUC.: A.B., Univ. Iowa, 1960; M.S., Univ. Pittsburgh, 1963; Ph.D., ibid., 1965. M. Jacqueline Y. Sessi, 15 April 1968. LEPID. SOC.: Ed. Com. (*News*), 1961-69; Region. Coord. (Ann. Summary), 1964-68; Ed. (Supplements), 1971- ; Ed. Com. (*Journal*), 1972- ; Sec. 1972- . OTHER: U.S. Dept. Agric., Plant Pest Control Div., Inspector, 1960-61; Carnegie Museum, Section of Insects & Spiders, Curatorial Asst., 1962-65; Catholic Univ. Amer., Asst. Prof., 1965-68; Carnegie Mus., Res. Assoc., 1965- ; Allyn Found. and Allyn Mus. Entomol., Curator, 1968-; Field Mus. Nat. Hist., Res. Assoc., 1971- ; J. Res. Lepid., Assoc. Ed., 1971- ; Bull. Allyn Mus. Entomol., Ed., 1971- ; Royal Entomol. Soc. Lond., Fellow; Soc. Syst. Zool.; Amer. Entomol. Soc.; Soc. Study Evolution. EXPED.: Mexico (1961, 66, 67, & 69). AUTH.: Experimental Entomology, with K. W. Cummins, N. A. Smith, and R. M. Fox, Reinhold, New York, 1965. MONOG.: Family Hesperiidae, *in* The Butterflies of Liberia, Mem Amer Entomol Soc (19), 1965. The Higher Classification, Phylogeny and Zoogeography of the Satyridae (Lepidoptera), Mem Amer Entomol Soc (24), 1968. CONTRIB.: About 30 scientific papers, published in Ann Carnegie Mus; Bull Allyn Mus; Bull Inst franc d'Afrique Noire; Diss Abstr; Entomol News; J Lepid Soc; J Res Lepid; Mem Amer Entomol Soc; Science; Trans Amer Entomol Soc.

MUNROE, Eugene Gordon (Ph.D.), Lepid. Sec., Biosyst. Res. Inst., Canada Dept. Agric., Ottawa, Ontario, CANADA. Systematics, biogeography (Pyraloidea, Rhopalocera, Saturniidae). 8 September 1919, Detroit, Michigan. EDUC.: B.Sc., Mc Gill Univ., 1940; M.Sc., ibid., 1941; Ph.D., Cornell Univ., 1947. M. Isobel Margaret Douglas, 1944. CH.: Janet Gordon, 1945; Donald Douglas, 1946; Susan Margaret, 1950; Elizabeth Anne, 1954. LEPID. SOC.: Charter Mem.; Nom. Com., 1951, 1952 (chm), 1955 (chm); Assoc. Ed. (*News/Journal*) 1951-53; Ed. Com. (*Journal*) 1955-65; Ed. Bd., 1964- ; Pres., 1959; Hon. Life Mem., 1972; V. Pres., 1973. OTHER: Privy Council Off. Canada, Sci. Secretariat, 1965-68; Sci. Advisor, 1965-67; Princ. Sci. Advisor & Head of Studies, 1967-68. MacDonald College, Quebec, Inst. Parasit., Lecturer & Res. Asst., 1947-50. Amer. Assoc. Adv. Sci.; Assoc. Tropical Biol.; Entomol. Soc. Amer.; Can. Soc. Zool.; Can. Entomol. Soc., Ed. 1958-61, Pres. 1964; Entomol. Soc. Quebec; Ottawa Field-Naturalists' Club; Roy. Entomol. Soc. Lond.; Roy. Soc. Canada, Fellow, Rapporteur, 1969-70 (Anim. Biol. Subj. Div.); Sigma Xi. PATRONYMS: *Ludia orinoptena monroei* [sic] Rothschild & Jordan, 1936; *Munroeia* Marion, 1954; *Munroeia* Amsel, 1956 [nec Marion] ; *Munroessa* Lange, 1956; *Munroeodes* Amsel, 1957 (replacement name for *Munroeia* Amsel). AUTH.: Fac. 13.1A, Scopariinae, Nymphulinae *in* Moths of America North of Mexico; Fac. 13.1B, Odontiinae, Glaphyriinae, ibid.; Fac. 13.1C, Evergestinae, ibid. CONTRIB.: More than 160 papers, dealing largely with the Pyralidae, and published in Ann Entomol Soc Amer; Amer Mus Novit; Bull Zool Nomenclature; Can Entomol; Entomol News; Entomologist; J Econ Entomol; J Lepid Soc; J NY Entomol Soc; Lepid News; Pacific Insects Monogr; Proc Entomol Soc Wash; Proc Int Cong Entomol; Rev Can Biol; Rev Fr Entomol; Science; Tijdsch Entomol; and others.

NEKRUTENKO, Yuri Pavlovich (Ph.D.), Senior Res. Staff, Ukrainian Res. Inst. Plant Protection, Kiev 127, Ukraine, USSR. Taxonomy. 20 April 1936, Kiev, Ukraine, USSR. EDUC.: Grad., Kiev State Univ., 1962; Ph.D., Ukrainian Acad. Agr. Sci., 1969. M. 1961, one dt. 2nd M., 1970. LEPID. SOC.:

M-a-l Ex. Council, 1969-71. OTHER: Lepid. Res. Foundation; Ukrainian Entomol. Soc.; Polish Entomol. Soc.; Assoc. Romana di Entomol. Served with Inst. Geol. Sci., Acad. Sci., Ukrainian S.S.R., 1962-67; Ukrainian Res. Inst. Plant Protection, 1967- . MONOG.: Phylogeny and Geographical Distribution of the Genus *Gonepteryx* (Pieridae), 1968. CONTRIB.: About 35 papers, published in Abh Ber Naturkundemus Goerlitz; Biol Nauki; Dopov Akad Nauk Ukr Rsr; Dtsch Entomol Z; J Res Lepid; J Lepid Soc; Nature (Lond); Paleontol Zh; Proc Int Cong Zool; Zool Zh; and others.

NEWCOMER, Erval Jackson, 1738 Naomi Place, N.E., Seattle, Washington, USA. Biology, distribution (Pacific Northwest Rhopalocera). 5 May 1890, Oregon, Illinois. EDUC.: A.B. (Entomol.), Stanford Univ., 1911. M. 1913, four ch. LEPID. SOC.: Ed. Com. (*News*) 1961-64, 1971-72; Ed. Bd. (*Journal*) 1965-69; Ed. *News*, 1964-70. OTHER: Lepid. Res. Foundation; U.S. Dept. Agr., Entomologist, 1914-55. CONTRIB.: Many papers and bulletins, mostly on fruit insect pests and Rhopalocera.

NICOLAY, Stanley S. (Col., U.S. Marine Corps, Ret.), Virginia Beach, Virginia, USA. Taxonomy (Lycaenidae, Hesperiidae). 14 March 1917, Colfax, Washington. EDUC.: B.S., Univ. Washington, 1940; Naval Officer and Aviator, May 1941. M. Lillian Emily Dozier, 31 March 1942. CH.: Michael D.; Stephen C.; Joseph J. LEPID. SOC.: Charter/Sustaining Mem.; Treas., 1969-74; Pres. 1976. PATRONYMS: *Calycopis nicolayi* Field, 1967. CONTRIB.: A number of papers describing new taxa of Hesperiidae and Lycaenidae, published in J Lepid Soc.

OBRAZTSOV, Nikolaĭ Sergeevich (Ph.D.) (1906-1966). (Obit., biblio. and portrait, *J. Lepid. Soc.* 20:255-265.) Microlepidoptera taxonomy (Ctenuchidae, Tortricidae). 18 August 1906, Rostov on the Don, RUSSIA; d. 6 May 1966, Sea Cliff,

New York, USA. EDUC.: Inst. Pedagogy, Nikolaev, Russia, 1934; Ph.D., Univ. Munich, Germany, 1951. M. Vera Nikolaëvna Obraztsova, 1934. CH.: One s. and one dt. LEPID. SOC.: Sustaining Mem.; M-a-l Ex. Council, 1955-57; Ed. Com. (*Journal*), 1958-63. OTHER: U.S. Nat. Mus. and Amer. Mus. Nat. Hist., Res. Assoc. CONTRIB.: Published 139 papers in English, French, German, Latin, Polish, Russian, and Ukrainian.

OGATA, Masami (D.M.Sc.), Osaka, JAPAN. Morphology (genitalia). 10 July 1922, Osaka, JAPAN. EDUC.: M.D., Kyoto Univ., 1946; D.M.Sc., ibid., 1953. M. 22 January 1959. CH.: Mari; Miyuki; Masato (2 dts., 1 s.). LEPID. SOC.: M-a-l Ex. Council, 1970-72. OTHER: Entomol. Soc. Amer.; Lepid. Soc. Japan; Entomol. Soc. Japan. PATRONYMS: *Callygris compositata* ab. "ogatai" Inoue, 1954. AUTH.: Icones Heterocerorum Japonicorum *in* Coloribus Naturalibus (Noctuidae & Agaristidae), 1958, Hoikusha, Osaka; 2nd ed., 1971. CONTRIB.: More than 30 papers published in Ann Entomol Soc Amer; J Lepid Soc; Trans Lepid Soc Japan.

PHILIP, Kenelm W. (Ph.D.), Assoc. Prof. Physics, Geophysical Inst., Univ. Alaska, College, Alaska, USA. Life histories, ecology (Arctic Rhopalocera). 6 November 1931, Staten Island, New York. EDUC.: B.S., Yale College, 1953; M.S., Yale Univ., 1958; Ph.D., ibid., 1963. M. Betty Anne Phillips, 26 July 1958. CH.: Mary Van Ness. LEPID. SOC.: Ed. Com. (*News*) 1970- ; Coord. Ann. Summary (Z-VIII), 1969-; V. Pres., 1971. OTHER: Amer. Astron. Soc.; Amer. Assoc. Adv. Sci.; Sigma Xi; N.Y. Acad. Sci; Alaska Conserv. Soc.; Nat. Assoc. Watch & Clock Coll.; Entomol. Soc. Amer.; Univ. Alaska, Inst. Arctic Biol., Res. Assoc.; Univ. Alaska Mus., Res. Assoc. CONTRIB.: J Lepid Soc.

POVOLNÝ, Dalibor (D.Sc.), Prof. Zool., Depart. Zool., Coll. Agr., Brno, CZECHOSLOVAKIA. Zygaena, *Lithocolletis* of central Europe. 13 November 1924, Třebíč, MORAVIA. EDUC.: Ing. Agr., Univ. Sch. Agr., Forestry & Vet. Med., Brno, 1949; Dr. techn., ibid., 1950; D.Sc., ibid., 1966. M. 2 September 1950. CH.: Drahomíra; Vladimír. LEPID. SOC.: V. Pres., 1957; M-a-l Ex. Council, 1961-63. OTHER: Czechoslovak Entomol. Soc., Presidium, Ed. Board, Sec. Gen.

Moravian Branch; Czechoslovak Zool. Sco.; Czechoslovak

Parasit. Soc.; Deutsche Entomol. Ges., Berlin; Wiener Entomol. Ges.; Soc. Syst. Zool. EXPED.: Afghanistan (1965, 66, 67 & 69). HON.: Univ. Agr. Brno, Medal De optime meritis; 50th Ann. Czechoslovak Republic, medal 1968. PATRONYMS: *Lycaena athamantis povolnyi* T.G. Howarth, (in press). AUTH.: The Members of the Genus *Zygaena* Fab. of Moravia and Silesia, Brno, 1946; The Sugar Beet Moth *Scrobipalpa ocellatella* (Boyd), Bratislava, 1960; The Study of the Sugar Beet Moth in Czechoslovak Republic, Moskva, 1958. CONTRIB.: About 125 papers published in Acta Entomol (Praha); Acta Entomol Mus Nat Pragae (Praha); Acta Faun Entomol Mus Nat Pragae (Praha); Acta Mus Moraviae (Brno); Acta Sci Nat Acad Sci Bohemoslovacae (Brno); Acta Sci Nat Moravo-Silesiacae (Brno); Acta Univ Agr Silvic; Ann Naturhist Mus Wien; Ark Zool; Bull Entomol Res; Deutsche Entomol Z; Entomol Tidskr; Folia Entomol (Brno); Folia Parasitol (Praha); Folia Zool (Brno); J Lepid Soc; Syst Zool.

POWELL, Jerry A. (Ph.D.), Prof. Entomol., Univ. California (Berkeley), USA. Biosystematics, biogeography, phylogeny & evolution (primarily microlepidoptera). 23 May 1933, Glendale, California. EDUC.: B. S. (Entomol.), Univ. Calif., Berkeley, 1955; Ph.D. (Entomol.), ibid., 1961. M. Frances Crisp, 1956. CH.: One s., two dts. LEPID. SOC.: Ed. (*Journal*) 1964-69; M-a-l Ex. Council 1973-75. OTHER: Anim. Behav. Soc.; Entomol. Soc. America; Can. Entomol. Soc.; Kansas Entomol. Soc.; Pacific Coast Entomol. Soc., V. Pres., 1963, Pres., 1964, Pub. Com. (chm), 1965-68, 1972; Pan-Pacific Entomol., Asst. Ed., 1957-58, 1960-62, Acting Ed., 1959-60. HON./AWARDS: National Sci. Foundation Grants (2) 1965-70; Smithsonian Institution, Fellowship, 1970-71; Calif. Acad. Sci., Fellow, 1971. CONTRIB.: About 100 papers published in Bull Entomol Soc Amer; Bull Mus Comp Zool; Can Entomol; Entomol News; Hilgardia; J Kans Entomol Soc; J Lepid Soc; J NY Entomol Soc; Lepid News; Pan-Pac Entomol; Proc Biol Soc Wash; Smithsonian Contr Zool; Univ Calif Publ Entomol; Wasmann J Biol.

REMINGTON, Charles Lee (Ph.D.), Dept. Biol., Yale Univ., New Haven, Connecticut, USA. Genetics, evolution, adult & immature biology and taxonomy. 1922, Reedville, Virginia. EDUC.: B.S., Principia Coll., Ill., 1943; Ph.D., Harvard Univ., 1948; Guggenheim Fellow, Oxford Univ., 1958-59. M. Jeanne Ehrenborg, 1946. CH.: Eric; Janna; Sheldon. LEPID. SOC.: Co-founder; first Treas. & Gen. Mgr., 1947-50; Co-editor, 1947; Ed.-in-Ch., 1948-63; Nom. Com., 1951 (chm); Ed. Bd., 1964- (chm); Hon. Life Mem., 1966; Pres., 1971. Received hand engraved Certificate of Merit, in recognition of his unceasing efforts in behalf of the Society, 28 Dec. 1957. OTHER: Sigma Xi; Royal Entomol. Soc. Lond.; Finish Entomol. Soc.; Soc. Study Evol.; Genetics Soc. Amer.; Ecol. Soc. Amer.; Zero Population Growth, V. Pres., 1970?; Council Pop. & Environ., Bd. Directors, 1971. PATRONYMS: *Erebia epiposodea remingtoni* Ehrlich, 1952; *Euphydryas editha remingtoni* Burdick, 1959. CONTRIB.: Many papers dealing with genetics and hybridization, published in Adv Genet; Ann Rev Entomol; Bull Brooklyn Entomol Soc; Entomol News; Evol Biol; J Lepid Soc; Lepid News; Postilla; Proc (XVI) Int Cong Zool; Psyche J Entomol; Trans Ill State Acad Sci; Yearbook Amer Philos Soc; and others.

RILEY, Norman Denbigh, 7 McKay Road, Wimbledon, London S.W. 20, ENGLAND. Taxonomy (Rhopalocera). 26 September 1890, London, ENGLAND. EDUC.: Dulwich College, 1904-09. M. Edith Vaughan, 6 May 1920. CH.: John; Pauline. MIL. SVC.: Comm. August 1914, served 1914-1918; promoted to Captain, 1916. LEPID. SOC.: Charter Mem.; M-a-l Ex. Council, 1951-52; V. Pres., 1954; Pres., 1958. OTHER: The Entomologist, Ed., 1924-1960; Roy. Entomol. Soc. Lond., Sec., 1926-29; Treas. 1939-40; Sec. 1941-50; Pres. 1951-52; Hon. Fellow 1957- ; Zool. Soc. Lond., former mem. of Council; V. Pres.; Int. Union Biol. Sci., former Sec. of Entomol. Sect.; Int. Entomol. Cong., Sec. Permanent Com. 1948-56; Amer. Mus. Nat. Hist., Corresp. Mem.; British Mus. (Nat. Hist.) Dept. Entomol., Keeper, 1932-1955; Soc. Promotion Nature Reserves, Sec., 1952-68, Treas. 1968-72; Int. Comm. Zool. Nomenclature, Commissioner, 1950-60; Philadelphia Acad. Sci. HON. MEM.: Entomol. Soc. France, Entomol. Soc. Belgium, Entomol. Soc. Quebec, So. Lond. Entomol. & N. H. Soc., Pres. 1924-25. AUTH.: Some British Moths, Penguin Books, 1944. Insects in Colour, Blandford, 1963. Field Guide to the Butterflies of

Europe, with L. G. Higgins, Collins, 1970. CONTRIB.: Many papers, published in numerous scientific journals; describing about 50 new species of Rhopalocera.

RINDGE, Frederick H. (Ph.D.), Curator Lepid., Dept. Entomol., Amer. Mus. Nat. Hist., New York, USA. Taxonomy, life histories (Geometridae). 1 January 1921, Los Angeles, California. EDUC.: B.S., Univ. Calif., Berkeley, 1942; Ph. D., ibid., 1949. M. Phyllis Jane Denton, 1943. CH.: Janet; Barbara; Marguerite. LEPID. SOC.: Charter Mem.; 1st Sec., 1951-54; Nom. Com., 1956 (chm), 1959; 1st V. Pres., 1960; Pres., 1965. OTHER: Los Angeles Co. Nat. Hist. Mus., Res. Assoc., 1973; Amer. Assoc. Adv. Sci., Fellow; Amer. Mus. Nat. Hist., Assoc. Benefactor, 1970; N. Y. Entomol. Soc.; Pacific Coast Entomol. Soc.; Sigma Xi; So. Calif. Acad. Sci., Fellow; Soc. Study Evolution. PATRONYMS: *Drepanulatrix rindgearia* Sperry, 1948; *Chaetophlepsis rindgei* Reinhard, 1952; *Pyrausta unifascialis rindgei* Munroe, 1957; *Aphelia alleniana rindgeorum* Obraztsov, 1959; *Lozotaenia rindgei* Obraztsov, 1962; *Agathymus rindgei* Freeman, 1964; *Udea itysalis rindgeorum* Munroe, 1966; *Gamelia rindgei* Lemaire, 1967; *Decinea rindgei* Freeman, "1968" (1969); *Nemoria rindgei* Ferguson, 1969; *Napeogenes ithra rindgei* Fox & Real, 1971; *Orgyia definita rindgei* Riotte, 1972; *Gesneria rindgeorum* Munroe, 1972. CONTRIB.: More than 70 papers, including 25 generic revisions, published in Amer Mus Novit; Ann Carnegie Mus; Bull Amer Mus Nat Hist; Bull Zool Nomenclature; J Lepid Soc; J NY Entomol Soc; Lepid News; Pan-Pac Entomol; Proc Calif Acad Sci.

RIOTTE, Jules Charles Emile (Rev.), Res. Assoc., Entomol.,

Roy. Ontario Mus., Toronto, CANADA. Holarctic Sphingidae and Notodontidae. 22 April 1901, Europe. EDUC.: L.Th., Wroclaw; M.A., Univ. Montréal. LEPID. SOC.: Sustaining Mem.; M-a-l Ex. Council, 1961-63. OTHER: Parish Priest, northern Ontario; Econom of Eparchy of Toronto, 1962- ; Roy. Ontario Mus. (Entomol.), Res. Assoc., 1962- ; Michigan Entomol. Soc., Toronto-Ontario Branch, Chairman, 1967- ; Toronto Entomol. Assoc., Occasional Pub., Chm. Ed. Board, 1970- ; Mem. other entomol. soc. in Canada, USA and Europe. CONTRIB.: About 40 papers, published in Beitr Entomol; Dtsch Entomol Z; Entomol Z; J Lepid Soc; J Res Lepid; Lepid News; Life Sci Contr Roy Ontario Mus; Mich Entomol; Ont

Field Biol; Proc Entomol Soc Ont; Mid-Cont Lepid Ser.

SARGENT, Theodore David (Ph.D.), Prof., Dept. Zool., Univ. Massachusetts, Amherst, USA. Behavior, ecology (especially *Catocala*). 25 October 1936, Peabody, Mass. EDUC.: B.S., Univ. Mass., 1958; M.S., Univ. Wisconsin, 1960; Ph.D., ibid., 1963. M. Katherine A. Doktor, 19 March 1966. CH.: David E.; Meryl A. LEPID. SOC.: Ed. *Journal*, 1972-74; V. Pres., 1976. OTHER: Amer. Assoc. Adv. Sci.; Amer. Ornith. Union; Anim. Behav. Soc.; British Ornith. Union; NY Entomol. Soc.; Northeastern Bird-Banding Assoc.; Soc. Study Evolution. AWARDS/HON.: Phi Eta Sigma, 1955; Phi Kappa Phi, 1958. AUTH.: Legion of Night: The Underwing Moths. 1976. U. Mass. Press. CONTRIB.: About thirty papers, published in Anim Behav; Auk; J Lepid Soc; J NY Entomol Soc; Nature (Lond); Science; and others.

SEVASTOPULO, Demetrius George, P.O. Box 5026, Mombasa, KENYA. Life histories, genetics. 18 September 1903, London, ENGLAND. EDUC.: Stanmore House Prep. Sch.; Haileybury Coll. M. 5 March 1940. CH.: George Demetrius, 1941; John Demetrius, 1942; Marigo Anne, 1951. LEPID. SOC.: V. Pres., 1967. OTHER: Roy. Entomol. Soc. Lond.; British Entomol. & Nat. Hist. Soc., Life Mem.; Bombay Nat. Hist. Soc., Life Mem.; Bengal Nat. Hist. Soc., Life Mem. PATRONYMS: *Leucoma sevastopuloi* Collenette, 1955; *Soritia sevastopuloi* Tremewan, 1959; *Sevastopuloa* (Lymantriidae) Collenette, 1960; *Rhodometra sevastopuloi* Carcasson, 1964. CONTRIB.: More than 300 papers published in Entomologist; Entomol Mon Mag; Entomol Rec & J Var; J Bengal Nat Hist Soc; J Bombay Nat Hist Soc; Trans Roy Entomol Soc Lond; Trans So Lond Entomol Soc; J East African Nat Hist Soc (Nature in E Africa); J Lepid Soc.

STALLINGS, Don Brigg (Juris Dr.), Caldwell, Kansas, USA. Racial speciation (Megathymidae). 4 August 1910, Oswego, Kansas. EDUC.: LL.B., Univ. Kansas, 1933; Juris Dr., ibid., 1968. M. Viola N. Turner, 30 January 1933. CH.: Don B. Jr.; John Reuben. LEPID. SOC.: Charter/Sustaining Mem.; 1st Coordinator, Ann. Summary (Reg. IV), 1947; Sec., Organization Comm., 1950; M-a-l Ex. Council, 1954-56; V. Pres., 1959; 1st V. Pres., 1963; Pres., 1966-67. OTHER: Yale Univ., Peabody Mus. Nat. Hist., Res. Affl. (Entomol.), 1963-. Central States Entomol. Soc.; Ecol. Soc. Amer.; Entomol. Soc. Amer.; Soc. Syst. Zool. PATRONYMS: *Celaenorrhinus stallingsi* Freeman, 1946; *Stallingsia* Freeman, 1958 (1959); *Calephelis stallingsi* McAlpine, 1971. CONTRIB.: Thirty-

five papers, largely on the Megathymidae, published in Bull So Calif Acad Sci; Can Entomol; Proc Denver Mus Nat Hist; Entomol News; J Lepid Soc; J Kans Entomol Soc; Lepid News; Pan-Pac Entomol.

SUOMALAINEN, Esko (Ph.D.), Prof., Inst. Genetics, The Univ., Helsinki, FINLAND. Genetics, Cytology, Lepidoptera of Scandinavia. 11 June 1910, Helsinki, FINLAND. EDUC.: M.Sc., Univ. Helsinki, 1934; Ph.D., ibid., 1940. M. Tellervo Tynni, 1943. CH.: Kaarina, 1944; Olavi, 1947. LEPID. SOC.: V. Pres., 1966. OTHER: Univ. Helsinki, Lect. genetics, 1941-48, Prof., 1948- , Dept. Head, 1949- ; Kaiser Wilhelm-Inst. Biol., Berlin-Dahlem, Invit. sci., 1942-43; Univ. Lund (Sweden), Visit. Prof., 1954; Academy of Finland, Sec., 1951-56; United Nations Educ., Sci. & Cultural Organization, Finnish Nat. Commis., V. Pres., 1957-65, Pres., 1965; Finn. Acad. Sci. & Letters, Sec. of Sci., 1951-64; Sec. Gen., 1964-70; Soc. Sci. Fenniae; Roy. Physiographic Soc. Lund; Roy. Danish Acad. Sci. & Letters; Amer. Assoc. Adv. Sci.; Amer. Soc. Human Genet.; Deutsche Ges. Genetik; Deutsche Ges. Vererbungswissenschaft; Entomol. Soc. Copenhagen; Entomol. Soc. Finland, Council Mem., 1937-62; V. Pres., 1948-62; Entomol. Soc. Lund; Finn. Biol. Soc. Vanamo, Council Mem., 1938-61; Sec., 1942-45; Finn. Lepid. Soc., Pres., 1955- ; Scandinavian Assoc. Genet., Council Mem., 1960- ; V. Pres., 1971- ; Scandinavian Soc. Cell Biol.; Soc. Biochemica, Biophysica et Microb. Fenn.; Soc. Genet. Fenn., V. Pres., 1944-65; Pres., 1965- ; Soc. Geogr. Fenn.; Soc. Gerontologica Fenn., Council Mem., 1958-59; Soc. pro Fauna et Flora Fenn. ED.: Ann Acad Sci Fenn A IV Biol; Ann Zool Fenn; Ann Bot Fenn. ED. BD.: Caryologia; Exp Cell Res; Hereditas; Luonnon Tutkija; Not Entomol; Valvoja; Z Zool Syst Evolutionsforsch. Cytologia (Tokyo), Standing Collab. CONTRIB.: Many papers, published in more than 30 scientific journals.

THORNE, Fred T., El Cajon, California, USA. Life histories of Nearctic Rhopalocera. 25 June 1909, Casper, Wyoming. EDUC.: San Diego State Coll., 1937-41. M. Dorothy Mary Argo, 1931. CH.: Valerie Barbara, 1932; Diana Patricia, 1939. LEPID. SOC.: Charter/Sustaining

Mem.; 1st V. Pres., 1958; M-a-l Ex. Council, 1967-69; Ed. Com. (*Journal*), 1957-69; Assoc. Ed. (*Journal*) (Especially for collectors) 1957-66; Ed. Com. (*News*), 1960, 1965-69; Regional Ed., Ann. Summary (Z-I), 1961-69; Program Com. (Pacific Slope Sec. meetings) 1954-56, 1960, 1965, 1969; Program Com. (Nat. meetings) 1953, 1962. OTHER: Entomol. Soc. Amer.; Los Entomologos; Lepid. Res. Foundation; Lorquin Entomol. Soc; San Diego Nat. Hist. Soc., Fellow. CONTRIB.: Bull So Calif Acad Sci; J Lepid Soc; J Res Lepid; Los Entomol; Mid-Cont Lepid Ser.

TILDEN, James Wilson (Ph.D.), Prof. Emeritus Entomol., San Jose State Coll., California, USA. Taxonomy (Rhopalocera western USA). 31 December 1904, Philo, Calif. EDUC.: A.B., San Jose State Coll., 1942; M.S., Stanford Univ., 1947; Ph.D., ibid., 1948. M. Hazel I. Miller, 19 June 1943. CH.: James W. Jr.; Bruce A.; Janice E. LEPID. SOC.: Charter Mem.; M-a-l Ex. Council, 1957-59; Ed. Bd. and Ed. (*News*), 1959-64; V. Pres., 1970. OTHER: Amer. Mus. Nat. Hist., Assoc. Mem.; Calif. Acad. Sci., Fellow, 1968; Calif. Tomorrow; Coleopterists' Soc.; Entomol. Soc. Amer.; Lepid. Res. Foundation, Charter; National Geog. Soc.; Pac. Coast Entomol. Soc., Pres., 1960; Soc. Syst. Zool; Western Soc. Naturalists. PATRONYMS: *Hesperia tildeni* Freeman, 1956; *Philotes enoptes tildeni* Langston, 1964. CO-AUTH.: Yosemite Butterflies, 1963. AUTH.: Butterflies of the San Francisco Bay Region, 1965. CONTRIB.: About 50 papers, published in Bull So Calif Acad Sci; Calif Vector Views; Coleopt Bull; Entomol News; J Lepid Soc; J Res Lepid; Lepid News; Microentomology (NY); Pan-Pac Entomol; Syst Zool; Turtox News; Wasmann J Biol.

TRAVASSOS, Lauro Pereira (Prof.) (1890-1970). (Obit. and portrait, *Atas Soc. Biol. Rio de Janeiro*, 15(1), 1971). 2 July 1890, Angra dos Reis, Rio de Janeiro, BRAZIL; d. 20 November 1970. LEPID. SOC.: Charter Mem.; Organization Com., 1950; Pres., 1955. OTHER: Inst. Oswaldo Cruz, Medical Zool. Div., Director; Nat. Veterinary Sch. Med.,

Univ. Sao Paulo, Cathedratic Prof., Parasitology; Univ.

Federal Dist. Sch. Sci., Prof. Zool.; Brazilian Acad. Sci.; National Acad. Med.; Soc. Biol. Rio de Janeiro, Founder; Soc. Entomol. Brazil; Brazilian Soc. Progress Sci.; Helminth. Soc. Washington; Zool. Soc. Lond.; Acad. Nazionale dei Lincei. HON.: Alfred Jouzikowski prize, National Acad. Med., 1968; hon. mention, Oswaldo Cruz prize, 1970. MONOG.: Revision of the family Trichostrongylidae Leiper, 1912, Monog. 1, Oswaldo Cruze Inst., 1937. Revision of the family Dicrocoelidae Odhner, 1911, Monog. 2, Oswaldo Cruz Inst., 1944. CONTRIB.: Published 440 papers on Zoology, ranging from protozoans to Arthropods, and including about 80 papers on Lepidoptera. Of these, 46 were on Arctiidae and 14 on Adelocephalidae, published in Ann Acad Bras Cienc; Arq Mus Nac Rio de J; Atas Soc Biol Rio de J; Bol Mus Nac Rio de J Zool; Bol Mus Paraense Emilio Goeldi Zool Nova Ser Zool; Mem Inst Oswaldo Cruz (Rio de J); Rev Bras Biol; Rev Entomol; Rodriguesia.

VÁZQUEZ GARCÍA, Leonila (Ph.D.), Inst. Biol., Lab. Entomol., National Autonomous Univ. Mexico,

MEXICO. Rhopalocera life histories. 17 January 1914, Jiménez, Chihuahua, MEXICO. EDUC.: M.S. (Biol.), Nat. Auton. Univ. Mexico, 1936; Ph.D., ibid. 1946. LEPID. SOC.: V. Pres., 1958, 1969. OTHER: Entomol. Soc. Amer., 1939- ; Mexican Soc. Nat. Hist., 1946-; Mexican Soc. Entomol., Founder 1952; Treas., 1952-63; Acad. Sci. Investig. Mexico, Founder 1959; Treas., 1959-61; Soc. Study Evolution, 1961- ; Entomol. Soc. Rome, 1965-68; Mexican Soc. Genet., Founder 1967; Assoc. Tropical Biol., 1968- . AWARDS/HON.: Univ. Merit Award, 1971, Nat. Auton. Univ. Mex. (for 35 years of academic work); Entomol. Merit Award, 1971, Mexican Soc. Entomol. PATRONYMS: *Synopsia leonilaria* Hoffmann, 1936; *Heliconius charitonius vazquezae* Comstock & Brown, 1950; *Anaea (Memphis) morvus leonila* Comstock, 1961. CONTRIB.: More than 40 papers describing an equal number of new taxa, published in Ann Entomol Soc Amer; Ann Inst Biol Univ Nac Auton Mex; Mexican Entomol Folio; Publ. Geophysics Inst., Univ. Mexico.

VOSS, Edward G. (Ph.D.), Prof. Botany & Curator Vascular Plants, Univ. Michigan Herbarium, Ann Arbor, Michigan, USA. Lepidoptera of Northern Michigan, Hesperiidae of world. 22 February 1929. EDUC.: B.A. (Biol.), Denison Univ., 1950; M.A. (Biol.), Univ. Mich., 1951; Ph.D., (Bot.), ibid., 1954. LEPID. SOC.: Charter/Life Mem.; Ed. Com. (*News*), 1959-62; 1st V. Pres., 1961. OTHER: Amer. Inst. Biol. Sci.; Amer. Soc. Plant Taxonomists; Intern. Assoc. Plant Taxon., Sec. Ed. Com., Intern. Code Bot. Nomen., 1969- ; Sec. Gen. Com. Bot. Nomen., 1969- ; Ohio Acad. Sci.; Soc. Syst. Zool.; Soc. Biblio. Nat. Hist.; Mich. Entomol. Soc.; Cranbrook Inst. Sci.; Hist. Soc. Mich.; Little Traverse Reg. Hist. Soc.; Isle Royale Nat. Hist. Assoc.; Mich. Acad. Sci.; Mich. Bot. Club, Ed., 1962- ; Mich. Natural Areas Council; Mich. Parks Assoc.; Mich. United Conserv. Clubs; Fed. Ontario Nat.; Assoc. Minn. Entomol.; Amer. Mus. Nat. Hist.; Nat. Geographic Soc.; Nat. Wildlife Fed.; Wilderness Soc.; Sierra Club; Nat. Audubon Soc. HON.: Phi Beta Kappa; Omicron Delta Kappa; Sigma Xi; Phi Kappa Phi; Eta Sigma Phi; Soc. Religion Higher Educ., Kent Fellow, 1952; Univ. Mich., Emma J. Cole Fellow (Bot.), 1953-54. CONTRIB.: Amer Midl Nat; Ann Entomol Soc Amer; Encycl Britannica; J Lepid Soc; Lepid News; Mich Entomol.

WARNECKE, Georg Heinrich Gerhard (Dr. h.c.) (1883-1962) (Obituary, *News Lepid. Soc.* 1:1/63; bibliography (282 titles), *Verh. Ver. naturw. Heimatforsch* 36:7-17) Zoogeography and migration. d. 20 September 1962, Hamburg, GERMANY. LEPID. SOC.: V. Pres., 1956.

WILLIAMS, Carrington Bonsor (Sc.D.), Kirkcudbright, SCOTLAND. Insect migration, ecology. 7 October 1889, Liverpool, ENGLAND. EDUC.: B.A., Univ. Cambridge, 1911; M.A., ibid., 1916; Res. Student, John Innes Horticultural Inst., 1911-16; Sc.D., Univ. Edinburgh, 1930. M. Ellen Margaret Bain, 1920. CH.: Three s. LEPID. SOC.: Nom. Com., 1951; V. Pres., 1969; Hon. Life Mem., 1959. OTHER: Roy. Entomol. Soc. Lond., Hon. past Pres.; British Ecol. Soc., Hon. past Pres.; Amateur Entomol. Soc. (Britain), past Pres.; Netherlands Entomol. Soc., Hon. Life Mem.; Entomol. Soc. Egypt, Hon. Life Mem. AUTH.: Migration of Butterflies, 1930; Insect Migration, 1958; Patterns in the Balance of Nature, 1964; Style and Vocabulary-Numerical Studies, 1970. CONTRIB.: Numerous papers, about 60 on insect migration, published in Ann Entomol Soc Amer; Bull Soc Entomol Egypt; Entomologist; J Bombay Nat Hist Soc; J East Africa Uganda Nat Hist Soc; J Lepid Soc; Lepid News; Nature (Lond); Proc Int Cong Zool; Proc Roy Soc Lond Sec B; Trans Roy Entomol Soc Lond.

ZIEGLER, John Benjamin (Ph.D.), Summit, New Jersey, USA. Systematics (Lycaenidae, Theclinae, Riodinidae). 2 January 1917, Rochester, New York. EDUC.: B.S., Univ. Rochester, 1939; M.S., Univ. Illinois, 1940; Ph.D., ibid., 1946. LEPID. SOC.: Sustaining Mem; Treas., 1951-52; Nom. Com., 1956, 1964, 1966. OTHER: Amer. Inst. Chem., Fellow; Lepid. Res. Foundation; N. Y. Acad. Sci. HON.: Sigma Xi; Alpha Chi Sigma; Phi Lambda Upsilon. CONTRIB.: Bull Brooklyn Entomol Soc; J Lepid Soc; J NY Entomol Soc; Lepid News.

FINANCIAL SPONSORS

Members

Adams, Christian	*New Jersey*
Ae, S. Albert	*JAPAN*
Agenjo, Ramon	*SPAIN*
Argicola, David V.	*Ohio*
Ahrenholz, David	*Minnesota*
Alderman, Atton L.	*Oregon*
Andersen, William A.	*Maryland*
Anderson, Lauren D.	*California*
Anderson, Richard A.	*New York*
Anhilger, Carl L.	*Wisconsin*
Armin, L. Clair	*California*
Armstrong, William E. & Josephine	*New York*
Arnold, Richard A.	*California*
Aronheim, Robert E.	*Connecticut*
Axtman, Michael*	*Texas*
Azevedo, John J., Jr.	*California*
Babcock, William F.	*Ohio*
Badger, F. Sidney, Jr.	*California*
Barksdale, Charles M.	*California*
Bellinger, Peter F.	*California*
Belmont, Robert A.	*California*
Blanchard, André & May Elise	*Texas*
Boughner, Jackson L.*	*Florida*
Bowe, John J.	*New Jersey*
Bracher, Ray W.*	*Indiana*
Brault, Loran D.*	*New York*
Brewer, Jo T.	*Massachusetts*
Bridges, Charles A., III	*New Mexico*
Bristol, Maurice L. (d. 1975)	*Illinois*
Broadway, Dale E.*	*Illinois*
Brower, Auburn E.	*Maine*
Brower, Lincoln P.	*Massachusetts*
Brown, Richard M.	*California*
Bruggemann, Paul F.*	*(Ontario) CANADA*
Bryson, Charles T.	*Mississippi*
Burns, John M.	*District of Columbia*
Cadbury, John W.	*New Jersey*
Calkins, Virgil F.	*Washington*
Cannon, Marvin S.	*Texas*

Carter, Nicholas G.	*Maryland*
Chandler, Beatrice M.	*Ohio*
Chermock, Ralph L.	*Alabama*
Cilek, Jim*	*Indiana*
Clarke, J. F. Gates	*District of Columbia*
Cooper, John T.	*California*
Covell, Charles V., Jr.	*Kentucky*
Cowan, Charles F.	*ENGLAND*
Dalpis, John R.	*Pennsylvania*
DeFoliart, Gene R.	*Wisconsin*
Detmar, Arthur O.	*Colorado*
DeWind, Joan M.	*Connecticut*
Dick, Martha	*Colorado*
Dimock, Thomas E.	*California*
Dockweiler, John E. S.	*California*
Dominick, Richard B. (d. 1976)	*South Carolina*
Dornfeld, Ernst J.	*Oregon*
Douglas, Malcolm G.	*Arkansas*
Draper, Jerome C., Jr.	*California*
Duncan, J. B.	*Utah*
Edgecomb, Homer E.	*California*
Ehrlich, Paul R.	*California*
Emmel, Thomas C.	*Florida*
Ettman, James K.	*Kentucky*
Eyer, John R.*	*New Mexico*
Fales, John H.	*Maryland*
Falk, Baldhard G.	*California*
Ferge, Leslie A.	*Wisconsin*
Ferguson, Douglas C.	*District of Columbia*
Fitter, James P.	*Oklahoma*
Flaschka, Herman A.	*Georgia*
Fowler, Timothy H.	*Massachusetts*
Franclemont, John G.	*New York*
Frederick, Albert C.	*Texas*
Froemel, E. A.	*Nebraska*
Funk, Richard S.	*Illinois*
Gage, Edward V.	*Texas*
Gardner, Karl R.	*Pennsylvania*
Gatrelle, Ronald R.	*South Carolina*
Gehring, George R.	(not listed)
Gertsch, W. J.	*Arizona*
Gibson, Loran D.	*Kentucky*
Gillette, Clyde F.	*Utah*
Glick, Perry A.	*Texas*
Good, Albert I.	*Pennsylvania*
Gorelick, Glenn A.	*California*
Gray, Richard E.	*New Hampshire*

Greenfield, Joseph C., Jr.	*North Carolina*
Gregg, Robert V.	*Kentucky*
Griepentrog, Elmer L.	*Oregon*
Groothuis, Dennis	*Illinois*
Haas, Theodor D.	*New York*
Habeck, Dale H.	*Florida*
Hansen, Kenneth C.	*Arizona*
Harrington, Don G.	*Texas*
Harris, Lucien, Jr.	*Georgia*
Harry, Jack L.	*Colorado*
Hedges, William A.	*Arizona*
Heineman, Bernard	*New York*
Heitzman, John R.	*Missouri*
Hessel, Sidney A. (d. 1974)	*Connecticut*
Hinshaw, Horton C., Jr. *	*California*
Holmes, James P.	*Massachusetts*
Howard, Maurice L.	*Colorado*
Hoying, Louis A.	*Ohio*
Hughes, Keith C.	*California*
Husher, Neil*	*Massachusetts*
Hutchinson, Kerry K.*	*California*
Imig, Robert*	*California*
Irwin, Roderick R.	*Illinois*
Johnson, Kurt	*New York*
Justice, John A.	*New Mexico*
Karpuleon, Fay H.	*Wisconsin*
Keji, Joseph A.	*New York*
Kendall, Roy O. & Connie	*Texas*
Kimball, Charles P.	*Massachusetts*
Kimura, Michio G.	*Washington*
Klopp, Wayne	*Florida*
Knowlton, Carroll B., Jr.	*New York*
Knudsen, J. P.	*North Carolina*
Kopp, Larry J.	*Pennsylvania*
Koyama, Bob	*California*
Kuehn, Roger M.	*Wisconsin*
Lawrence, Donald A. & Mrs.	*Illinois*
Leeuw, Irwin	*Illinois*
Leser, James F.	*Arizona*
Letsinger, Rae	*Missouri*
Leuschner, Ronald	*California*
Levatich, Julius*	*New York*
Lienesch, John F.	*Washington*
Lipes, Jack E.*	*Maryland*
Littahorsky, Anton	*Missouri*
Lombardini, John B.	*Texas*
Makielski, Sally K.	*Louisiana*

Martin, Leland L. *Ohio*
Martin, Lloyd M. & Dorothy *Arizona*
Mason, John W. *Maryland*
Mather, Bryant & Katherine *Mississippi*
Maudsley, James R. *Georgia*
Maxwell, Nancy N. *Texas*
May, Robert L. *New Jersey*
McAlpine, Wilbur S. & Mrs.* *Michigan*
McBurney, John T. *California*
McElvare, Rowland R. *North Carolina*
McGowen, J. H., Jr.* *California*
McGuffin, W. C. *(Ontario) CANADA*
McHenry, Paddy *California*
McLean, L. N. *New York*
Melton, Lee J., III *Maryland*
Merritt, James R. *Kentucky*
Merz, David C. *New York*
Miller, Thomas A. *Maryland*
Mitchell, Robert T. *Maryland*
Monk, Harry C. *Tennessee*
Munroe, Eugene G. *(Ontario) CANADA*
Nelson, Robert P. *(British Columbia) CANADA*
Nielsen, Mogens C. *Michigan*
Nielson, Gregory *Illinois*
O'Brien, Michael *New Jersey*
Osmundson, Sharpe H. *California*
Parshall, David K. *Ohio*
dos Passos, Cyril F. *New Jersey*
Patterson, Donald *California*
Perkins, Edwin M., Jr. *California*
Perkins, Owen A. *Michigan*
Persky, Barry S. *New York*
Pfeiler, Edward J., Jr. *MEXICO*
Philip, Kenelm W. *Alaska*
Phillips, Lyle V.* *Utah*
Phillips, W. Levi *Utah*
Phillipson, Donald E. *Colorado*
Pierce, Mark W.* *Illinois*
Powell, Llewellyn* *Virginia*
Prentiss, Jack B. *Texas*
Preston, Floyd W. & June *Kansas*
Proshold, Fred *North Dakota*
Reimer, Jonathan *Indiana*
Richter, Max L. *New York*
Roberds, Joseph *California*
Roeller, Herbert * *Texas*
Rogers, Wesley H.* *Washington*

Rolfe, William S.	*New York*
Roman, Nicholas A.	*Oregon*
Rozman, Robert S.	*Maryland*
Rupert, Laurence R.	*New York*
Rupprecht, Jerome	*Pennsylvania*
Rutkowski, Frank E.	*New York*
Sanford, Robert (Mrs.)*	*Michigan*
Schedivy, Milton E.	*New Jersey*
Schmitt, August	*New York*
Schoene, William J.	*California*
Schoenfeld, Morris	*New York*
Schrader, L. G.*	*Iowa*
Sekerman, Charles A.	*California*
Selman, Charles L.*	*Ohio*
Sheppard, Arthur C.	*(Quebec) CANADA*
Shields, James E.	*Indiana*
Shields, Oakley	*California*
Shull, Ernest M.	*Indiana*
Sieker, William E.	*Wisconsin*
Sielski, Lester B.	*Indiana*
Sleeter, Ronald T.	*Illinois*
Smith, David S.	*Texas*
Smith, Eugene S.	*Connecticut*
Smith, Michael J.	*California*
Smith, Richard H., Jr.	*Virginia*
Snocker, Franklin D.	*Nebraska*
Spelman, M.	*Florida*
Stallings, Don B. & Viola	*Kansas*
Stanford, Ray E.	*Colorado*
Steele, Wayne J.	*California*
Stein, George L.	*Florida*
Sternitzky, Robert F.*	*Arizona*
Steward, John P.	*California*
Stewart, Robert H., Jr.	*California*
Swank, Charles K.*	*Illinois*
Taylor, Milton D.	*South Carolina*
Taylor, Orley R., Jr. & Toni	*Kansas*
Thomas, Roy R. & Mrs.	*New York*
Thorne, Fred T.	*California*
Tilden, James W. & Hazel	*California*
Tindale, Norman B.	*California*
Toalson, George & Nan	*Texas*
Toliver, Michael E.	*Illinois*
Travis, Irven*	*Pennsylvania*
Treat, Asher E.	*New Jersey*
Vázquez, G., Leonila	*MEXICO*
Vernon, John B.	*Texas*

Voss, Edward G.	*Michigan*
Wagner, Warren H. Jr.	*Michigan*
Was, Howard	*Wisconsin*
Wells, Ralph E.	*California*
Werschkow, George	*Pennsylvania*
Whittaker, Robert H.	*New York*
Wiley, James R.*	*Illinois*
Williams, Benjamin D. III	*Massachusetts*
Wilson, Kent H.	*Oklahoma*
Wind, Clo M.	*California*
Winter, William D. Jr.	*Massachusetts*
Woodley, Robert E.	*Washington*
Wright, David A.*	*New Jersey*
Yu, C. K.*	*FORMOSA*
Ziegler, J. Benjamin	*New Jersey*

Dealers

(see Advertisements following pages)

Australian Entomological Magazine	*AUSTRALIA*
Queensland Butterfly Company	*AUSTRALIA*
The Butterfly Company	*New York*
K. H. Chen	*(Formosa) TAIWAN*
E. W. Classey, Ltd.	*ENGLAND*
Entomological Reprint Specialists	*California*
Muh Sheng Entomological Center	*(Formosa) TAIWAN*
Insect Lore Products	*California*
Lumawig-Ocampo Enterprise	*PHILIPPINES*
Sciences Nat	*FRANCE*
Miguel Serrano	*CENTRAL AMERICA*
Taiwan Novelty Company, Ltd.	*(Formosa) TAIWAN*
Clo Wind Company	*California*

* not shown in the 1976 membership list

Commercial Firms

ENTOMOLOGICAL BOOKS & LITERATURE, PRINTERS & PUBLISHERS

AUSTRALIAN ENTOMOLOGICAL MAGAZINE
14 Chisholm Street
Greenwich, N.S.W. 2065, AUSTRALIA

E. W. CLASSEY, LTD.
Park Road, Faringdon, Oxon, SN7 7DR, ENGLAND
(*see Advertisement next page*)

ENTOMOLOGICAL REPRINT SPECIALISTS
P.O. Box 77224, Dockweiler Station, Los Angeles, California 90007, USA (*see Advertisement next page*)

SCIENCES NAT
2, Rue André Mellenne, Venette F-60200, Compiegne, FRANCE. SUPPLIES & EQUIPMENT (wide variety selected around the world and our own made) BOOKS (current and antiquarian in all languages) LEPIDOPTERA SPECIMENS (living and dead) NOTE: *Importation of live material requires appropriate government permits.*

LEPIDOPTERA, OTHER BIOLOGICAL SPECIMENS AND MATERIALS

THE BUTTERFLY COMPANY
51-17 Rockaway Beach Blvd., Far Rockaway, New York 11691 USA (*see Advertisement preceding page*)

CLO WIND COMPANY
827 Congress Avenue, Pacific Grove, California 93950 USA. ENTOMOLOGICAL EQUIPMENT Insect SPECIMENS from all parts of the world.

MUH SHENG ENTOMOLOGICAL CENTER
3 Ton Rong Road, Puli, Taiwan 545, REPUBLIC OF CHINA. PURCHASE butterflies and beetles from India, South Pacific Islands, South America and Africa. SELL Formosan butterflies, beetles and their art work products.

K. H. CHEN
P.O. Box 17-90, Taipei, TAIWAN (Formosa)

QUEENSLAND BUTTERFLY COMPANY
Long Road, Nth. Tamborine, Queensland 4272, AUSTRALIA. Worldwide LEPIDOPTERA, especially Australian and New Guinea. Most specimens from Australian region *ex pupae*. Write for catalogue. Wanted to contact WHOLESALE SUPPLIERS of A-1 material.

LUMAWIG-OCAMPO ENTERPRISE
P. O. Box 2684, Manila, PHILIPPINES
PRESERVED Philippine biological specimens. Land and marine animals including insects, especially LEPIDOPTERA and COLEOPTERA, mostly identified. Individual, museum, and institutional orders welcome. Write for list.

MIGUEL SERRANO
Avenida Alfredo Espino y Calle "B", Colonia, Campestre, San Salvador, El Salvador, CENTRAL AMERICA.

TAIWAN NOVELTY COMPANY, LTD.
P.O. Box 860, Taipei, TAIWAN (Formosa)

INSECT LORE PRODUCTS
P.O. Box 1535, Shafter, California 93263 USA

INDICES

Information and data contained in the *News/Journal* published by the Society, during its first quarter-century of service to the world community of lepidopterists, are here indexed under three categories: author, subject, and scientific names; each is arranged in alphabetical sequence.

The Author index gives the surname first followed by given names or initials mostly. Then follows the reference citation, consisting of the volume and page number(s). If more than one paper was published by the author, the references are given in numerical sequence by volume number but only the first page of each reference is cited.

The Subject index is not a listing of published titles, but rather an index to the subject matter contained in these published papers. It is believed that such information will provide a better research tool than simply listing the numerous published titles. In most instances inclusive page references are given in this index.

The index of Scientific Names gives the species and genus in that order. Subspecies follow below but the genus is not repeated. Because generic names are frequently changed, it was thought arrangement by species would be more useful to the student. In those instances where the same species has been recorded for different genera, these too are listed in alphabetical order. No attempt has been made to correct misspellings; it was thought this was best left to the researcher. Following each species are the reference citations, again arranged in numerical sequence by volume number. As pointed out in the Introduction and in the text (p. 30), the letter S preceding a reference citation means one of the Supplement series. In volume 2-supplement (Field Season Summary, but not part of the summary per se) is a list of species and their larval foodplants by J. G. Franclemont, these are cited in the index as "2:S-viii." New taxa are shown in **BOLD FACE** type, including family, genus, species and subspecies which are new.

A feature of this index, not found in most indices of this kind, is the indication of published illustrations. Simply by referring to the scientific names, the reader can determine what illustrations (I), if any, were published pertaining to the species. Symbols used to denote illustrations are: (IA) adult, (IC) components - head, antennae, wing, etc., (IF) larval foodplant, (IG) genitalia, (IH) habitat, and (II) immatures. In addition, locally published larval foodplants are shown immediately below the biological species, indented four spaces. These features are demonstrated in the following examples from the index:

ABRAHAMI, *Carystoides (IA, IG) S2:34* [The new species, *abrahami* is described in Supplement 2, page 34- and the author illustrates both the adult or imago and its genitalia.]

anguina, Dasylophia 2:S-viii
    Robinia pseudoacacia [Here a larval foodplant is recorded. The reference cited is not one of the Supplement series but supplement to volume 2 as explained above.]

anicia, Euphydryas 2:92; 6:105; 8:3; 12:165; 18:228; 25:246
anicia alena 22:160
anicia eurytion 10:96; 14:184
anicia helvia 19:59 [This example lists three subspecies of *Euphydryas anicia* and reference citations.]

Authors

ABBOTT, Charles H. 16:229
ABBOT, John 2:28
ACKERMANN, Otto 2:31
ADAMS, M. S. 17:159, 22:231
AE, Shigeru Albert 10:9, 143, 11:207, 14:5, 15:175, 17:163
ALBRIGHT, Ray 14:158
AMSEL, H. G. 19:117
ANDERSEN, William A. 14:62, 15:99
ANTHONY, G. S. 23:103
APTER, R. L. 19:195
ARBOGAST, R. T. 10:112
ARNHOLD, F. R. 6:99
ARNOLD, A. 22:173
ARNOLD, R. A. 22:173, 23:257
AUSTIN, Edward J. 10:55
AUSTIN, George T. 10:55

BADGER, F. Sidney 12:41
BAILEY, E. G. 4:14
BAKER, James H. 7:15, 13:217
BAUER, David L. 2:81, 3:25, 7:146, 8:129, 9:7, 49, 12:97, 13:165, 207, 14:148, 15:91, 21:212
BAUER, W. R. 20:83, 21:235
BEALL, Geoffrey 5:37, 6:69, 7:41
BEEBE, Ralph 3:35, 7:28, 8:26, 9:142
BEIRNE, Bryan P. 4:25, 5:27
BELKIN, J. N. 6:61
BELL, Ernest L. 3:36, 59
BELLINGER, Peter F. 7:27, 170, 8:102, 9:23, 83, 161, 216, 10:58, 121, 179, 223, 11:170, 12:36, 40, 13:15, 35, 253, 14:175
BENNETT, D. 22:253
BERTONI, M. S. 22:231
BERUBE, D. E. 24:220
BIRD, Charles D. 10:107
BLANCHARD, André 17:234, 18:42, 20:247, 22:133, 23:274, 24:249
BLEVINS, T. B. 3:26, 48
BOCK, Theodore 3:74, 4:73, 5:70
BOLTE, K. 25:266
BORODIN, D. N. 3:29
BOSCOE, William F. 16:174

BOUTON, David W. 16:128, 18:157
BOWDEN, S. R. 12:7, 18:91, 19:170, 20:189, 25:6
BOWER, H. M. 15:64, 17:36
BRAUN, Annette F. 4:3
BREWER, J. 20:45, 235
BRONSKILL, Joan F. 15:102
BROOKS, James C. 16:198
BROU, V. A. 24:239
BROWER, A. E. 1:19, 57, 63, 6:37, 12:47, 14:78
BROWER, Jane Van Zandt 8:125, 15:23
BROWER, Lincoln P. 8:125, 12:101, 103, 15:23
BROWN, F. Martin 2:77, 101, 3:13, 20, 27, 84, 105, 4:10, 15, 39, 45, 51, 63, 73, 76, 5:4, 43, 63, 64, 112, 6:67, 8:64, 9:124, 185, 10:135, 140, 209, 11:1, 153, 14:156, 15:107, 16:239, 18:43, 19:197, 20:71, 239, 21:129, 22:57, 77, 192, 23:175, 24:307, 25:152, 176, 240, 298
BROWN, K. S., Jr. 19:107, 171, 21:77, 145, 22:147
BROWN, Leland R. 7:148
BROWN, R. M. 21:113, 23:95
BROWNLEE, R. G. 23:279
BRUGGEMANN, Paul F. 2:13, 3:49
BRUSSARD, P. F. 23:279, 25:22
BUCHHOLZ, Otto 4:62, 10:139
BUCKETT, J. S. 20:83, 21:235, 22:42
BURDICK, William N. 10:81, 11:22, 12:165
BURNS, J. M. 19:195, 20:207
BUSH, Guy L. 24:135
BUTLER, Robert B. 11:137, 18:116
BYERS, George W. 13:96, 25:124

CALLAGHAN, C. J. 24:228
CANNON, M. S. 25:150
CAPPS, Hahn W. 17:31
CARCASSON, R. H. 23:134, 24:73, 25:169
CARDÉ, R. T. 19:69
CARNEY, Dan Jr. 15:115
CARY, Margaret M. 2:86, 3:78, 5:20, 9:135, 136, 11:107, 24:267
CHAMBERS, Donald S. 16:200, 17:24
CHASE, Hazel 4:13
CHERMOCK, Ottilie D. 8:25
CHERMOCK, Ralph L. 3:96, 4:99, 5:101, 6:32, 7:102
CHINA, W. E. 13:216, 17:104, 253
CHVALA, Milan 17:169

CLENCH, Harry K. 1:1, 3, 8, 27, 30, 38, 2:6, 105, 108, 3:15, 4:14, 8:93, 94, 172, 9:105, 10:15, 151, 161, 11:18, 44, 57, 214, 12:56, 125, 15:226, 16:44, 17:40, 18:189, 19:223, 239, 20:65, 21:53, 141, 181, 22:125, 227, 23:134, 24:3, 56, 117, 240, 25:80
CLARK, Austin H. 4:13, 62, 6:42
CLARKE, C. A. 9:46, 10:47, 11:201, 24:229
COBB, Robert 3:49
COLE, L. R. 13:1
COMMON, I. F. B. 13:57
COMSTOCK, John Adams 11:4, 99, 12:127, 13:67, 16:247, 17:28, 18:32, 19:243, 20:213, 21:27, 23:15, 105
COMSTOCK, W. P. 4:30
CONCELLO, J. A. 24:227
CONWAY, Patrick J. 8:28, 10:112, 21:258
COOK, Carl 2:22
COUCHMAN, L. E. 23:280
COVELL, Charles V., Jr. 11:137, 14:237, 16:197, 18:111, 22:110, 23:278, 24:51
COWAN, C. F. 23:210
CROSS, Frank C. 3:74
CROWE, Charles R. 17:245, 20:1, 21:121
CUNNINGHAM, H. B. 24:271
CURTIS, Christopher C. 16:174
CUSHING, C. R. 24:305

DARLINGTON, P. J., Jr. 8:104
DAVIDSON, Susan E. 14:227
DAVIDSON, W. M. 3:64, 12:36, 17:225
DAVIS, Donald Ray 23:137, 25:168, 247
DE LA TORRE y Callejas, S. L. 2:86, 3:16, 65, 4:72, 11:178, 14:81
DE FOLIART, Gene R. 10:91
DENMARK, Harold A. 10:150, 18:1
DEXTER, Ralph W. 2:86, 11:206
DIAKONOFF, A. 2:79, 3:62, 4:48, 50, 5:36, 6:28, 80, 7:24, 8:67, 169, 9:41, 10:76, 11:55, 14:241, 15:130, 17:47, 19:102, 20:206, 255, 24:70
DICKSON, C. G. C. 10:37
DIMOCK, T. 22:146
DOMINICK, R. B. 25:84
DONAHUE, Julian P. 16:131
DORNFELD, E. J. 13:183
DOS PASSOS, Cyril F. 2:27, 41, 59, 3:19, 41, 42, 61, 4:38, 6:79, 7:26, 8:170, 10:29, 213, 11:176, 12:119, 191, 13:212, 14:24, 243, 15:209, 16:45, 17:103, 19:190, 192, 21:

1, 23:115, 24:26
DOWNEY, John C. 15:34, 16:235, 19:165, 20:129
DOYLE, J. F. III 22:122, 24:212
DRUMMOND, B. A. 24:135
DURDEN, Christopher J. 14:230, 19:186

EBNER, James A. 14:157, 25:73
EBERT, Heinz S-3:1-48
EDWARDS, Charles R. 25:84
EFF, J. Donald 1:91, 2:91, 3:26, 90, 4:93, 5:91, 7:86, 9:15, 22, 10:102, 16:137, 147, 22:159
EHLE, George 11:38, 12:15
EHRLICH, Paul R. 2:92, 8:100, 10:160, 11:103, 155, 12:52, 173, 13:31, 99, 14:169, 227, 15:63, 249, 16:20, 55, 24: 224, 297
EICHLIN, T. D. 24:271
EISNER, C. 11:55
ELDER, William H. 15:95
ELLIS, S. L. 22:159, 23:62
ELY, Rachel 8:29
EMMEL, John F. 16:23, 136, 17:7, 243, 18:65, 21:41, 114, 211, 23:161, 22:46
EMMEL, Thomas C. 15:197, 16:23, 136, 234, 17:7, 243, 20: 123, 21:41, 114, 211, 22:46, 23:161, 165, 279, 24:135, 25:12, 20
EMSLEY, M. G. 24:25
ESAKI, Teiso 3:62
ESCALANTE, T. 18:85
EVANS, W. H. (Brig.) 9:218
EVANS, William H. 2:42, 3:48, 74, 106, 4:70, 5:12, 6:100, 106, 8:10, 9:118, 10:168, 12:95, 13:93, 17:129, 19:237
EYER, John R. 17:237, 20:232

FAGER, Edward W. 6:44
FARNSWORTH, D. 15:105
FENDER, Kenneth M. 7:15
FERGUSON, Douglas C. 4:11, 5:53, 10:175, 16:209
FERRIS, Clifford D. 23:271, 24:202, 256, 306, 25:44
FIELD, William D. 11:12
FISCHER, Roland L. 15:121
FISHER, Mike 25:44
FLASHKA, H. 23:43
FLEMING, R. C. 19:52
FLOYD, J. 23:43
FORBES, Alexander 2:94

FORBES, W. T. M. 1:44, 46, 2:10, 24, 36, 60, 76, 88, 100, 110, 3:11, 24, 36, 59, 72, 4:16, 36, 60, 84, 5:16, 52, 59, 80, 126, 6:36, 67, 97, 109, 8:1, 167, 9:213, 10:35, 11:13, 147
FORD, E. B. 25:155
FORSTER, Walter 15:57, 116
FOX, Richard M. 15:23, 21:7
FRANCLEMONT, John G. 3:5, 5:6, 11:147, 21:11, 23:10
FRANKIE, G. W. 23:278
FRECHIN, Donald P. 5:66
FRECHIN, D. 23:115
FREDERICK, A. G. 12:207
FREEMAN, H. A. 5:95, 7:90, 9:196, 12:81, 13:89, 14:58, 17:89, 18:75, 19:83, 20:181, 227, 21:115, 205, 278, 24:68, 247, S-1:1-58, S-2:1-62
FREEMAN, T. N. 2:63, 3:101, 4:106, 5:41, 110, 7:118, 10: 137, 11:27, 97, 24:86, 272
FRYXELL, Thomas 8:103
FULLER, Stanley V. 16:59
FULLER, Woodson C. 15:34
FULTON, MacDonald 7:28
FUNK, R. S. 20:216, 22:63

GAGE, E. V. 24:270
GARDINER, B. O. C. 15:235, 21:138, 198
GATRELLE, Ronald R. 25:143, 145
GILBERT, L. E. 23:177, 278, 24:297
GILBERT, William M. 14:36
GILHAM, Nicholas 8:100, 146
GILMORE, Richard M. 17:102
GLICK, P. A. 19:129
GOBEIL, R. E. 19:181
GOLIGER, Melvin 10:45
GOODWIN, George H. 12:123
GORELICK, Glenn A. 22:111, 23:31, 280, S-5:1-41
GOZMÁNY, Lancelot A. 2:93, 3:26, 43, 75
GREY, L. Paul 2:86, 5:31, 10:17, 119, 11:171, 13:14, 173, 245, 16:81, 17:129, 18:58
GRAY, P. H. H. 3:80, 104, 5:35, 67, 72, 6:73, 7:5, 47, 127, 8:88, 102, 140, 145, 9:58, 11:54, 14:63
GRAY, Richard E. 25:65, 108
GREENE, G. L. 24:47
GREENFIELD, Joseph C., Jr. 25:278
GRIEWISCH, Louis 7:54
GROOTHUIS, D. 19:221

GUPPY, Richard 4:13, 73, 5:69, 6:43, 7:43, 56, 8:101, 9:15, 10:169, 11:227, 13:101, 170, 14:240, 16:64, 18:185, 21:122
GYVES, M. T. 23:18

HAMMOND, H. E. 14:67
HANNEMANN, H. J. 22:124
HARDESTY, R. 19:221
HARDWICK, D. F. 20:29, 21:22, 22:65, 23:26, 24:151, 228, 234, 282, 25:1, 109, 177, 181, 274, 287
HARDY, George A. 13:70, 78
HARRIS, Lucien Jr. 4:43, 6:76, 8:153
HEINEMAN, Bernard F. 7:167, 9:80
HEITZMAN, Richard 15:233, 16:249, 17:44, 18:64, 19:77, 179, 20:177, 23:105
HEMMING, Francis 9:144, 10:110, 162, 11:46
HENNE, Christopher 2:42, 11:169, 20:213, 21:27, 23:15
HENSEL, H. 19:242, 20:28
HERMAN, C. 15:43
HESSEL, J. H. 19:55, 20:242, 21:112, 23:278
HESSEL, Sidney A. 5:102, 6:34, 7:106, 8:57, 9:142, 10:200, 17:43, 18:27, 22:241, 23:254, 24:105
HESSELBARTH, Gerhard 4:50
HIGGINS, L. G. 12:161, 13:104
HINTON, H. E. 6:1
HODGES, Ronald W. 12:205, 15:81
HOFFMEYER, Skat 10:36
HOGUE, Charles L. 12:17, 209, 16:58
HOLLAND, R. 21:210, 23:33
HOLZMAN, Richard W. 15:191
HOPF, Alice L. 8:123
HOPFINGER, John C. 1:90, 3:87, 4:92, 5:88, 7:80, 15:124
HORN, D. J. 23:25
HOVANITZ, William 2:60, 11:10, 15:251, 23:279
HOWE, William H. 12:25, 130, 200, 13:26, 78, 18:26, 19:33, 20:215, 21:39, 24:217
HOYING, L. A. 22:118
HUBBARD, J. P. 19:231
HUBBELL, Stephen P. 11:37, 14:209
HUBER, R. L. 19:76, 20:34
HYNES, Vonta P. 3:64

IRWIN, Roderick R. 20:156, 22:21, 209, 24:143, 25:83
IWASE, Tarô 4:13, 43, 5:52, 6:43, 7:45, 166, 8:27, 95, 9:13

JABLONSKI, R. J. 8:28
JACKSON, T. H. E. 13:121
JAE, R. J. 24:233
JEWETT, Stanley G., Jr. 13:171
JOHNSON, F. M. 20:207
JOHNSON, John W. 12:17
JOHNSON, Kurt 25:216
JOHNSTON, Edward C. 4:27
JORDAN, Karl 7:3
JUMALON, Julian N. 12:115, 13:19, 14:155, 18:101, 21:49, 24:303, 306

KADDOU, Ibrahim K. 25:53
KARALUS, Karl E., Jr. 11:54
KARSNER, Katherine Hunn 17:242
KEIPER, R. R. 23:1, 205
KEJI, Joseph A. 5:69, 17:158
KENDALL, C. A. 25:29
KENDALL, Roy O. 11:168, 224, 225, 13:221, 14:176, 15:73, 16:201, 17:21, 18:129, 163, 190, 19:1, 105, 20:35, 229, 24:59, 266, 301, 25:29
KENNEDY, C. J. 18:41
KETTLEWELL, H. B. D. 17:173
KIMBALL, Charles P. 4:47, 9:16, 10:112, 11:21
KING, E. W. 23:191
IRIAKOFF, S. G. 2:3, 15, 39, 49, 61, 80, 103, 3:4, 14, 67, 6:7, 10:207, 17:1, 33
ISTNER, David H. 6:73
LOTS, Alexander B. 2:51, 3:25, 6:77, 5:24, 120, 10:203, 12:1
NOWLTON, George F. 7:11, 55, 8:17
NUDSEN, John P. 7:27, 8:141, 9:11
OLYER, John M. 17:178, 19:212, 20:13, 217, 21:107, 22:211, 23:77, 24:61, 125
RIVDA, W. V. 21:31, 22:191
UBO, Kaiya 14:203
UZUYA, Takeshi 13:175

AFONTAINE, J. D. 24:83, 25:299
AMBREMONT, Edward N. 17:148, 19:47, 22:182
ANE, M. C. 21:279
ANGER, T. W. 9:203, 11:162
ANGSTON, Robert L. 8:11, 9:18, 11:230, 17:201, 19:95, 138, 22:89, 23:49, 149
ATHAM, Roy 7:172, 8:27, 10:111

LEARNED, Elmer T. 8:44
LEECH, Hugh B. 5:6, 21:74, 24:189
LEES, Eric 17:105
LEGGE, A. H. 19:57
LEHMAN, Robert D. 25:150
LEMS, Kees 4:72, 6:42
LENGYEL, G. 3:43
LENNOX, Donald J. 10:53
LEWIS, G. G. 19:115
LIGHTY, Phillip M. 25:238
LINDQUIST, O. H. 21:15
LINDSAY, S. A. 25:64
LIPES, Jack E. 15:56, 114
LOELIGER, R. 4:61
LORKOVIĆ, A. 15:43, 16:5, 105

MAC LEOD, E. G. 9:54
MAC NEILL, C. D. 21:204
MACY, Ralph W. 7:25, 12:199
MAEKI, Kodo 7:36, 11;8, 13:193, 14:37, 127, 179
MAKINO, Sajiro 7:7
MALICKY, H. 24:190
MANLEY, Thomas R. 20:122, 25:146, 234
MARKS, L. S. 11:158, 23:18
MARSDEN, D. 22:121
MARTIN, Leland L. 16:59
MARTIN, Lloyd M. 1:88, 2:105, 3:86, 4:86, 5:82, 7:67, 25:215
MASTERS, John H. 21:120, 22:26, 158, 189, 252, 23:104, 155, 199, 251, 24:10, 15, 22, 203, 25:19, 84, 86, 149, 150, 256, 297
MATHER, Bryant 6:42, 74, 7:13, 8:102, 131, 9:67, 119, 10: 160, 204, 13:18, 71, 19:139, 20:102, 186, 245, 21:59, 24:176, 25:147
MATHER, Katharine 13:71
MATTONI, R. H. T. 2:104, 8:8, 11:5
MATTOON, S. O. 25:247
MC ALPINE, Wilbur S. 14:209
MC CORKLE, D. V. 23:279
MC DONALD, W. A. 6:61
MC DUNNOUGH, J. 2:76
MC ELVARE, Rowland R. 4:49, 6:53, 10:90, 20:91, 254
MC FARLAND, Noel 13:236, 15:172, 18:201, 19:82, 25:144
MC GUFFIN, W. C. 24:54, 25:262, 266
MC HENRY, Paddy B. 13:66, 16:104
MEINERS, Edwin P. 2:40, 43, 55, 62, 66, 83, 3:51, 5:11,

10:163
MERRITT, James R. 6:101, 8:49, 9:146, 203
MIELKE, O. H. H. 21:77, 145, 22:147
MILLER, J. Y. 24:244, 292
MILLER, Lee D. 15:97, 22:125, 23:201, 24:13, 120, 244, 292, 25:12
MILLER, William E. 14:231, 20:251, 25:284
MOECK, Arthur H. 4:47, 11:44, 16:81, 17:129
MOIZ, S. A. 21:127, 22:115, 183, 23:107, 24:19
MONROE, Burt L., Jr. 7:53
MONROE, R. S. 21:185, 243
MOUCHA, Josef 7:55, 11:45, 13:246, 15:74, 132, 16:145, 192, 17:38, 169, 18:109, 19:81, 23:189
MULLER, Joseph 8:29, 11:15, 12:179, 13:17, 27, 182, 14:159, 177, 20:118, 22:180, 23:48, 249, 24:1
MUNROE, Eugene G. 1:95, 3:3, 97, 4:44, 102, 5:7, 29, 55, 68, 81, 6:89, 7:16, 66, 9:140, 13:241, 14:1, 169, 24:287, 25:142, 185
MUNSHI, G. H. 21:127, 22:115, 183, 23:107, 24:19
MUSPRATT, Vera Molesworth 8:143
MYERS, J. 23:196

NABOKOV, Vladimir 4:75, 6:35, 41, 7:49
NEKRUTENKO, Y. P. 24:213, 218
NEWCOMER, E. J. 13:64, 16:67, 17:117, 171, 18:47, 19:161, 20:253, 21:71, 279
NICOLAY, S. S. 23:127, 25:87, S-4:1-39
NIELSEN, M. C. 12:37, 201, 14:57, 18:233, 20:41, 243, 24:42
NOSEK, Josef 9:199

OBRAZTSOV, Nicholas S. 7:128, 8:50, 135, 10:153, 176, 177, 11:178, 12:55, 14:112, 15:127, 16:175, 237
O'BYRNE, Harold I. 2:22
OGATA, Masami 11:216
OKADA, Yoshio 3:79, 11:216
OKAGAKI, Hiromu 11:216
OLIVER, Charles G. 20:44, 24:77, 25:143
OLSON, W. L. 16:136
OPLER, Paul A. 15:145, 16:193, 20:212, 21:113, 212, 22:89, 188, 25:115, 194
OWEN, D. F. 17:159, 25:271

PALKUTI, G. A. 25:150
PALMER, H. B. 22:211
PATTERSON, Donald 9:147, 13:229
PATTERSON, George F. 25:222
PAYNE, J. A. 23:191
PEASE, Roger W. 3:106
PEASE, Roger W., Jr. 9:4, 14:89, 17:27, 231, 22:197
PECHUMAN, L. L. 25:82
PENNY, George Henry 9:79
PERKINS, E. M. 20:103, 172, 21:213, 22:53
PERKINS, O. A. 22:119
PERKINS, S. F. 20:103, 172, 21:213, 22:53
PETERS, J. V. 25:114
PETERSEN, B. 18:165
PHILIP, Kenelm W. 15:56
PHILLIPS, Leonard S. 9:143, 10:44, 13:73, 15:101, 19:104, 20:124
PHILLIPSON, D. E. 21:261
PLATT, Austin P. 23:97, 109, 25:278
PLISKE, Thomas E. 11:42, 14:209, 25:294
PLOMLEY, J. M. 16:246
POVOLNÝ, Dalibor 9:199, 24:6
POWELL, Jerry A. 12:26, 13:229, 14:112, 15:145, 18:192, 241, 19:225, 21:211, 22:225, 23:106, 147, 203, 211, S-6:1-67
PRICE, Homer F. 15:93, 23:186
PRIDDLE, T. R. 20:119, 21:249
PRIESTAF, R. C. 24:226
PROCACCINI, D. S. 23:18
PRONIN, George F. 6:93, 8:121, 9:137, 18:35
PYLE, R. M. 22:172, 196

QUELCH, C. S. 6:44

RÁCZ, G. 21:241
RAHN, R. A. 23:273
RANDLE, Worth S. 7:119, 133
RAWSON, George W. 3:28, 5:70, 7:27, 9:133, 15:237, 16:250, 17:225, 22:27, 240
REICHEL, Johannes 4:74
REINTHAL, Walfried J. 16:201, 17:35
REMINGTON, Charles L. 1:1, 4, 9, 13, 15, 16, 17, 25, 26, 28, 31, 33, 37, 40, 42, 43, 49, 51, 52, 61, 64, 73, 74, 75, 76, 84, 87, 93, 94, 96, 97, 2:1, 2, 5, 14, 16, 25, 26, 27, 32, 36, 37, 42, 44, 45, 48, 49, 50, 62, 64, 66,

68, 73, 77, 80, 84, 85, 102, 103, 4:1, 13, 15, 16, 31, 37, 38, 42, 49, 52, 62, 68, 71, 73, 84, 107, 5:9, 10, 17, 42, 45, 46, 57, 60, 72, 6:40, 71, 78, 102, 111, 121, 7: 24, 26, 32, 57, 65, 122, 139, 8:9, 30, 31, 47, 48, 49, 75, 76, 104, 146, 163, 170, 171, 173, 9:4, 21, 22, 77, 82, 104, 132, 157, 158, 10:36, 56, 57, 80, 101, 109, 178, 192, 197, 240, 11:177, 200, 12:14, 94, 96, 129, 175, 205, 240, 13:15, 30, 33, 151, 193, 14:37, 127, 179, 201, 15: 125, 16:22, 58, 149, 150, 17:49, 63, 99

REMINGTON, Jeanne E. 1:39, 56, 69, 81, 98, 2:7, 14, 17, 42, 82, 98, 108, 3:53, 107, 10:74, 17:65

REMINGTON, P. Sheldon 2:91, 3:94, 4:97, 5:97, 6:103, 7:93, 9:17, 190, 13:15, 15:65, 16:71

RICKARD, M. A. 21:248, 281, 22:75

RILEY, N. D. 3:67, 10:193, 12:209, 13:11, 18:237

RINDGE, Frederick H. 5:1, 7:1, 126, 8:46, 12:208, 19:193, 20:250, 21:140, 24:208, 25:143

RIOTTE, J. C. E. 11:35, 62, 12:124, 15:92, 16:62, 243, 18: 84, 89, 21:8, 33, 135, 252, 22:40, 23:101, 24:227

RITTERBUSH, Philip C. 17:229

RODECK, Hugo G. 4:46

ROEVER, Kilian 16:1

ROGERS, Wesley H. 16:63, 17:171

ROGERS, W. P. 2:86

ROSS, Gary N. 17:148, 18:11, 19:47, 21:185

ROZMAN, R. S. 24:81

RUCKES, H. 19:190

RUPERT, Laurence R. 5:53

RUTIMEYER, E. 23:255

RUTKOWSKI, F. E. 3:48

RUTKOWSKI, Frank 20:43, 25:126, 137

RYDON, A. 18:51

SABROSKY, Curtis W. 11:29

SALA, Frank P. 4:71, 12:17, 13:65, 21:211

SAMUELSON, G. Allan 8:91

SARGENT, T. D. 23:1, 261, 24:105

SCHMID, F. 24:88

SCHOENHERR, W. H. 2:22

SCHWEHR, David Wm. 25:139

SCOTT, F. W. 22:254

SCOTT, J. A. 22:159, 237

SEELEY, C. 20:47, 189

SETTE, Oscar E. 13:238, 21:211

SEVASTOPULO, D. G. 7:172, 8:26, 9:79, 13:24, 18:73, 170, 25:80, 146, 293

SHAPIRO, Arthur M. 16:129, 199, 19:215, 22:157, 179, 23:189, 265, 24:224
SHEPARD, Jon H. 16:46, 19:94, 21:114, 211
SHEPPARD, A. C. 13:77
SHEPPARD, P. M. 9:46, 10:47, 11:201, 19:227, 24:229
SHIELDS, Oakley 17:111, 21:211
SHININGER, F. S. 22:53
SHIRÔZU, Takashi 2:103, 15:245
SHOUMATOFF, Nicholas 4:7, 6:64, 7:38
SHULL, Ernest M. 6:68, 11:167, 229, 16:143, 17:110, 224, 18:159, 22:20
SIBATANI, Atuhiro 11:216
SICHER, Harry 6:107, 108, 7:53, 16:98
SIEKER, William E. 15:124, 21:275
SIMMONS, Robert S. 10:157, 15:99, 17:107
SINGER, M. C. 23:279
SHULL, E. M. 23:186
SLANSKY, Frank, Jr. 25:294
SMALL, Gordon B., Jr. 16:195, 23:127
SMALLEY, S. B. 3:25
SMITH, Claude I. 2:34
SMITH, Marion E. 7:123, 8:13, 9:12, 11:221
SMITH, Richard S. 14:239
SORENSEN, J. T. 21:57, 258, 23:155
SPANGLER, Paul J. 11:205
SPENCER, O. D. 25:247
STAFFORD, C. W. 10:44
STALLINGS, Don B. 1:92, 3:93, 4:38, 96, 8:77, 10:1, 109, 11:113, 12:93, 13:99, 204, 15:19, 16:55, 17:81, 18:45, 20:163, 23:243
STALLINGS, Viola N. 13:204, 15:19, 17:8, 18:45, 20:163, 23:243
STANFORD, Ray E. 24:256
STARÝ, Petr 25:53
STEINHAUSER, Stephen R. 25:294
STEYSKAL, George C. 24:38, 25:264
STONER, Emerson A. 8:103
STRAATMAN, Raymond 9:74, 16:99, 161, 23:69, 25:58
STRALEY, G. B. 23:76
STRUBLE, George R. 6:107
SULLIVAN, J. Bolling 25:295
SUOMALAINEN, Esko 6:57
SWEADNER, Walter R. 2:80
SWEET, H. 23:278
SYME, P. D. 15:108, 113

TASKER, R. R. 24:226
TAYLOR, Barbara 19:189
TAYLOR, Robert 19:189
TEALE, E. W. 9:143
THAXTER, E. Lincoln 11:43
THOMAS, Edward S. 6:92, 10:46
THOMAS, G. M. 20:235
THORNE, Alvin L. 14:242
THORNE, Fred 8:9, 10:80, 172, 11:168, 15:200, 16:61, 23:278.
TILDEN, J. W. 5:70, 6:95, 8:5, 10:113, 11:43, 151, 12:33, 203, 16:199, 18:214, 19:53, 22:187, 23:280, 25:293
TINDALE, Norman B. 15:195
TOLIVER, Michael E. 22:177, 25:213, 246
TRAVASSOS, Lauro P. 3:32, 9:150
TREAT, Asher E. 8:23, 9:55, 69, 10:87, 21:169, 254
TRINNELL, J. R. 21:15
TURNER, J. R. 8:77, 10:1, 109, 11:113, 12:93, 13:99, 204, 15:19, 16:55, 17:81, 18:45, 20:163
TURNER, Richard D. 17:227

VALLETTA, Anthony 15:109
VAN SOMEREN, V. G. L. 9:127, 13:121
VAN SON, G. 9:1, 81, 12:6, 18:195
VÁRI, L. 25:211
VIETTE, Pierre E. L. 3:77, 5:9, 7:166, 9:61
VOSS, Edward G. 4:46, 70, 7:54, 10:18, 17:181

WAGNER, W. H., JR. 9:151, 10:18
WALLIS, J. B. 2:1
WEAST, Robert D. 13:213, 16:61
WEBER, B. H. 3:86
WELLING, Eduardo C. 12:118, 131, 170, 13:62, 17:37, 18:229, 239, 20:95, 201
WALLEY, Paul E. S. 17:39
WHITTAKER, R. H. 7:9
WIELGUS, R. S. 22:111
WILCKE, Hermann 3:7
WILCOX, LeRoy 5:71
WILKINSON, R. S. 20:12
WILLIAMS, C. B. 3:17, 39, 63, 24:157
WILLIAMS, R. N. 21:185
WILSON, Kent H. 11:17
WILTSHIRE, E. P. 10:116, 201, 13:79, 16:47
WOBUS, R. A. 20:123
WOHLFAHRT, Theodor A. 6:13

WOOD, J. B. 23:264, 25:83
WOODCOCK, H. E. 3:25, 6:108
WOODLEY, R. E. 23:113
WREN, George R. 9:12
WRIGHT, David Allan 21:255, 24:209, 25:68
WYATT, Alex K. 2:22, 10:111, 17:100, 18:118, 21:125
WYATT, Colin W. 4:72, 10:214, 11:36, 47, 142, 15:1, 18:78

YOUNG, Allen M. 22:185, 25:223
YOUNG, Frank N. 9:204, 10:46

ZEIGER, C. F. 18:31
ZIEGLER, J. Benjamin 7:33, 132, 10:89, 14:19, 18:85
ZIEMER, S. E. 2:34, 4:12

Subjects

A.A.A.S. Meeting in Chicago, report on 2:2
Abbot, John - some original paintings 4:25-26
Abbot & Smith - authorship of names 12:191-192
Aberrant *Colias* from Minnesota 22:158
Aberrant *Heliconius charitonius* 21:248
Aberrations 2:6
Aberration, *Phyciodes tharos* 25:143-144
Aberration, *Vanessa cardui*, experimental 22:146
Acanthopteroctetes, genus review & new species 23:137-147
Acentropus niveus in Massachusetts 8:23-25; 9:69-73
Actias luna: effects of photoperiod on
pupal diapause 21:255-258; 24:209-212
emergence structures employed by 21:249-252
Addendum to mites from noctuid moths 21:254
Additional notes: life history of *Neoterpes edwardsata;*
rearing and preserving larvae 18:32-34; 19:233-236
Additions to: A.M.N.H. collections 21:180
Membership list 4:24, 36, 84, 110
Adelpha bredowii in Oregon 11:44; 12:199-200
Adelpha, notes on two rare Mexican and related Central
American species 24:292-297
Adelocephala quadrilineata life history 15:233-234
Adopaea lineola in Ontario 16:62; in Pennsylvania 12:15-16
AEgiale hesperiaris, biology 11:114
Ae's larvae of interspecific hybrids in Japan 18:170
Affinities of the ITHOMIINAE and the SATYRINAE 24:297-300
Afghanistan Expedition, 1956 (Amsel, *et al.*) 15:128-129
Afghanistan Rhopalocera 15:1-18
African Lepidoptera, protective resemblance 13:121-150, 256
AGATHYMINI, new tribe 12:93-94
Agathymus, a divergent group in Baja and southern
California 23:243-249
Agathymus aryxna, fluid recycling in 20:242
Agathymus fieldi, new species from Mexico 14:58-62
Agathymus, four new species from Texas 18:171-185
Agathymus, new genus 12:82-84
Agathymus in Texas 14:58-62
Agathymus, larval behavior 23:243-249
Agathymus mariae micheneri new subspecies from Mex. 15:19
Agathymus, new species from Texas 20:181-185
Agenjo - Faunula Lepidopterologica Almeriense 9:160
Agenjo, Ramon (Personalia) 5:10

Agressive encounter, *Catocala cara* and *Polygonia interrogationis* 19:52
Agraulis vanillae overwintering in Missouri 17:227-228
Status of in Missouri and Kansas 19:33-34
Alaska Highway, collecting 6:103-106
Alaskan butterflies 7:123-126
Albinic female of *Pieris sisymbrii* from Oregon 21:121
Albinistic *Limenitis* in Illinois 2:22
Albino male *Eurema lisa* 18:64
Allan's book on larval foodplants 5:60
Allen, J. A., Colorado itinerary in 1871 10:209-212
Alsophila in Quebec 8:140
Alsophila pometaria and *Operophtera bruceata* females 7:127
Amateur research in Lepidoptera 14:1-4
Amathusia phidippus, abundance of in Cebu, Phil. 18:101-104
Amendments to Constitution 4:38, 5:111, 6:53, 8:48, 9:35-38, 12:157-158, 17:63-64, 20:197-200, 29:67-72
American moths in Britain 4:62
American white butterflies and English foodplants 25:6-12
Amsterdam symposium on classification 6:1-31
Anaea andria, marked specimen 3:65
Anaea monograph announced 10:178
Anepia, a new species from Sierra Nevada of Calif. 21:235
Anisota virginiensis, flight pattern of the male 25:84-85
Annaphila arvalis in British Columbia 13:78
Annaphila field notes 4:71
Annaphila pseudoastrologa, larvae and habits 19:237-239
Annotated list of butterflies collected in British Honduras in 1961 18:11-26
Announcement of meetings: 6:91; 9:184; 10:80, 136; 20:55
Annual meetings:
A.A.A.S., 1949 3:103
E.S.A., 1949 3:77, 103
Lepid. Society 4:37; 5:1-3; 5:42- 5:120; 22:64
Anomalous foreleg in *Dodonidia helmsii* 23:201-202
Another U.S. record for *Oeneis macounii* 19:125
Antennae of Lycaenidae 11:13-14
Anteos clorinde nivifera in Colorado 14:156
Antheraea eucalypti, neutron irradiation in 25:238-240
Antheraea polyphemus variant 10:45-46
Anthocharis, correct spelling 23:210
Anthocaris flight habits 8:10
Anthocaris in Mississippi 6:42
Anthocaris midea field notes 6:99-102
Anthocaris, new larval foodplant records 21:212
Anthocaris on hilltops 5:70

Anthocaris reakirtii marking and retrieving 9:118
Anthocharis sara notes 6:101, 106
Antigonus pulverulenta foodplant 15:73
Antillea Higgins, new genus 12:164
Ants associated with Lycaenid larvae, new aspects 24:190
Ants, specimens damaged by 25:83
Apantesis figurata forms reared 13:182-183
Aphania infida figured 11:28
Aphania salicaceana new species 11:27-28
Apodemia mormo complex in western USA, revised 15:145-171
Apodemia mormo dialeuca new subspecies from Baja California 15:167-168
Apodemia mormo tuolumnensis new subspecies from California 15:159-160
Apparatus for measuring maculation patterns 23:18-22
Apparatus for photographing mounted specimens 19:212-214
Application of automation in Rhopalocera research 20:1-12
Appointments: Asst. Treasurer 5:111; Assoc. Editor 5:72
Arachnis zuni larvae abundant 13:236-237
Arcas, review of genus and description of new species 25:87
Arctic butterfly collecting 11:47-53
Arctic Lepidoptera of Baker Lake, N.W.T., CANADA 2:63-65
Arctiidae, striated band of thorax 11:147-150
Argynninae, classification reviewed 11:171-176
Arizona butterflies, some comments on 19:107-115
Arizona Rhopalocera, new records 8:25
Articles of organization, Lepidopterists' Society 1:2
Assistant Treasurer appointed 5:111
Associate Editor appointed 5:72
Association between lycaenid larvae and ants 24:190-202
Association of Minnesota Entomologists 21:140
Asterocampa in Florida 12:36
Asymmetry in *Speyeria* 13:14-15
Atlantea Higgins, new genus 12:162
Atrytone arogos, life history of 20:177-181
Atrytone dukesii in Michigan 11:42
Atrytonopsis hianna, ecology and behavior 19:215-221
Attracting butterflies with bait 6:32-33
Attracting moths by light 4:47
Attraction of butterflies to light 18:73-74
Authorship of *Polites mystic* 20:239-242
Authorship of *Melitaea rubicunda* 23:175-176
Automation, application of in Rhopalocera research 20:1-12
Automeris io, gynandromorph 17:43-44; 18:27-31
Automeris io, mosaic gynandromorphs 25:234-238

Automeris zephyria, field notes on 19:82
Australian Rhopalocera, oviposition 16:99-103
Austrian Lepidoptera collecting 3:7-9

Back issues, Lepidopterists' News 4:24, 36
Backyard collecting 12:130
Baiting for moths 3:26
Bait trap, lightweight collapsible 23:97-101
Baja California butterflies 12:26-32
Baja California Lepidoptera 12:26-32; 13:229-235
Barnes, William reminiscences 17:117
Beginning of the butterfly season 19:239-241
Behavioral adaptations of cryptic moths:
  bark-like species 23:109; resembling dead leaves 23:205
Behavioral notes on *Marpesia petreus* 20:122
Behr, Dr. H. H., letters from to Herman Strecker 22:57-62
Bellinger, P. F. personalia 6:78
Bilateral gynandromorph of *Automeris io* 18:27-31
Biographies:
  Abbot, John 2:28-30, 108
  Barnes, William 3:53-54
  de Joannis, Joseph 3:77
  Dyar, H. G. 1:30
  Edwards, W. H. 1:8, 2:54
  Edwards, Henry 2:7
  French, G. H. 2:40
  Grote, A. R. 2:17
  Holland, W. J. 1:98
  Hübner, Jacob 3:107-108
  Hulst, G. D. 2:66
  Kusnezov, N. J. 3:29-31
  Murtfeldt, Mary E. 2:83
  Packard, A. S., Jr. 1:39; 2:54
  Reakirt, Tryon 18:211-214
  Riley, C. V. 1:56; 2:54
  Scudder, S. H. 1:17-18; 2:54
  Skinner, Henry 1:69
  Smith, J. B. 1:81
  Strecker, F. H. H. 2:98
Biological & systematic considerations, *Epinotia* 25:115
Biological observations on *Callophrys viridis* 21:113-114
Biological notes on three *Pyrginae* in New York 22:179-180
Biological studies of the genus *Ethmia* in Calif. S6:1-67
Biology and description, *Exoteleia nepheos* 21:15-21
Biology and hostplant associations: *Ornithoptera periamus*
  urvilleanus 23:69-76; *Ornithoptera victoriae* 23:69-76

Biology and host preferences of *Vanessa tameamea* 22:111-114
Biology and lab culture of *Chlosyne lacinia* 24:135-142
Biology and seasonal history of
Polydorus aristolochiae 24:19-22
Biology of *Ocnerogyia amanda* 25:53-57
Biology of *Papilio indra kaibabensis* 21:41-49
Bionomic notes on *Haeterini* and *Biini* in Venezuela 24:15-18
Biosystematics of *Callophrys* in California S5:1-41
Bird collection to American Museum of Natural History 14:82
Black light, a new portable 21:112
Board of specialists 1:13-14, 86, 100; 2:10, 48; 3:2; 4:2
Bog collecting in Maine 10:119-120
Boisduval authorship of *Papilio* names 16:45-46
Boisduval & LeConte: dates of publication 13:212
names proposed by 12:121-122
Bolivian Cordillera Real butterfly fauna 15:116-120
Boloria astarte compared with *Boloria distincta* 11:142-146
Boloria distincta, observations on 11:142-146
Boloria epithore complex, a new race 20:103-117
Boloria eunomia, flight period of 19:184-185
Boloria eunomia in Wisconsin 25:149
Boloria eunomia laddi in Colorado 25:84
Boloria frigga in Michigan 11:37-38
Boloria freija new to Wisconsin 22:252
Boloria hegemone houri new subsp. from Afghanistan 15:8-9
Boloria in central Colorado 8:64-66
Boloria montinus on the mount Washington Carriage rd. 2:86
Boloria polaris in northern British Columbia 23:113-115
Boloria selene in: Oregon 14:158
Pacific Northwest 17:171-172
Washington 15:115
Boloria titania, habitat in Ontario 14:230
Boloria toddi, flight habits 19:104; in Md. 9:54; 10:160
Bomolocha, a new species from Florida 21:125-126
Book notices, reviews (see LITERATURE REVIEWS/NOTICES)
Brazilian "*Cercyonis*" 25:12-19
Brazilian Lepidoptera collecting 13:79-88
Breeding cage 3:26
Brephidium exilis yucateca new subspecies 24:3-6
British Honduras, butterflies collected in 1961 18:11-26
Brown, Colorado Butterflies. Part I. Satyridae 9:21
Brownsville, Texas butterfly records 15:56, 114
Bruggemann, Paul F. (Personalia) 5:10
Bryk, Felix (1882-1957), obituary 11:55
Buchholz collection acquired by American Museum 12:208
Buchholz, Otto (1874-1958, obituary 13:27-29

Buck-moths (*Hemileuca*) collecting of 2:81-82
Burdick collection to University of Colorado 16:246
Butterflies:
  attracted to light in Gujarat State, India 18:159-163
  from Coahulila, Mexico 22:227-231
  from Honduras, recent collections 21:185-197
  higher classification 11:103-106
  in Colorado, an extraordinary swarm of 22:172
  in northern Pennsylvania 25:222
  new to Wisconsin 22:252
  north of Mexico - Society project 2:77-78
  of Contra Costa County, California 22:89-107
  of Eastern Brazil S3:1-48
  of Formosa in colour (Shirôzu), reviewed 14:243
  of Kent Island, New Brunswick 19:181-183
  of Knob Lake, Northern Quebec 5:7-9
  of Madison County, Illinois 24:305
  of Michigan, addenda to list of 22:119-120
  of northeastern Indiana, list of uncommon 23:186-188
  of Roosevelt County, New Mexico 25:213-214
  of Yakima County, Washington 18:217-228; 20:253-254
Butterfly and moth plastic mounts 2:31
Butterfly:
  collecting at the time of a solar eclipse 18:109-110
  collecting, field techniques 4:10, 73
  faunas, forest-prairie transitions 25:216-221
  hunting in central Labrador 19:242-243
  migrations in Mexico during 1963-1964 18:229-230
  migration near Italy 2:94
  records for Ontario and for Canada 21:135-137
  records from central California 19:171-176
  season, the beginning of 19:239-241
  traps, notes on use of in East Africa 18:51-58

Caenurgina caerulea, possible mimicry of 22:122-124
Caenurgina caerulea, mimicry & distribution of 19:165-170
Calephelis borealis life history 7:119-122, 133-138
Calephelis virginiensis, a colony at Houston, Texas 21:281
California Academy of Sci. centennial vol. reviewed 13:35
California Rhopalocera, ecology 16:23-24, 136
  field observations 17:7-20, 111-116
Calinaga, systematic position 12:173
Callimorpha equitalis, aggregation in Himalayas 17:224
Callophrys comstocki in Nevada 24:228; range ext. 21:114

Callophrys eryphon occurrence in Michigan 20:41-42
Callophrys fotis bayensis, new subspecies from San Francisco 23:95-96
Callophrys iroides larval foodplants 22:225-226
Callophrys macfarlandi, observations on in Sandia Mountains, N. M. 19:177-179
Callophrys (Mitoura) johnsoni in California 16:193-194
Callophrys (Sandia) macfarlandi life history 16:55-57
Callophrys viridis, biological observations on 21:113-114
Callophrys xami, possible record in California 20:212
Callophrys xami life history 18:85-89
Callosamia promethea, color variant of in N. Y. 22:185-186
Calosaturnia meridionalis name changed 12:17
Canadian butterflies 5:7-9, 41-42
Canadian species of *Lithocolletis* feeding on *Salix* and *Populus* 24:272-281
Cape Cod, exotic Lepidoptera after hurricanes 10:112
Caradrina morpheus new to North America 13:77
Carbon tetrachloride danger 5:6
Care of a collection and library 2:27
Cargolia with transfer of species 18:111-115
Caria domitianus and *ino* and description of a new subspecies 21:53-57
Caria ino melicerta at light 24:13-15
Carpenter, G. D. Hale (Personalia) 5:10
Carterocephalus palaemon in California 11:43
Catalogues and checklists of Nearctic Rhopalocera 10:29-34, 213-214
Categories of classification 2:90
Caterpillar cranium terminology, review of Snodgrass' paper 1:82
Caterpillar survives long immersion 4:12
Catocala allusa on fir logs in Vancouver Island 14:240
Catocala cara and *Polygonia interrogationis*, aggressive encounter 19:52
Catocala, collecting methods 1:19-20
Catocala connubialis form "*broweri*", new form 14:177-178
Catocala, continuous var. in related species of 22:231-236
Catocala feigning death 4:46
Catocala fraxini emergence and longevity 15:235-236
Catocala, highest catch at bait 12:201-202
Catocala ilia, evolution of melanism 17:159-162
Catocala of southern New England: Abundance and seasonal occurrence 24:105-117
Catocala parta forms reared 13:182-183
Catocala, suggestion regarding hindwing diversity among 23:261-264

Catopsilia migration in India 6:68-69
Causes of death encountered during rearing of *Danaus plexippus* 20:235-238
Celerio galii intermedia in Ontario 11:35-36
Celerio lineata larvae, food for Indians 21:144
Celerio lineata outbreaks in Utah 7:11-12
Celluloid, use of in spreading 2:86
Century of progress in the natural sciences, reviewed 13:35
Cephise, a new record for Mexico 24:68-69
Ceramidia viridis from California 22:187
Ceratonyx arizonensis, larva and pupa 21:11-14
Ceratonyx satanaria, identity of 21:11-14
Cercyonis meadi group, a new subspecies 23:161-164
Cercyonis pegala nephele at fluorescent light 20:12
Cercyonis pegala pegala occurrence in Mississippi and variation 20:186-188
Cercyonis pegala pegala variations in N. J. 22:180-182
Cercyonis stephensi, type locality of, revisited 19:161-164
Cercyonis, taxonomy, distribution and biology characteristics 23:165-175
Change of name, Lepidopterists' News 12:240
Charaxidinae, chromosomes 14:179-201
Chemoreceptors, *Danaus gilippus berenice* 23:196-198
Chilo erianthalis, new species from Louisiana 17:31-32
Chlaenogramma jasminearum in Pennsylvania 17:102
Chlosyne fulvia life history and habits of 22:237-240
Chlosyne gloriosa new species from Mexico 13:165-169
Chlosyne hoffmanni manchada new subspecies 13:207-211
early stages 21:71-73
Chlosyne lacinia and *californica*, larval notes 16:61
Chlosyne lacinia, biology and lab culture of 24:135-142
Chlosyne riobalsensis new species from Mexico 14:148-154
Chlosyne rosita browni new subspecies from Mex. 14:148-154
Chlorocresol method for field collecting 15:195-197
Chromatography, study of fluorescent pigments 22:27-40
Chromosome numbers of Japanese Rhopalocera 7:36-38
Chromosome numbers in taxonomy 11:8-9
Chromosomes of N. A. Rhopalocera 14:37-57, 127-147, 179-201
Chromosomes of Papilionidae 13:193-203
Chromosomes of *Utetheisa ornatrix* complex 17:231-233
Chrysalids, gregarious habit of *Nymphalis antiopa* 23:273-274
Cisseps fulvicollis, biology in laboratory 17:65-80
Cisthene subjecta on Long Island 10:111
Citheronia speculchralis, larva of from New Jersey 23:25
Classification of the butterflies by A.H. Clark, a rev. 2:73
Class study of wings of Lepidoptera 6:61-63
Clench, Harry K. (Personalia) 5:40
Cocoon with two pupae 3:106

Code of zoological nomenclature proposals 6:88
Coenonympha inornata in Maine 24:202
Coenonympha inornata phantasma new subspecies in Colorado 10:81-86
Coenonympha tullia in New York 17:158
Coenonympha tullia yellow forms in Oregon 21:1-7
Coenyropsis new genus 12:6
Coleopterists' Bulletin 1:60; 3:16
Colias activity in early winter in Pennsylvania 16:129-130
Colias cesonia therapis, seasonal variation in Venezuela 23:251-253
Colias cesonia, two variant females 22:253
Colias croceus, genetics of female dimorphism 15:43-55
Colias discal spot genetics 8:163-166
Colias erate poliographus genetics 10:143-149
Colias eurytheme: a melanic female 20:215-216; aberrant from Minnesota 22:158; electrophoretic variation in esterases 20:207-211; X *interior* hybrids 10:9-14
Colias eye colors 2:92
Colias in Canada 2:13
Colias nastes in the U.S.A. 16:46
Colias new genes, "Whitish" & "Blonde" 7:139-145
Colias philodice: effect of X-irradiation on larvae 23:257-260; new pale male 8:76; unusual variant in N.H. 25:108
Collecting buck-moths 2:81-82
Collecting: butterflies in Colorado 5:63; in California, winter 2:34; in Kentucky, early spring 2:22; Ithomiidae with Heliotrope 22:108-110; Lepidoptera by airplane, review of 19:129-137; Lepidoptera, discouragements and rewards 17:43-44; localities of Dr. Edward Palmer 21:129-134; methods for *Catocala* 1:19-20; rearing, and preserving larvae of Macrolepidoptera, notes on 18:201-210; rules for U.S. National Parks 4:15, 49; season summary 1:49, 50, 72, 87-96; *Speyeria egleis secreta* 2:91-92; sphingids and other moths on the Mississippi Gulf Coast 19:189-190; Sphingidae in Jamaica 2:86; Sphingidae in Venezuela 3:78; trip in Yukon and Alaska 19:57-62; trip notes 1:10, 22, 42, 58, 84; trip to Negros, Philippines 24:306; trip to northern Queensland, Australia 22:121-122
Collection and Library, care of 2:27; comments on 2:55
Collection, care and maintenance 24:51-54
Collection of Hesperiidae from Honduras 21:243-247
Collector's travels in U.S.A. 15:109-113
Colony of *Calephelis virginiensis* at Houston, Texas 21:281
Colony of *Thymelicus lineola* at Edmundston, N.B. 20:28
Colorado Butterflies, Part III, reviewed 11:57-60
Colorado butterflies & skippers, field data 22:159-171

Colorado butterfly collecting 5:63
Colored illustrations of the butterflies of Japan, rev.12:56
Colored illustrations, insects of Japan, reviewed 12:56
Common, I.F.B., (Personalia) 6:78
Communal roosting in *Colias* and *Phoebis* 24:117-120
Communal roosting in *Pieris rapae* 25:144-145
Comstock, William Phillips (1880-1956), obituary 13:30
Confusion between *Lethe creola* and *L. portlandia* 25:145-146
Congress of Entomology announcement 9:134
Congress of Entomology, 8th International 1:72
Connecticut butterfly collecting 10:55, 200
Constitution & By-laws 10:197-200
Constitution, proposed amendments 5:111; 12:157-158
  amendments approved 6:53
Considerations sur le male d'Ornithoptera allottei et sur la phylogenie des Ornithopteres 24:88-105
Consul panariste in Venezuela 25:19
Cook, John Hawley (1878-1958), obituary 12:207
Copulation of Rhopalocera 23:105-106
Corrected list of n.e. Arkansas butterflies 21:277-278
Corrections and addenda: 1:Index; 2:Index; 3:Index; 4:15,22 52; 5:Index-ii; 6:53, 73; 7:56, 198; 9:40, 218, 240; 10: 239; 13:253, 256; 18:115-116, 163; 19:170; 20:254;23:280
Coryphista meadi, host-plant finding by odor 24:220-224
Courtship and mating between *Chlosyne neumoegeni* and *C. californica* 24:226
Cover artist, drawing: 10:174; 11:96; 12:160; 15:33; 16:66; 17:49
Crambidia allegheniensis synonymized 8:93-94
Crambidia pura new for Canada 24:227
Cramer & Stoll, names proposed by 12:195-198
Cresote holders 3:49
Critics & criticisms 2:41
Crymodes exulis distribution 4:52
Ctenuchidae, priority of name 3:32
Cuba, additions to list of butterflies 4:72
Cuban butterflies, list 3:65; additions 4:72
Curating collections, creosote 3:49
"Curso de Entomologia" from Argentina 4:68
Cutworm rearing cage 15:105-106
Cynthia-Cecropia mating 9:142
Czechoslovakia museum, butterfly exposition 11:45

Danainae chromosomes 14:127-147
Danaus and *Papilio* in the Niagara Peninsula 4:14
Danaus berenice larval sex differences 8:123-124
Danaus gilippus berenice, distribution of foodplant chemoreceptors 23:196-198

Danaus gilippus in Mississippi 9:67-68; in Ohio 14:36
Danaus plexippus attacking red-winged blackbird 25:294
causes of death during rearing 20:235-238; in Gulf States 9:119-124; in Iowa and Virginia 10:44; in Japan 8:27; notes for 1949 4:45-46; parasites 2:92; some experimental color aberrations 20:45-47
Danaus in Niagara Peninsula 4:14
Daritis early stages 15:95-96
Dates of publication of Lepidoptera by H. Strecker 18:43-44
Davidson, W.M. (1887-1961), obituary 16:250
reminiscences 17:117
Day-flying Pericopidae new to Texas and the U.S. 24:301-303
Dealers in literature 3:49, 84
Death of netted butterflies 9:80
Decodes aneuretus new species from California 14:121-122
Decodes bicolor new species from California 14:122-123
Decodes johnstoni new species from California 14:119-121
Decodes montanus new species from California 14:118-119
Decodes new genus of Tortricidae 14:113-114
Definitions of common Latin terms 2:50
Definitions of terms 2:38
Denmark, butterfly collecting in 11:162-166
DeRabié paintings of Lepidoptera 5:55-57
Describing new species (editorial) 5:46
Description and habits of larvae of *Annaphila pseudoastrologa* 19:237-239
Description of the male of *Lithophane gausapata* 22:42-45
Descriptions of larvae of two eastern species of *Lygris* 25:262-264
Devastation of Yucatecan forests - insect abundance and local climates 20:201-205
Diakonoff, A. (Personalia) 5:10
Diapause of Lepidoptera in hot, arid climates 10:201-203
Diapausing larvae, care of 4:70
Diapause, termination of in saturniids 20:47-48
Dichogaster early stages, a correction 18:164
Dimorphism in larvae 9:79
Dimorphism in *Pieris napi oleracea*, environmental regulation 24:77-81
Dione moneta poeyii, ecology and natural dispersal in Texas 23:177-185
Discovery and observations of *Boloria eunomia* in Michigan 18:233-237
Distribution and geographical differentiation of *Marpesia eleuchea* 25:185-193
Distribution and hosts of five *Philotes* in Calif. 19:95-102
Distribution of *Plebeius saepiolus*, *Lycaena mariposa*, and *Hesperia comma* on Vancouver Island 18:185-188

Distribution, butterflies Contra Costa Co. Calif. 22:89-107
Distributional notes on the genus *Mestra* in N.A. 24:203-208
Distributional oddities 12:203-204
Distributions, describing of 11:21
Diurnals attracted to mercury vapor lamp 23:264
Diurnal Lepidoptera attracted to light, more observations 19:179-180
Djakonov, Aleksandr M. (1888-1956), obituary 11:55-56
Dodonidia helmsii, an anomalous foreleg in 23:201-202
Dragonfly attacks *Limenitis* defending its territory 25:146
Drepana pumping at water 11:18-21
Dryas julia delia life history 22:75-76
Drying and relaxing specimens 17:178-180
Dry Tortugas, Lepidoptera records 17:225-227
Dues 5:42, 111
Dyspteris pumping at water 11:18-21

Eana (Ablabia) idahoensis new species from Idaho 16:191
Eana (Tortricidae) in North America, revised 16:175-192
Eana (Ablabia) subargentana n. sp. from N.A. 16:183-187
Eacles imperialis gynandromorph 20:118
Early spring collecting in Kentucky 2:22
Early stages of *Chlosyne hoffmanni manchada* 21:71-73
Early stages of Lepidoptera 11:4
Early stages of *Sphinx sequoiae engelhardti* 21:27-31
Echinargus isola in Illinois 10:112
Ecological & behavioral notes, *Hesperia metea & Atrytonopsis hianna* 19:215-221
Ecological races 2:39
Ecology of natural dispersal - *Dione moneta poeyii* 23:177
Ecology of Rhopalocera in California 16:24-44, 136
Ecpantheria scribonia new foodplant record 24:227
Ecuadorian butterfly notes 4:67
Editorial 5:46; 18:64
Editorial and printing changes 10:80, 113, 240
Editorial announcing new section for the *NEWS* 9:146-147, 203
Editorial moves 11:200
Editorial organization changes proposed 17:63-64
Editor's report for 1952 8:9
Edwards, William Henry: reminiscences published 5:45
    specimens in two midwestern collections 20:156-162
    unnoticed paper by 10:135
    visit to Coalburg home of 3:61-62
Effect of barometric pressure on eclosion of *Pieris rapae* 22:211-225
Effect of certain environmental factors & chemicals on markings of *Pieris rapae* 20:13-27
Effects of photoperiod on initiation of pupal diapause in

Actias luna 21:255-258
Effects of vitamins on the development of *Nymphalis antiopa* 21:241-242
Effects of X-irradiation on larvae of *Colias philodice* 23:257-260
Egg-laying of captive Lepidoptera 13:24
Eggs of Theclini, collecting in winter 13:175-181
Ehrlich collection to AMNH 14:82
Ehrlich, P. R. (Personalia) 6:78
Electrophoretic variation in esterases of *Colias eurytheme* 20:207-211
Electrostatic charges and larval growth, 18th Century experiments 17:229-230
Embedment of specimens in clear polyester castings 21:107
Encyclopédie methodique, names proposed in 12:119-120
Entomological publications (see LITERATURE REVIEWS/NOTICES)
Entomological Society of America 1:73
Entomological Society of Canada 4:67
Entomologists' Gazette 7:58
Environmental factors and chemicals, effect on markings of *Pieris rapae* 20:13-27; 23:77
Environmental variations in *Euphydryas anicia eurytion* 21:261-270
Epionotia, biological and systematic coosiderations 25:115
Erebia disa mancinus in Washington, a correction 19:94
Erebia discoidalis in Wisconsin 15:95-96; probable second U. S. record 19:76
Erebia in Europe 13:104
Erebus and *Thysania* in Connecticut 4:13
Erebus odora and *Thysania zenobia* in Wisconsin 2:34
Erebus odora: in Alaska 11:205; in Connecticut 4:13; in Ohio 2:86; migrations 13:65-66; rearing attempted in New Jersey 13:17-18
Eressa javanica and allied forms 8:135-139
Erora, distribution and taxonomic notes 16:1-4
Erora laeta: at high altitude 6:34; in Michigan and Pennsylvania 10:18-24, 161-162; in North Carolina 25:295-296;in New Hampshire 14:239-240
Errata 1:Index; 2:Index; 3:Index; 4:15, 22, 52; 5:Index-ii; 6:53, 73; 7:56, 198; 9:40, 218, 240; 20:254; 23:280
Erynnis pacuvius foodplant 9:15
Erynnis zarucco larval foodplant in Louisiana 14:176
Epimecis, draft key to species 5:59-60
Especially for Field Collectors: 12:47-51, 127-132, 201-204 13:19-26, 79-103, 175-183, 229-237; 14:67-80, 159-160, 203-207, 237-240; 15:65-72, 109-113, 195-199, 237-244; 16:64-70, 137-146, 201-203, 249-250; 17:43-46, 111-116, 178-180, 243-245; 18:51-58, 59-64

Ethmia species in Pennsylvania 11:44-45
Euchloe ausonides mayi in Ontario 22:40-41
range extension 24:226
Euchloe olympia and *ausonides* in Ontario 15:113-114
Euchloe olympia field notes 6:99-102; host plant 25:64
Eumaeus atala rediscovery and early stages 15:237-244
in Florida Keys 20:216
Eunica tatilia in Kansas 12:25
Euphydryas anicia eurytion environmental variations 21:261
Euphydryas anicia in Oklahoma 25:246
Euphydryas editha bayensis life history 12:33-35
Euphydryas editha remingtoni new subspecies 12:168-169
Euphydryas pattern chart 12:166
Euphydryas phaeton in Indiana 10:46
new larval foodplant 23:48
Euphydryas subspeciation in western states 13:171-173
Eupithecia revision 3:108
Eupragia hospita new species from Florida 23:23-25
Euptychia mitchellii in New Jersey 20:43-44
life history 14:209-226
localities 12:41
Euptychia areolata distribution and variation in Mississippi 19:139-160
Eurema chamberlaini first U.S. record 18:31
Eurema daira daira in Mississippi 10:204-206
Eurema dina helios new record from Florida 16:246
Eurema hecabe hibernating behavior 6:43
Eurema lisa swarm of under mercury vapor lamp 23:264
Eurema nicippe in Connecticut 10:200
pupal coloration 12:95-96
Eurema, notes on 2:51-53
Eurema salome in Texas 23:188
Euristrymon ontario ontario distribution 16:195-197
Duristrymon ontario, some records of 25:80-82
Euristrymon polingi in Texas 16:201-203
European Lepidoptera reared in Ohio 4:13
subspeciation 5:27-28
Eustis, Henry W., obituary 6:77
Eustixia pupula larval foodplants 22:157-158
Eutachyptera psidii early stages 11:99-102; 13:24
Eutricopis nexilis life history 24:151-156
Evaluation of an insect light trap designed to separate beetles and moths 18:1-10
Evans, A catalogue of the American Hesperiidae 9:160
Evans, Brigadier W.H., obituary and bibliography 10:101, 193-199
Everes comyntas larval foodplant in Texas 22:122
Evolutionary and biological significance of multiple pairing

in Lepidoptera 22:197
"Excelsior" complex 12:131-132
Exoteleia, a new Nearctic species on pine 21:9-11
Exoteleia nepheos biology and immature stages 21:15-21
Exotic Lepidoptera captured in California 22:187
Experimental color aberrations in *Danaus plexippus* 20:45-47

Fagitana littera reared 7:166
Feeding habits of *Oeneis jutta reducta* 24:306-307
Feigning death by moths 6:42, 108
Felder & Felder, names proposed by 12:193-194
Ferguson collection to Smithsonian Institution 25:168
Fertility, fecundity, etc., of *Cisseps fulvicollis* 17:65-80
Field notes on *Automeris zephyria* and larvae of *Hemihyalea edwardsi* in New Mexico 19:82
Field notes on three skippers in Texas 24:212
Field notes and subspecific status of *Oeneis melissa* in northern Quebec 23:103-104
Field season summary for 1948 2:Supplement
Field & technique notes: 4:10, 73; 5:69-70; 6:42-44, 73,100 107-108; 7:27-28, 53-56, 172; 8:26-29, 94, 101-103, 140, 146; 13:78, 164, 170, 183; 14:36, 57, 62, 156, 158, 230, 242; 15:56, 64, 73, 92, 101, 108, 113-114, 115; 16:44, 46, 59-63, 98, 128, 136, 174, 198-202, 246; 17:102, 110, 158, 224, 245
First records:
Boloria eunomia in Wisconsin 25:149
Eurema chamberlaini in the U.S. 18:31
Thorybes bathyllus in Minnesota 20:34
Thymelicus lineola in Connecticut 19:195-196
two butterflies in Wisconsin 22:252
Flight habits of *Boloria toddi* 19:104; *Morpho theseus justitiae* 20:95-101
Flight pattern of *Anisota virginiensis* 25:84-85
Flight period of *Boloria eunomia* 19:184-185
Florida: check list near completion 10:109; collecting 3:25 collecting guide 9:204-212; "Florida" of Hübner 4:62 Lepidoptera checklist 7:65; 9:22
Fluid recycling in *Agathymus aryxna* 20:242
Fluorescence of butterfly wing colors 13:73-77
Fluorescent light, *Cercyonis pegala nephele* at 20:12
Fluorescent pigments in Lepidoptera 22:27; 22:240
Flyways of butterflies 4:13, 46
Flyways of *Papilio glaucus* 2:92
Foodplant preferences in *Lycaena helloides* complex 17:24-26
Foodplants, Allan's book on 5:60
Foodplants of *Syssphinx heiligbrodti* in Texas 18:42
Foodplants of Texas Lepidoptera 13:221-228

Ford, *MOTHS* 9:213-215
Forest-prairie transitions and the admixture of butterfly faunas 25:216-221
Formal organization for the Lepidopterists' Society 4:1, 38
Formosa, butterflies of (Shirozu), reviewed 14:243
Forster & Wohlfahrt, Schmetterlinge Mitteleuropas 6:79-80
Forsyth, Marguerite S., obituary 6:76-77
Fossil Lepidoptera 1:43
Freeman, H.A. collection of Megathymidae to AMNH 24:208
Freeze-drying caterpillars, a simplified method 23:43-48
Fulvia, new genus 12:162-163
Further comments on possible mimicry of *Caenurgina caerulea* 21:122-124
Further distribution records and taxonomic notes on *Philotes rita* 21:141-142

Galleria mellonella rearing cage 15:102-104
Gaspé Peninsula collecting trip 5:53-54
Generic notes on two hairstreaks new to the U.S. 24:56-59
Genetics and taxonomy 11:10-12
Genetics of female dimorphism in *Colias croceus* 15:43-55
Genetics, *Pieris napi-bryoniae* group 16:5-19, 125-127
Genitalia function and terminology 11:5-7, 216-220
Genitalia of ...Tortricidae ...of the British Islands 15:121
Genitalic structure 10:17; terminology 8:67-74, 167-168, 169
Genus concept in Rhopalocera 3:3-4
Geographic subspeciation in Lepidoptera, Symposium 5:17-35
Geographic variability in *Speyeria* (Moeck), reviewed 13:245
Geometridae of Baker County, Oregon 13:217-220
Geometrinae: new species from Africa 25:169-176
Georgia butterflies, notes 4:43-44, 62
German Lepidoptera, books on reviewed 11:62, 176
Gerould, Harvard, & Jones collections to Yale University 8: 47, 104, 173
Gerould, John H., ninetieth birthday 12:190
Glassine envelopes, papering Lepidoptera in 25:65-68
Glaucina mayelisaria, the female of 20:250
Glaucina, new species from Texas 20:247-250
Glaucopsyche in California, review of 23:149-154
Glaucopsyche lygdamus, a record from Missouri 18:41-42
Glaucopsyche piasus, generic review and description of new subspecies 25:240-246
Gloveri crosses 3:106
Godart & Latreille, names proposed by 12:119-120
Gonadectomy and transplantation in sex races of *Lymantria dispar* 7:7-8
Goenepteryx aspasia, comments on forms described by Shu-iti Murayama 24:213-217

Grammar of the name *Heliothis* 25:264-266
Graphium marcellus in Mississippi, var. of 24:176-189
Graphium philolaus new to Texas and North America 24:233
Gregarious roosting in tropical butterflies of the genus
Morpho 25:223-234
Guia de naturalista sudamericanos, reviewed 12:56
Guia para el entomologio principiante (Hayward), rev. 16:22
Gynandromorph (bilateral) of *Pieris rapae* 24:224
Gynandromorph of *Automeris io* 17:43-44
two mosaics 25:234-238
Gynandromorphic *Eacles imperialis* 20:118
Phaeoura mexicanaria 23:274-275
Gyops (Noctuidae), life history and taxonomy 16:47-54

Hairstreak genera, redefinition of North American 14:19-23
Hairstreaks new to Texas and the U.S. (three) 24:59-61
Hairstreaks new to the U.S., generic notes on two 24:56-59
Hand-pairing *Limenitis* butterflies, simple technique 23:109
techniques 10:47-53
Haploa clymene, hibernating larvae on bridges 24:271
Harrisina mass rearing technique 8:11-12
Hasora chromus, migration in India 17:110
Heliconius antiochus salvinii, note on 23:199-201
Heliconius besckei, use of color for species recog. 24:25
Heliconius charitonius aberrant 21:248
Heliconius hecale and *Xanthocles* in Venezuela 23:104-105
Heliconius males and female pupae 8:102
Heliolonche carolus life history 23:26-30
Heliolonche pictipennis life history 25:287-292
Heliopetes domicella distribution 5:70
Heliopetes sublinea, on the status of 21:205-206
Heliothiinae types sought 6:53
Heliothis, grammar of the name 25:264-266
Heliothis oregonica life history 25:1-6
Helophilour moths of Hungarian Moorlands 2:93-94
Hemihyalea edwardsi in New Mexico, larvae of 19:82
Hemileuca maia: field observations 5:71; in Ohio 10:46
in Wisconsin 8:29
Memming, Francis retirement 12:36
Herculia phaezalis life history 13:67-69
Hesperia metea, ecological and behavioral notes 19:215-221
Hesperia pawnee in Michigan 12:37-40; 14:57
Hesperia tildeni new species from California 9:196-198
Hesperiidae catalogue announcement 9:160; 10:110
Hesperiidae larval foodplants in Texas 13:221-228;20:229-232
Hesperiidae new records from Mexico 20:226-228
Hesperiidae of St. Louis area 9:190-195
Hesperiidae, three new species from Mexico 21:115-119

Hesperioidea chromosomes 14:37-57
Hesperioidea taxonomic notes 14:24-36
Hesperocharis longstaffi rediscovered in Venezuela 22:189
Hessel, S.A. appointment at Peabody Museum 12:185
Hestia leuconoe 12:115-117
Heteropterus morpheus in Netherlands 4:72
Hibernating:
butterfly larvae, method for overwintering 19:187
larvae of *Haploa clymene* 24:271
Nymphalidae on Vancouver Island 4:13-14
Hilltops and butterflies: 6:101-102; 7:38-44; 8:141-145; 9:127-134; 16:64-66
Hinton, H.E. (Personalia) 5:10
Hispaniolan Lepidoptera, paintings 5:55-57
Hispaniola new *Phlegethontius* form 11:107-112
History of Lepidopterology: Hungary 3:43; Missouri 3:51-52
Holarctic butterfly subspeciation 5:24-27
Holomelina aurantiaca buchholzi new subspecies from Virginia 17:100; correction 18:118
Holomelina Nearctic species, taxonomic notes and partial key 19:69-76
Honduras, recent collections of butterflies from 21:185-197
Hopfinger, John Carl (1888-1961) obituary and localities 16:147-150
Host plant finding by odor in adult *Coryphista meadi* 24:220
Host plant: identification 1:25; 4:2
records from British Columbia 8:101
specificity of *Polydorus aristolochiae* 21:127-128
How to know the butterflies (Ehrlich & Ehrlich), rev. 14:201
Hübner's "Florida" 4:62; 5:6
Hulstina inconspicua life history 17:28-30
Hulst, Smith, & Sperry collections to AMNH 8:30, 46
Humidity, darkness, and gold spots possible factors in pupal duration 18:230-232
Humidity effects in *Pieris rapae* 8:88-90
Humidity tests with *Papilio* pupae 5:67
Hungarian lepidopterology:
fauna of Carpathian basin 3:75-76
short history 3:43
National Lepidoptera Collection saved 10:192
Hyalophora isolating mechanisms 13:213-216
Hybridization experiments with *Papilio* 13:151-164
in Australia 16:161
in Japan 17:163
Hybrids:
among species of *Hyalophora* 25:68-73
between *Colias eurytheme* X *Colias interior* 10:9-14
between *Papilio xuthus* *P. polyxenes-machaon* 14:5-18

Hybrids of *Agathymus* 13:204-206; of *Limenitis* 24:270
Hypaurotis crysalus: evening mating in Colorado 16:200
foodplant 13:164; multiple capture at light 19:55-56
Hyperaeschra georgica larva described 10:203-204
Hypermnestra helios ariana new subspecies from Afghanistan 15:2-3
Hyphantria cunea in Europe 7:55
Hyponephele shivacola new species from Afghanistan 15:16

Identity of *Ceratonyx satanaria* 21:11-14
Identification of host plants 4:2; parasites 4:2
Immature stages of *Scopula ancellata* 24:54-55
Immature stages, preservation of 1:31, 46
Importance of preserving natural habitats 21:275-276
Incisalia:
augustinus life history and host plant 7:33-35
foodplants 10:89-90
eryphon life history 13:70
fotis schryveri bionomic notes and life hist. 24:256-266
henrici in Manitoba and Minnesota 22:26
mossii habitata in British Columbia 13:101-103
Incubation apparatus for larvae and pupae 5:35
Indian butterfly records 16:131-135, 143-145
India, *Hasora* and *Callimorpha* swarms 17:110, 224
Indo-Australian collecting methods 9:74-76
Inexpensive apparatus for photographing specimens 19:212-214
Ingham, Charles Henry (1904-1957), obituary 11:169-170
Ingham collection, part acquired by Yale 12:94
Inoue, Check List of the Lepidoptera of Japan Part I 9:159
Insect net bags, use of nylon 2:86
Insecticides, curating collections 2:36
International aid 2:61
International Code of Zoological Nomenclature, rev. 15:125
International Commission on Nomenclature, notices 17:104,253
Intergeneric mating 23:115
Interspecific hybridization between *Limenitis arthemis astyanax & L. archippus* 25:278-284; mating of *Lycaena* 6:108
pairing in Lycaenidae 16:235-237
Iowa butterfly records 15:97-98
Iraq Lepidoptera (Wiltshire), reviewed 13:33
Irradiation (neutron) in *Anthereae eucalypti* 25:238-240
Irwin collection to Illinois Natural History Survey 25:83-84
Isolating mechanisms of *Hyalophora* 13:213-216
Isolation influence in Geometridae 6:93-94; 7:172
Itame ribearia variation in larval colour patterns 25:266
Ithome edax new species from Texas 15:88-89
Ithome ferax new species from Florida 15:89-90
Ithome genus revised 15:81-90

Ithome lassula new species from Florida 15:89
Ithomiinae and Styrinae, affinities of 24:297-300
Ithomiidae collecting with Heliotrope 22:108-110
Ithomiidae: Tribes Tithoreini & Melinaeini, checklist of 15:25-33
Itineraries of the Wheeler survey naturalists 20:71-82

Japan, collecting mountain butterflies in 14:203-207
Japanese: butterfly life histories 8:95-100
  Lepidoptera check list reviewed 11:177
  Lepidoptera observations 6:78
Jones, Frank Morton ninetieth birthday 12:190
Jordan, H.E. Karl notice of death 12:198
Junonia coenia in Mississippi, variation 21:59-71

KAAD mixture for preserving larvae 2:14
Kalshoven, Plagen van de Cultuurgewassen in Indonesie 6:80
Karanasa pamira kotandari new subspecies from Afghanistan 15:11-12
Katahdin, collecting on 12:47-51
Kent Island, New Brunswick butterflies on 19:181-183
Key to the genera of Psaphidini, a new genus and species from North America 20:83-91
Killing: agent, ethyl acetate 10:150-151
  butterflies, methods 3:26, 49
  methods for Lepidoptera 4:73
Kirby, W. F. rare 1858 paper 16:104
Kiriakoff, S. G. (Personalia) 6:78
Klages collection to Cornell University 6:36
Klots, A. B. (Personalia) 6:78
Klots' Field Guide inaccuracies 6:41
Knob Lake field notes on butterflies 5:7-9
Komárek & Tykač, Atlas Motylu 9:158
Kricogonia castalia migration in northern Mexico 25:124-125
Kricogonia migration of in southern Texas 19:223-224

Labels and pins 4:31, 52
Label printing in the field 13:96-98
Laboratory techniques for *Cisseps fulvicollis* 17:65-80
Lacey, H.G. unpublished field notes on Lepidoptera 25:29-44
Lambremont, The Butterflies and Skippers of Louisiana 9:158
Larva and pupa of *Orthosia hibisci quinquefasciata* 20:213
Larva feeding on aphid exudations 4:52
Larvae and pupae, incubating 5:35
Larvae of *Celario lineata* as food for Indians 21:144
Larval: behavior of *Agathymus* in Baja and so. Calif. 23:243
  colour patterns of *Itame ribearia*, var. in 25:266-270
  description of *Polydorus aristolochiae* 23:107-108

prolegs structure for classification 6:1-6, 7-12
versus imaginal evolutionary rates 14:101-103
Larval foodplants:
for *Callophrys iroides* 22:225-226
for *Euchloe olympia* for northern populations 25:64
for *Eustixia pupula* 22:157-158
for *Everes comyntas* in Texas 22:122
for Loochooan butterflies 18:105-109
for
for sphingids in Canada 21:252-254
for Texas Hesperiidae 19:1-33; 20:35-41; 20:229-232
for Texas Rhopalocera 18:129-157
identification 1:25; 3:2; 4:2
Latin terms, definitions of 2:50
Leaf-mining Lepidoptera, book reviewed 11:177-178
Leaf-mining Lepidoptera, rearing methods 4:3-6
LeConte, names proposed by 12:121-122
Legna perditalis foodplant 8:27
Lepidoptera:
as research tools 17:65-80, 173-177
associated with pig carrion 23:191-195
biology, open for study 2:37
classification based on tympanic structures 17:1-6
from southern Florida, unusual species 25:294
hibernation 2:13
in the Carnegie Museum 2:80
measurements 4:51-52, 75-76
of the Central Brazil Plateau -
Preliminary list of Rhopalocera 21:77-106, 145-168
Nymphalidae, Libytheidae 21:77-106
Lycaenidae, Pieridae, Papilionidae, Hesperridae 21:145-168
New genera, species, and subspecies of Hesperiidae 22:1-20
Partial list for the Belo Horizonte area, showing the "Blend Zone" 22:147
of New York, Part IV (Forbes), reviewed 17:40-42
ovipositing on plants toxic to larvae 18:104
parasite identification 2:53
Lepidoptera of Iraq (Wiltshire), reviewed 13:33
Lepidoptera, Rhopaloceres-Heteroceres dates of publication by Herman Strecker 18:43-44
Lépidoptères de l'Amérique septentrionale, names proposed in 12:121-122
Lepidopterists' Society: formal organization 4:1, 38
NEWS, back issues 4:24, 36; review (1948) 3:1
Lepidopterists' Societies, foreign 1:72
Lepidopterology in Argentina 2:42

Lepidopterous larvae associated with aphids 4:13

Leptotes marina in Illinois 10:112

Lerema ancillaris new to Texas and the U. S. 24:266

Lerodea dysaules first U. S. record 18:214-216

Lethe anthedon borealis a note on 25:256-261

Lethe creola and *L. portlandia,* confusion between 25:145-146

Lethe creola, notes on with designation of lectotype 24:143

Letters from Dr. H. H. Behr to Herman Strecker 22:57-62

Letters to the editor 3:19-20, 67-68; 4:49

Leucophlebia lineata brunnea synonymy 21:8

Libytheana bachmanii: date of publication 12:123

- migrating in Mexico 6:44; migration in Arizona 3:28
- migration in southern Texas 19:223-224

Libytheinae chromosomes 14:179-201

Life history of:

- *Atrytone arogos* 20:177-181
- *Chlosyne fulvia* 22:237-240
- *Dryas julia delia* 22:75-76
- *Eutricopis nexilis* 24:151-156
- *Heliolonche carolus* 23:26-30
- *Heliolonche pictipennis* 25:287-292
- *Heliothis oregonica* 25:1-6
- *Incisalia fotis schryveri* 24:256-266
- *Lithophane subtilis* 23:15-18
- North American Lepidoptera 2:34
- *Ornithoptera alexandrae* 25:58-64
- *Orthosia hibisci quinquefasciata* 20:213-215
- *Papilio indra martini* 22:46-52
- *Papilio indra minori* 18:65-73
- *Papilio oregonius* 22:53-56
- *Problema byssus* 19:77-81
- *Pyrrhia exprimens* 24:234-239
- *Schinia cupes deserticola* 25:109-114
- *Schinia felicitata* 21:22-26
- *Schinia florida* 24:282-287
- *Schinia ligeae* 25:274-277
- *Schinia niveicosta* 20:29-33
- *Schinia separata* 25:177-180
- *Schinia walsingham* 25:181-185
- studies 1:9, 31
- studies of Japanese Rhopalocera 15:245-248

Lights attracting butterflies 16:128, 131-135, 174, 234

Lights attracting butterflies in Central America 17:37-38

Light traps:

- designed to separate beetles and moths 18:1-10
- for insects 3:63-64; 13:24, 57-61
- review of design principles with description 22:65-75
- weather-resistant for collecting Lepidoptera 18:79-83

with multiple-reflectors 5:72
Limenitis:
arthemis astyanax hybridization with 25:278-284
astyanax on Nantucket 9:16
hybrid 7:53; hybrid from Arizona 8:129-130
lorquini complex, a review of 20:172-176
weidemeyerii complex revision & description of a new subspecies 21:213-234
Lindsey, A. W. (1894-1963), obituary & biblio. 17:181-190
List of members 6:125-141
Literature for sale 3:20
Literature Reviews/Notices:
A CENTURY OF PROGRESS IN THE NATURAL SCIENCES 13:35
Allan, *LARVAL FOODPLANTS* 5:60
Amsel, *MICROLEPIDOPTERA PALAEARCTICA* 19:117-125
Amsel, *MICROLEPIDOPTERA VENEZOLANA. ÜBER DIE VON PATER CORNELIUS VOGL IN CARACAS UND MARACAY GEFANGENEN KLEINSCHMETTERLING. Part I* 12:55
Avinoff & Sweadner, *THE KARANASA BUTTERFLIES* 7:16-23
Barcant, *BUTTERFLIES OF TRINIDAD AND TOBAGO* 25:297
Beaufoy, *BUTTERFLY LIVES* 3:37
Beirne, *BRITISH PYRALID AND PLUME MOTHS* 8:49
Bourgogne, *LES LÉPIDOPTÉRES* 5:60
Bowman, *AN ANNOTATED LIST OF THE LEPID. OF ALBERTA* 5:60
Braun, *ERGEBNISSE DER DEUTSCHEN AFGHANISTAN-EXPEDITIONS 1956 DER LANDESSAMMLUNGEN FÜR NATURKUNDE IN KARLSRUHE* 15:128-129
Brewer, *WINGS IN THE MEADOW* 22:188
Brown, *COLORADO BUTTERFLIES. Part III. LIBYTHEIDAE, RIODINIDAE & LYCAENIDAE* 11:57-60
Chopard, *LE MIMÉTISME* 5:61
Clark, *BUTTERFLIES OF THE DISTRICT OF COLUMBIA* 2:108
Clark & Clark, *BUTTERFLIES OF VIRGINIA* 6:71-72
Clark, *CLASSIFICATION OF THE BUTTERFLIES, WITH THE ALLOCATION OF THE GENERA OCCURRING IN NORTH AMERICA NORTH OF MEXICO* 2:73
Clarke, *MICROLEPIDOPTERA OF JUAN FERNANDEZ IS.* 19:164
Collins & Weast, *WILD SILK MOTHS OF THE UNITED STATES, SATURNIINAE, EXPERIMENTAL STUDIES & OBSERVATIONS OF NATURAL LIVING HABITS AND RELATIONSHIP* 16:58
Comstock, *BUTTERFLIES OF CALIFORNIA* 1:38
Comstock, *LEPIDOPTERA OF AMERICAN SAMOA* 21:21
Corbet & Pendlebury, *THE BUTTERFLIES OF THE MALAY PENINSULA*
Cowan, *ANNOTATIONES RHOPALOCEROLOGICAE* 22:187, 24:307
Crotch, *A SILKMOTH REARER'S HANDBOOK* 10:178
CURSO DE ENTOMOLOGIA - ARGENTINA 5:62
Curtis, *A SHORT HISTORY, THE BROWN-TAIL MOTH* *24:228*

Davis, *A REVISION OF THE MOTHS OF THE SUBFAMILY PRODOXINAE* 23:203-204
del Ponte, *CURSO DE ENTOMOLOGIA. III. MORFOLOGIA INTERNA* 5:62
Derksen & Scheiding-Gollner, *INDEX LITTERATURAE ENTOMOLOGICAE* 21:144, 23:188
Derxen & Scheiding, *INDEX LITTERATURAE ENTOMOLOGICAE* 19:62
Diakonoff, *A REVISION OF THE FAMILY CERACIDAE* 5:61
Diakonoff, *MICROLEPIDOPTERA OF NEW GUINEA. I.* 7:128
Diakonoff, *MICROLEPIDOPTERA OF NEW GUINEA. III.* 10:176
Diakonoff, *MICROLEPIDOPTERA OF THE PHILIPPINE IS.* 23:106
Doring, *ZUR MORPHOLOGIE DER SCHMETTERLINGSELER* 10:56
dos Passos, *ENTOMOLOGICAL REMINISCENCES OF WILLIAM H. EDWARDS* 5:45
dos Passos & Grey, *SYSTEMATIC CATALOGUE OF SPEYERIA* 2:5
Duane & Tyler, *OPERATION SATURNID* 5:62
Edwards, *BUTTERFLIES OF NORTH AMERICA* 1:75
Ehrlich & Ehrlich, *HOW TO KNOW THE BUTTERFLIES* 14:201
Englehardt, *NORTH AMERICAN AEGERIIDAE* 1:15
ENTOMOLOGISCHES NACHRICHTENBLATT - SWISS/AUSTRIAN 4:62
ENTOMOLOGISCHE VEREENIGING IN INDONESIE successor to *NEDERLANDSCH-INDISCHE ENTOMOLOGISCHE VEREENIGING* 4:50
ENTOMOPS 20:28
Ferguson, *A REVISION OF THE MOTHS OF THE SUBFAMILY GEOMETRINAE OF AMERICA* 23:242
Ferguson, *THE LEPIDOPTERA OF NOVA SCOTIA. Part I, MACROLEPIDOPTERA* 8:170
Ferris, *A KEY TO THE BUTTERFLIES OF WYOMING* 25:176
Fichter, *INSECT PESTS* 21:143-144
Field, *A MANUAL OF THE BUTTERFLIES AND SKIPPERS OF KANSAS* 3:102
Fontes & Parodis, *GUIA DE NATURALISTA SUDAMERICANOS* 12:56
Forbes, *LEPIDOPTERA OF NEW YORK* 1:63
Forbes, *THE LEPID. OF N.Y. & NEIBHBORING STATES, II* 3:5
Forbes, *et al.*, ... Part IV 17:40-42
Ford, *BUTTERFLIES* 1:3
Forster & Wohlfahrt, *DIE SCHMETTERLINGE MITTELEUROPAS* 7:26; 8:170, 11:176
Freeman, T.N., *et al.*, *BUDWORM SIBLINGS IN CANADA* 7:57
Freeman, H.A., *ECOLOGICAL & SYSTEMATIC STUDY OF THE HESPERIOIDEA OF TEXAS*
French, *THE BUTTERFLIES OF THE EASTERN U. S.* 2:55
Fulmek, *PARASITINSEKTEN DER BLATTMINIERER EUROPAS* 16:237
Garth, *BUTTERFLIES OF GRAND CANYON NATIONAL PARK* 5:61
Gifford, *A LIST OF THE BUTTERFLIES OF MALAWI* 21:7-8
Harris, *THE BUTTERFLIES OF GEORGIA* Revised 4:42
Harris, *THE ENGLISH LEPIDOPTERA OR THE AURELIAN'S POCKET*

COMPANION 25:148
Hayward, *GUIA PARA EL ENTOMOLOGOGO PRINCIPIANTE* 16:22
Hayward, *RHOPALOCERA OF ARGENTINA. Vol. I.* 3:66
Hemming, *THE GENERIC NAMES OF THE BUTTERFLIES AND THEIR TYPE-SPECIES* 21:274
Hemming, *THE GENERIC NAMES OF THE HOLARCTIC BUTTERFLIES Vol. I* 3:44
Herbulot, *ATLAS DES LÉPIDOPTÈRES DE FRANCE - III, HÉTÉROCÈRES* 3:68
Herbulot, *ATLAS DES LÉPIDOPTÈRES DE FRANCE, BELGIQUE, SUISSE, ITALIE DU NORD HÉTÉROCÈRES, Fasc II* 2:88
Hering, *BESTIMMUNGSTABELLEN DER BLATTMINEN VON EUROPA EINSCHLIESSLICH DES MITTELMEERBECKENS UND DER KANARISCHEN INSELN* 11:177-178
Hering, *BIOLOGIE DER SCHMETTERLINGE* 2:95
Higgins, *A REVISION OF THE MELITAEINE GENUS CHLOSYNE AND ALLIED SPECIES* 18:266
Higgins & Riley, *A FIELD GUIDE TO THE BUTTERFLIES OF BRITAIN AND EUROPE* 25:152
Hodges, *THE MOTHS OF AMERICA NORTH OF MEXICO, Fascile 21 SPHINGOIDEA* 25:198
Hoffmeyer, *DE DANSKE SPINDERE* 5:62
Hoffmeyer, *DE DANSKE UGLER* 5:62
Hogue & Truxal, *THE INSECT REALM* 25:152
Holland, *BUTTERFLY BOOK* 1:27
Holland, *MOTH BOOK* 1:51, 23:241
Hrubý, *PRODROMUS LEPIDOPTERORUM SLOVACIAE* 19:81
IDEA - New German 4:50
Igarashi, *et al.*, *CONTRIB. TO THE INSECT FAUNA OF NEPAL* 21:40
Inoue, *CHECK LIST OF THE LEPIDOPTERA OF JAPAN, II* 11:177
INTERNATIONAL CODE OF ZOOLOGICAL NOMENCLATURE ADOPTED BY THE XV INTERNATIONAL CONGRESS OF ZOOLOGY 15:125-126
Janse, *ON THE TYPES OF SOUTH AFRICAN MICROLEPIDOPTERA DESCRIBED BY THE LATE EDWARDMEYRICK AND PRESERVED IN THE TRANSVAAL MUSEUM, PRETORIA, AND THE SOUTH AFRICAN MUSEUM, CAPE TOWN Vol. 1 - TINEIDAE* 23:147-148
Jones, *CHECK LIST OF BRITISH COLUMBIA MACROLEPID.* 6:79
Kalshoven, *DE PLAGEN VAN DE CULTUURGEWASSEN IN INDONESIA* 4:50
Klots, *A FIELD GUIDE TO THE BUTTERFLIES OF NORTH AMERICA East of the Great Plains* 4:107-108
Klots & Clench, *NEW SPECIES OF STRYMON FROM GEORGIA* 7:26
Kozhantshikov, *FAUNA SSSR. NASEKOMYE CHESHUEKRYLYE. Vol. III, Part 2.* 10:177-178
Koch, *FALTER BEI TAG UND NACHT. AUS DEM LEBEN UNSERER SCHMETTERLINGE* 11:62
Koch, *WIR BESTIMMEN SCHMETTERLINGE. Part I, TAGFALTER*

DEUTSCHLANDS 11:62
Part II, BAEREN, SPINNER, SCHWAERMER UND BOHRER DEUTSCHLANDS 18:84
Part III, EULEN DEUTSCHLANDS 18:84
Part IV, SPANNER DEUTSCHLANDS 18:84

Krogerus, *INVESTIGATIONS ON THE LEPIDOPTERA OF NEWFOUNDLAND, I. Macrolepidoptera* 10:175-176

Krzywicky, *PIERIDAE - PAPILIONIDAE* 16:192

Langer, *NORDENS DAGSOMMERFUGLE (The Butterflies of Scandinavia)* 12:205-206

LEPIDOPTERA, Danish Journal, Lepidopterologisk Forening in Copenhagen 20:27

Lhomme, *CATALOGUE DES LÉPIDOPTÈRES DE FRANCE ET DE BELGIQUE: 2 vols.* 3:38

MacKay, *LARVAE OF THE NORTH AMERICAN TORTRICINAE* 18:228

MacKay, *NORTH AMERICAN AEGERIIDAE:* Revision based on late instar larvae 23:126

MacNeill, *THE SKIPPERS OF THE GENUS HESPERIA IN WESTERN NORTH AMERICA* 18:240

Macy & Shepard, *BUTTERFLIES* 3:44

Manley & Allcard, *FIELD GUIDE TO THE BUTTERFLIES AND BURNETS OF SPAIN* 25:299

Mann, *CACTUS-FEEDING INSECTS AND MITES* 23:241-242

Mather & Mather, *THE BUTTERFLIES OF MISSISSIPPI* 13:33-35

McDunnough, *REVISION OF NORTH AMERICAN EUPITHECIA* 3:108

McDunnough, *SPECIES OF THE GENUS HYDRIOMENA OCCURRING IN AMERICA NORTH OF MEXICO* 8:171

McGuffin, *GUIDE TO THE GEOMETRIDAE OF CANADA. I. SUBFAMILY STERRHINAE* 21:282, 22:110

Michener, *SATURNIIDAE OF WESTERN HEMISPHERE* 6:109-111

Miller, *HIGHER CLASSIFICATION, PHYLOGENY & ZOOGEOGRAPHY OF THE SATYRIIDAE* 23:276-277

Mitchell & Zim, *BUTTERFLIES AND MOTHS* 19:225

Moeck, *GEOGRAPHIC VARIABILITY IN SPEYERIA* 13:245-246

Morris (Editor), *THE DYNAMICS OF EPIDEMIC SPRUCE BUDWORM POPULATIONS* 18:116

Moucha, *BEAUTIFUL BUTTERFLIES* 20:56

Moucha, *MOTÝLI* 16:150

Newman, *BUTTERFLY FARMER* 8:49

Niculescu, *FAUNA REPUBLICII POPULARE ROMANE, vol. XI, fasc. 6 & 7* 20:33

Obenberger, *ENTOMOLOGIE* 19:90

Osthelder, *DIE SCHMETTERLINGE SUCBAYERNS UND DER ANGRENZENDEN NORDLICHEN KALKALPEN* 10:56

Patočka, *DIE TANNENSCHMETTERLINGE DER SLOWAKEI* 15:74

Peterson, *LARVAE OF INSECTS, Part I, Lepidoptera and Hymenoptera* 3:37

Phinhey, *BUTTERFLIES OF RHODESIA* 7:25

Pierce & Metcalfe, *THE GENITALIA OF THE GROUP TORTRICIDÆ OF THE LEPIDOPTERA OF THE BRITISH ISLANDS* 15:127-128
Ponec, *NAŠE MOTÝLE* 15:74
Prentice, *FOREST LEPIDOPTERA OF CANADA RECORDED BY THE FOREST INSECT SURVEY* 18:210
Rao & Chuan, *PESTS OF HEVEA PLANTATIONS IN MALAYA* 20:206
Reiprich, *MOTÝLE SLOVENSKA. OBLAST SLOVENSKÉHO RAJA* 13:246
Rindge, *REVISION OF THE GENERA MELANOLOPHIA, PHEROTESIA, AND MELANOTESIA* 19:46
Ross, *INSECTS CLOSE UP* 10:57
Schmid & Endicott, *MARIPOSAS DE VENEZUELA* 25:86
Schwarz, *MOTÝLI DENNÍ* 5:62
Shaffer, *REVISION OF THE PEORIINAE AND ANERASTIINAE OF AMERICA NORTH OF MEXICO* 23:280
Shapiro, *BUTTERFLIES OF THE DELAWARE VALLEY* 20:244
Shirôzu, *BUTTERFLIES OF FORMOSA IN COLOUR* 14:243
Shirôzu, *EARLY STAGES OF JAPANESE BUTTERFLIES* 19:211
Smith, *THE NATURAL HISTORY OF THE RARER LEPIDOPTEROUS INSECTS OF GEORGIA* 2:43
Sthr & Cook, *REVISION OF THE GENUS MALACOSOMA IN NORTH AMERICA*
Stokoe & Stovin, *THE CATERPILLARS OF BRITISH MOTHS, Including the Eggs, Chrysalids and Foodplants.* 3:38
THE CATERPILLARS OF THE BRITISH BUTTERFLIES ... 3:38
Takeuchi, *COLORED ILLUSTRATIONS OF THE INSECTS OF JAPAN* 12:56
Tietz, *THE LEPIDOPTERA OF PENNSYLVANIA* 8:172
Tilden, *BUTTERFLIES OF THE SAN FRANCISCO BAY REGION* 19:138
TRANSACTIONS OF THE SHIKOKU ENTOMOLOGICAL SOCIETY 4:71
Turner, *PHYLOGENY AND CLASSIFICATION OF THE LEPID.* 2:65
Tuxen, *TAXONOMIST'S GLOSSARY OF GENITALIA IN INSECTS* 12:125-126
Viette, *LÉPIDOPTÈRES HOMONEURES* 5:62
Villiard, *MOTHS AND HOW TO REAR THEM* 23:270, 25:221
Warner & Bijok, *BUTTERFLIES AND MOTHS* 20:122
Werny, *UNTERSUCHUNGEN ÜBER DIE SYSTEMATIK DER TRIBUS THYATIRINI, MACROTHYATIRINI, HABROSYNINI UND TETHEINI* 23:101-102
Williams, *THE MIGRATION OF BUTTERFLIES* 3:32
Wiltshire, *THE LEPIDOPTERA OF IRAQ* 13:33
Wolfsberger, *DIE MACROLEPIDOPTEREN-FAUNA DES GARDASEEGEBIETES* 21:120
Yagi, *STUDIES ON THE COMPOUND EYES OF LEPIDOPTERA* 8:146
Yokoyama, *COLORED ILLUSTRATIONS OF THE BUTTERFLIES OF JAPAN* 12:56, 19:46
Zagulajev, *FAUNA OF THE USSR, LEPIDOPTERA. Vol. 4, Part*

2, *TINEIDAE, Part 2. SUBFAMILY NEMAPOGONINAE* 19:102
TINEIDAE, Section 3, TINEINAE 15:130-132
Zimmerman, *INSECTS OF HAWAII* 21:106

Lithocolletis, Canadian species feeding on *Salix* and *Populus* 24:272-281
Lithophane, two new species from California 23:10-14
Lithophane subtilis life history notes 23:15-18
Life insects, regulations for shipping 3:13
Lizard as moth predator 3:64
Locality label printer 12:129
Loeliger, Robert (Personalia) 6:78
Lonomia cynira rearing larvae of 21:138-140
Loochooan butterflies, recent foodplant records 18:105-109
Lorquin's localities "Sonora" and "Utah" 21:271-274
Louisiana butterfly records 17:21-24, 148-158; 20:102
Lucas authorship of *Papilio* names 16:45-46
Lycaeides argyrognomon in Wisconsin 7:54
Lycaeides argyrognomon sublivens 6:35-36
Lycaena:
aeolus new species from Afghanistan 15:17
helloides complex, foodplants in Colorado 17:24-24
interspecific mating 6:108
nivalis browni, Washington population 15:107-108
phlaeas albinos in Connecticut and Ohio 16:59-60
phlaeas and *L. hypophlaeas* spot genetics 8:125-129
phlaeas experimental sexual dimorphism 17:105-106
thoë in Mississippi 8:102
Lycaenidae:
antennal characters 11:13-15
chromosomes 14:127-147
interspecific pairing 16:235-237
larvae, aberrant feeders 7:45-46
of South Africa, larval structures 10:37-43
synonymy and systematic position of some species from Texas 20:65-70
Lycaenopsis pseudargiolus in light trap 14:242
Lyela amirica new species from Afghanistan 15:16-17
Lycomorpha and *Propyria* transferred to Lithosiidae 11:149
Lygris larval descriptions for two eastern species 25:262

Mackenzie River butterfly collecting 11:47-53
Macroglossa balteata date of publication 12:123
Macrolepidoptera collecting, rearing, preserving larvae 18: 201-210; 19:233-236
Maculation patterns, apparatus for measuring 23:18-22
Madagascar Lepidoptera 9:61-66
Mailing dates, *NEWS* 4:24; 5:Index-ii

Maine bog collecting 10:119-120
Maine butterfly season 14:78-80
Malay Butterflies, correction 3:62
Malay Peninsula, butterflies of, reviewed 11:60-62
Manitoba, Rhopalocera of White Shell Reserve 10:107-108
Marked Lepidoptera recovered 7:65
Marking migrant Monarch butterflies 4:49
Marking North American Lepidoptera 5:9
Marking techniques for population studies 14:227-229
Marpesia petreus behavioral notes 20:122
Martin & Truxal, N.A. Lepidoptera in the LACM 9:158-159
Maryland, new butterfly records 10:157-159, 160; 15:99-101 ; 17:107-109
Mating:
- a notable intergenetic observation 23:115
- behavior in butterflies, some aspects of 22:125-132 23:271-272
- flight of butterflies with mimetic females & non-mimetic males 24:72; 25:80
- time of Lepidoptera 18:35-41

Mayan names for Lepidoptera 12:118
McDunnough, James Halliday (1877-1962): note on 1:7; obituary & bibliography 16:209-228; erratum in obituary 17: 99; reminiscences 17:117
Mead, T. L. itinerary in 1871 9:185-190
Measurements and Lepidoptera 4:51-52, 75-76
Measurements for population analysis 6:13-27
Meetings, Pacific Slope Section (1968) 22:132; (1969) 24:73
Megalura petreus in Colorado 12:200
Megathymidae:
- chromosomes 14:37-57
- from Mexico & the U.S., new species and new subspecies 20:163-172
- generic revision 12:81-92
- higher classification 11:130-135
- notes on five 18:45-47
- of Mexico 11:113-130
- research in Mexico 15:65-72
- suprageneric categories 12:93-94

Megathymus:
- *aryxna* lectotype designation 11:29-34
- *baueri* new species and relatives 8:77-87
- *beulahae* new species 11:121-125
- *estellae* new species 11:119-121
- foodplant relationships 12:184-185
- *gayleae* new species from Texas 17:81:85
- *harrisi*, collecting technique 11:54
- *hazelae* new species 11:126-129, 135

juliae new species 11:124-127
life history in Georgia 8:153-162
polingi notes 10:109
remingtoni new species 11:117-119
streckeri biological notes 12:186-190
ursus notes 10:1-8
violae new species 10:1-8
yuccae biological notes 12:175-184
yuccae coloradensis and *streckeri* partial courtship 22: 177-178
yuccae complex in Texas 17:87-88, 89-99
yuccae in North Carolina 11:137-141
yuccae louiseae new subsp. from Texas 17:95-97
yuccae reinthali new subspecies from Texas 17:91-93
yuccae reubeni new subspecies from Texas 17:87-88
yuccae, two new subspecies from Texas 19:83-90
yuccae wilsonorum new subspecies 11:128-130

Melanic:
aberration of *Papilio cresphontes* 24:239
Danaus, Colias, Speyeria 10:111-112
female of *Colias eurytheme* 20:215-216
moths of the genus *Opostega* 20:232-234

Melanism in *Catocala ilia* 17:159-162

Melitaea:
callina, true identity 19:197-211
kuchi new species from Afghanistan 15:7-8
nycteis in Pennsylvania 11:38-41
pola, true identity 19:197-211
rubicunda, authorship 23:175-176
shandura apsara new subspecies from Afghanistan 15:6-7
theona bolli larval foodplant 15:56

Melitaeine, new genera 12:161-164

Membership list, announcements & additions: 1:61-62; 2:100; 3:12, 24, 36, 60, 72, 84, 114; 4:24, 36, 84, 110; 5:16, 52, 126; 6:56, 125-141; 7:32, 64, 132, 181-198; 8:56,120 152, 181-198; 9:40, 104, 184, 220-237; 10:80, 136, 192, 240; 11:96, 200

Mestra, distributional notes on the genus in N.A. 24:203-208

Meteorologic influences on California Rhopalocera 17:7-20

Method for overwintering hibernating larvae 19:187-188

Mexican butterfly migrations 16:249-250

Mexican expeditions 15:65-72, 197-199

Mexican Megathymidae reviewed 11:113-130

Mexican satyrid at light 18:116

Mexican tropics, butterfly collecting in 12:127-129

Michigan, *Boloria frigga* and *Atrytone dukesii* 11:37-38, 42

Michigan Lepidoptera, notes 4:14, 46; sampled by trap 7:28

Microlepidoptera preparation for spreading 12:205

Microlepidoptera subspeciation 5:29-31
Microlepidoptera venezolana, reviewed 12:55
Microscope for studying scales 10:160
Microtia dymas imperialis 12:98-100
Midge-flies biting butterflies 15:91-92
  biting lycaenids 16:20-22
Migrant butterflies outside North America 3:39-40
Migrant Lepidoptera in North America 3:17-18, 28, 37
Migration in Atlas Mountains of Morocco 4:72
Migration in Mexico 6:44
Migration in Switzerland 4:61-62
Migration of:
  an American butterfly, first record 6:42
  Ascia monuste 7:13-14
  butterflies 16:63, 229-233, 249-250; 17:110
  butterflies in India 11:229
  butterflies in Japan and Korea 4:43
  butterflies in Virginia 10:44
  butterflies in Yucatán 13:62-64
  Catopsilia in India 6:68-69
  Danaus plexippus 5:37-40; 6:69-70; 7:28
  Erebus odora 13:65-66
  Kricogonia castalia in northern Mexico 25:124-125
  Kricogonia lyside in Mexico 18:26
  Libytheana bachmanii 3:28
  Libytheana and *Kricogonia* in southern Texas 19:223-224
  Nymphalis californica 7:9-10, 15
  Phoebis sennae eubule in Alabama & Mississippi 22:182
  Vanessa cardui in late summer and fall of 1965 20:123
  Vanessa cardui in Montana and Wyoming 21:39-40
  Vanessa cardui in Ohio 22:118-119
  Vanessa cardui in Utah 8:17-22
  Vanessa cardui with special ref. to N.A. 24:157-175
  Vanessa carye 7:53
Mildew, prevention of 2:86
Mimicry and distribtion of *Caenurgina caerulea* 19:165-170
Mimicry and the Monarch butterfly 19:227-230
Mimicry in Africa 13:121-150, 256
Mindanao butterflies 12:115-117
Minute moths, a new technique for spreading 19:115-116
Minutes of annual meetings: 6:121-124; 7:1-2; 8:5-7; 9:17-18; 10:74-75, 172-174; 11:158-161, 230-232; 12:52-54,209 212; 13:31-32, 238-240; 15:63-64, 200-203, 249-252; 18: 117-118
  Pacific Slope Section: 9:18-20, 147-150; 20:49-55; 21: 210-212; 23:278
Miscellaneous notes: 1:7, 21, 24, 34, 36, 45, 48, 50, 58,60 70, 82, 83, 86, 99; 2:67, 74, 78, 86, 88, 92, 100; 3:12,

24, 35, 36, 49, 51-52, 60, 64, 65, 72, 80, 84, 106; 4:13 -15, 36, 47, 52, 60, 67, 68, 72, 73, 110; 5:4, 111, 126; 7:32, 52, 58, 64, 65, 147; 8:9, 45, 48, 56, 120, 140,145 151, 152, 162, 166, 167, 174; 9:22
Mississippi, butterflies of , reviewed 13:33-35
Mississippi butterfly checklist, 1st supplement 13:71-72
Missouri Lepidoptera collecting 10:163-168
Mites from noctuid moths 21:169-179; addendum 21:254
Mitoura gryneus, colony in Kansas 21:120
Mitoura gryneus on Swamp White Cedar 9:4-6
Mitoura:
  hesseli in New Hampshire 17:27
  johnsoni in Oregon and California 13:183
  nelsoni muiri identification 6:95-96
  spinetorum host records 12:14
Mnemonic auricyanea in Pennsylvania 10:161-162
Monarch butterfly:
  eaten by birds 18:165-169
  in northern Saskatchewan 21:31-32
  migration in winter 6:69-70
  migration study 5:37-40
  mimicry 19:227-230
  observations in mid-town and lower Manhattan,NYC 23:254
  overwintering breeding colony in Arizona 22:63-64
Moroccan Lepidoptera collecting 10:214-222
Morpho gregarious roosting 25:223-234
Morpho theseus justitiae flight habits 20: 95-101
Morphinae chromosomes 14:127-147
Moth Ear Mite distribution 9:55-58
Moths: attracting to light 4:47; 19:179-180; California winter moths 2:42
Motýle Slovenska (Reiprich), reviewed 13:246
Motýli (Moucha), reviewed 16:150
Multiple capture of *Caria ino melicerta* at light 24:13-15
  Hypaurotis crysalus at light 19:55-56
Museum visits in Europe 7:167-170
Myelois neophanes to *Apomyelois bistriatella* 17:39

Names of butterflies in eastern USA 6:74-76
Naše Motýle (Ponec), reviewed 15:74
National Parks (U.S.) collecting rules 4:15, 49
Natural habitats, importance of preserving now 21:275-276
Nomenclature of the specific complex 2:15-16
Nearctic Butterflies, The 3:13, 20, 27, 62, 84, 105; 4:73
Nearctic Lycaenidae, a revised synonymic catalogue 24:26-38
Nearctic Rhopalocera, notes for synonymic list 17:103-104
Neoterpes edwardsata, notes on life history 18:32-34
Neptis hylas, sounds produced by 22:254

Net-traps, notes on the use of at Palawan, Phil. 24:303-304
Neutron irradiation in *Antheraea eucalypti* 25:238-240
New:
- and corrected butterfly records for Ontario 21:135-137
- and scarce captures for New Jersey 24:1-3
- *Bomolocha* from Florida 21:125-126
- classification of the Lepidoptera, review of 1:33-34
- *Callithomia* from Bolivia 24:22-24
- Canadian records for *Thymelicus lineola* 20:243-244
- distributional data on three n.e. U.S. butterflies 23:265
- distribution records for Ark., La., Tex. 18:190-191
- foodplant record for *Ecpantheria scribonia* 24:227
- genus (*Symbiopsis*) of hairstreak from Central and South America S4:1-39
- hairstreak for the U.S. 18:189-190
- heliothid moth from southwestern U.S. 20:91-94
- Hesperiidae records for Mexico 20:226-228
- Journals (see LITERATURE REVIEWS/NOTICES)
- larval foodplant for *Euphydryas phaeton* 23:48
- larval foodplant records for *Anthocaris* 21:212
- lepidopterous leaf-miners associated with *Quercus agrifolia* 25:194-211
- members 1:86, 100; 2:48, 60, 76; 12:80, 160
- Mexican Hesperiidae 24:120-124
- Michigan butterfly records 24:42-47
- moths from Texas 22:133-145
- name for *Papilio zetes westwood* 25:142
- Nearctic species of *Exotelei* on pine 21:9-11
- *Papilio* from Colombia and a new sphingid from New Guinea 23:255-257
- *Perisama* from Bolivia 24:10-13
- publications 12:157-158
- or unusual butterfly records from Florida 24:240-244
- race and discussion of the *Boloria epithore* complex 20:103-117
- records for New Jersey 24:1-3
- Secretary 11:96
- society and new journal 21:140
- state records for La. butterflies and skippers 19:47-52

New species of:
- Adelid moths 23:211-240
- African Geometrinae 25:169-176
- *Agathymus* from Texas 18:171-185; 20:181-185
- *Anepia* from the Sierra Nevada of California 21:235-240
- *Apodemia* from Texas 18:75-77
- *Cameraria* on bur oak in Manitoba 24:86-87
- *Cargolia* from Mexico 18:111-115
- *Eupragia* from Florida 23:23-25

Glaucina from Texas 20:247-250
Gnorimoschemini from the Lesser Antilles 24:6-10
Hesperiidae from Mexico S2:1-62
Lithophane from California 23:10-14
moth destructive to pine cones in Mexico 20:251-253
Phaneta refusana 25:284-287
Phycitinae from Texas 24:249-255
Piruna from Texas 24:247-249
Syllepis poey with a key to known species 24:287-291

New subspecies of:
Brephidium exilis from Yucatan 24:3-6
Callophrys fotis from San Francisco Bay area 23:95-96
Cercyonis meadi group 23:161-164
Glaucopsyche piasus 25:240-246
Gonepteryx rhamni from Tian-Shan Mountains, USSR 24:218
Marpesia eleuchea 25:185-193
Megathymidae from Mexico and the U.S. 20:163-172
Pyrrhopyge creon from Panama 23:127-130

Newfoundland butterflies, notes on 23:33-42
New Mexico, butterfly collecting in north 14:237-238
New Mexico, butterflies of the Pinos Altos Mts. 19:231-232
New sphingid from New Guinea 23:255-257
New sphinx moth from the West Indies 24:267-270
New sphinx moth record for the U.S. 24:217
New technique for spreading minute moths 19:115-116
Noctasota curiosa synonymized 11:215
Noctuid light trap 22:65-75
Noctuid moths swarming in Kansas 13:26
Nocturnal copulation of Rhopalocera 23:105-106
Nocturnal moths feeding in daylight 5:69

Nomenclature:
controversy 3:19-20, 67, 104; 4:16, 68
in zoology 7:32, 122
notes 2:46; 10:110, 162, 237; taxonomic 11:29-34, 46 170
notice 9:215; 13:216
of *Papilio ajax* 4:110
rules pamphlet 9:132

Nominations for Lepidopterists' Society offices 4:38; 5:80; 6:53, 88; 7:126; 8:120; 9:184; 10:136; 11:96; 12:158
Nordens ragsommerfugle, reviewed 12:205-206
N. A. butterflies described by Linnaeus in 10th edition Systema nature, 1758 22:77-88
N. A. season summary, preparations for 1948 2:25
North Carolina Entomological Society formed 10:90
Northern Canadian butterflies 5:41-42
Northwestern U.S.A., butterfly collecting 16:67-70

Notes on:
Caria domitianus and *ino* with description of a new subspecies 21:53-57
Edwards specimens in two midwestern collections 20:156
five Megathymidae 18:45-47
larval foodplants of some sphingids in Ontario 21:252
Pyrgus communis and *P. albescens* 19:91-94
some south Florida Lepidoptera 25:137-142
the genus *Cephise* with a new record for Mexico 24:68-69
Trogonoptera trojana of Palawan-Philippines 21:49-52
two rare Mexican *Adelpha* and related C.A. spcies 24:292
uncommon moths in central and southern Ontario 21:33-39
Notices by members 1:1-2, 11, 12, 23, 35, 37, 47-48, 59, 71, 72, 85, 99; 2:11, 23, 35, 47, 59, 75, 87, 99, 109; 3:12 23, 35, 59, 71, 83, 113, 114; 4:23, 35, 59, 83, 109; 5: 15, 51, 79, 125; 6:54-55, 87, 120; 7:31, 63, 65, 131, 180; 8:55, 119, 151, 174; 9:39, 103, 183, 219; 10:79, 135, 191, 237; 11:94-95, 198, 199, 245; 12:79, 159
Novara report, dates of publication 12:193-194
Nova Scotia, rearing notes from 4:12
Number of butterflies in Middle East in one day 10:116-118
Number of butterflies in India in one day 11:167; 16:143-145
Number of butterfly species in one day 9:77-78, 145, 203-204
Nymphalinae chromosomes 14:179-201
Nymphalis antiopa:
effects of vitamins on the development of 21:241-242
gregarious habit of chrysalids 23:273-274
migration 9:143
Nymphalis californica:
in Illinois and Iowa 20:124
in Pennsylvania and California 8:94, 103
migration in Northwest 13:64
migration in Washington 16:63
pupae in unusual site 6:107
Nympahlis j-album at light 15:101; in Maryland 14:62
Nymphalis milberti habits 9:15-16; larval mortality 5:69
Nyphanda fusca larva an ant-guest 6:43
Obituaries (see OBITUARIES/BIOGRAPHIES/BIBLIOG. of text)
Observations on:
Arkansas Rhopalocera, and list from n.e. Ark. 21:206-209
biology of *Ocnerogyia amanda* 25:53-57
Callophrys macfarlandi in the Sandia Mountains, New Mexico 19:177-179
forest *Oeneis* 23:155-161
Oeneis macounii in Manitoba and Minnesota 21:258-260
Ochlodes snowi at blacklight in Colorado 16:234
Ochlodes yuma, taxonomy and distribution 11:151-154
Ocnerogyia amanda pest of *Ficus* in Iraq 25:53-57

Occurrence of *Callophrys eryphon* in Michigan 20:41-42
Occurrence of certain Rhopalocera in Oregon & Wash. 18:47-50
Œneis:
caryi, type figured 11:22
chryxus in Wisconsin 2:34
chryxus valerata new race 11:22-26
daura, new locality 3:25
field observations on 23:155-161
jutta in Wisconsin 25:150
jutta reducta feeding habits 24:306-307
macounii, another U.S. record 19:125
macounii in Manitoba and Minnesota 21:258-260
melissa field notes and subspecific status in northern Quebec 23:103-104
nevadensis hilltopping 16:64-66
Oenomaus ortygnus, first U.S. record 18:189-190
Oklahoma butterfly records requested 6:55
Ontario butterfly records 15:92, 108, 113-114; 16:62, 243
Ontogeny of Insects (Czech. symposium vol.) rev. 17:38
Opostega, melanic specimens of 20:232-234
Opostegidae of North America, pictorial key 17:237-242
Organization Committee report 4:38
Orders of insects 3:45-47
Oregon Geometridae 13:217-220
Oregon Rhopalocera, field observations 17:111-116, 171, 245
Ornithoptera alexandrae life history 25:58-64
Ornithoptera periamus urvilleanus biology and larval foodplants 23:69-76
Ornithoptera victoriae biology and larval foodplants 23:69
Orthosia hibisci quinquefasciata, larva and pupa 20:213-215
Overwintering of Monarchs, a breeding colony in Ariz. 22:63
Ovipositing by Rhopalocera on toxic foodplants 16:99-103
Oviposition in captivity 2:74; 3:48
Oviposition observations 5:69
Oviposition of Rhopalocera in the lab, technique 2:74

Pacidara venustissima synonymized 11:214-215
Packard, A.S., arrangements for monog. of bombycines 11:206
Paintings of Lepidoptera by DeRabié 5:55-57
Pairing in Lepidoptera, evolutionary & biological significance of multiple 22:197-209
Palatability of model and mimic butterflies to mice 15:23
Palmer's collecting localities in southern Utah and northwestern Arizona 21:129-134
Panoquina ocola mass flight 9:79
Panthiades m-album early stages and Penn. notes 15:226-232
Papaipema mulleri new species from New Jersey 10:139-140
Papaipema, guide to collecting larvae 8:57-63

Papilio:
additions to Brown's analysis 5:16
aegeus X *fuscus* hybrids & foodplants 16:161-174
ajax nomenclature of 4:110
alexanor hazarajatica new subsp. from Afghanistan 15:2
antimachus, systematic position of 24:224
aristodemus ponceanus, observations on 25:126-136
brevicauda field notes 4:11-12
cresphontes melanic aberration 24:239
demoleus from California 22:187
demodocus pupal color in relation to season 25:271-274
flight pattern 2:143
flyways 3:25, 62
foot note to Forbes' comment 5:52
glaucus, flyways of 2:92
glaucus forms 3:48
glaucus group, foodplant specificity 12:103-114
glaucus melanic male 14:157-158
glaucus mosaic melanic 16:98
glaucus size in Mississippi 8:131-134
glaucus venation aberration 18:157-158
glaucus X *eurymedon* hybrids 11:201-205
glaucus X *rutulus* hybridization 9:46-48
gothica, is it a good species 24:229-233
hippocrates a distinct species 13:151-164
humidity tests with pupae 5:67
hybridization experiments 13:151-164
idaeoides 12:115-117
indra kaibabensis biology of 21:41-49
indra kaibabensis new subsp. from Grand Canyon 9:49-54
indra martini population biology & life history 22:46-52
indra minori 16:137-142
in the Niagara Peninsula 4:14
machaon aliaska, aberration 11:36
machaon complex in Arizona 9:7-10
machaon hudsonianus melanic 12:124
multicaudatus and *rutulus* 11:17, 224
multicaudatus nomenclature 12:101-102
multicaudatus reared 9:12
of the Americas 4:39-41, 63-67
oregonius life history 22:53-56
palamedes foodplants in Georgia 16:198
philolaus new for North America 24:233
polytes X *protenor* and *polytes* X *helenus* hybridized in Japan 17:163-169
polyxenes asterius, effects of x-irradiation on larvae 22:173-177
protenor X *helenus* and *bianor* X *maackii* in Japan 15:175

species names, authorship 16:45-56
xuthus X *polyxenes-machaon* group hybrids 14:5-18
zelicaon, 2-year pupa 3:64
zetes westwood, new name for 25:142
Papilionidae chromosomes 13:193-203
papilionidae higher classification 14:169-175
Papilionidae of Manus Island, New Guinea, notes on 25:73-80
Papilionoidea, Nearctic, taxonomic notes 15:209-225
Parabasis pratti notes 10:15-17
Paralasa asura new species from Afghanistan 15:13
Pararge eversmanni shiva new subsp. from Afghanistan 15:9
Parasite of *Danaus plexippus* 2:92
Parasites of Lepidoptera, identification of 2:53; 3:2; 4:2
Parasites of Lepidoptera in Illinois 3:48
Parasitinsekten der Blattminierer Europas, reviewed 16:237
Paris Museum type specimens 5:9
Parnassius infra-subspecific names 10:140-142
Parnassius re-discovered in France 4:15
Parnassius tianschanicus shiva new subsp. from Afghanistan 15:4
Partial courtship between *Megathymus yuccae coloradensis* & *M. streckeri* 22:177-178
Patricia demylus gemellus, additional records 25:149
Peale's Lepidoptera Americana 12:101-102
Pennsylvania Lepidoptera records 10:161-162
Perichares philetas dolores symbiotic relationship with Maxillaria orchids 25:20-22
Periodical, new Japanese entomological Journal 14:175
Pero macdunnoughi foodplants 10:168
Personalia 3:11, 42, 64, 77, 103; 4:9, 31, 48, 69; 5:10, 40; 6:78; 7:23, 166
Peters, A Provisional Check-list of the Butterflies of the Ethiopian Region 9:81-82
Phaeoura mexicanaria gynandromorph 23:274-275
Phalaenidae ovipositing in California 3:74
Phaloesia saucia new to Texas and the U.S. 24:301-303
Phaneta refusana, identity of with description of a new species 25:284-287
Philatelic Lepidoptera 8:13-16; 9:12; 11:221-224; taxonomic list 22:241-252
Philippine butterflies baited 13:19-23
Philotes distribution and hosts in California 19:95-102
Philotes in central coastal California 17:201-223
Philotes of N.A.: Synonymic list & distribution 23:49-62
Philotes rita dist. records & taxonomic notes 21:141-142
Philotes taxonomy and distribution 8:8
Phlegethontius afflictus figured 11:107-112
Phlegethontius johanni new species 11:107-112

Phlegethontius quinquemaculatus impaled 3:74
Phocides polybius foodplant 15:114
Phoebis agarithe migrating in Mexico 6:44
Phoebis in Florida, notes on 2:22
Phoebis sennae eubule, mass flight in Ala. & Miss. 22:182
Pholus fasciatus in New York 11:43
Photography apparatus described 7:148-166
Photography of types 3:41, 42
Photographing mounted specimens 19:212-214
Photomicrography, inexpensive apparatus 24:61-68
Photoperiod effects in *Colias* 11:207-214
Photoperiod, effect on the termination of pupal diapause in *Actias luna* 24:209-212
Phyciodes batesii in Miss., probability of occurrence 20:245
Phyciodes behavior in Missouri 2:22
Phyciodes tharos, melanic aberration 25:143-144
Phycitinae, observations and description of two new species from Texas 24:249-255
Phylogeny & classification of the Lepidoptera, summary 2:65
Pieridae: American butterflies and English foodplants 25:6
Pieridae chromosomes 14:37-57
Pieridae - Papilionidae (Krzywicky), reviewed 16:192
Pieris:
- *beckeri*, recent range extension in Wyoming 19:221-222
- hybrids, sex ration 20:189-196
- *protodice & Urbanus dorantes* in southern Fla. 24:244-247
- *protodice*, an extreme phenotype 23:189-190
- *napi-bryoniae* group, evolutinary biology 16:5-19, 105
- *napi*, form "sulphurea" for experimental purposes 18:91
- *napi macdunnoughii* named 8:75
- *rapae*, bilateral gynandromorph 24:224
- *rapae*, effect of barometric pressure and other factors on eclosion 22:211-225
- *rapae*, effects of environmental factors on 23:77-94
- *rapae*, some properties of cuticular materials 20:217-225
- *rapae*, specialized case of communal roosting 25:144-145
- *rapae*, variations in markings induced during the pupal stage 24:125-135
- *rapae*, winter survival at The Pas, Manitoba 22:191
- sexual mosaics 12:7-13
- *sisymbrii*, an albinic female from Oregon 21:121
- *virginiensis* in Michigan and Massachusetts 10:18-28
- *virginiensis* new to Wisconsin 22:252

Pig carrion, Lepidoptera associated with 23:191-195
Pine Tip Moth, new species from Caribbean region 14:231-236
Pinos Altos Mts., N. M., some butterflies of 19:231-232
Pins and moisture 4:49
Plastic mounts for collections 2:31

Platychasma and *Cyphanta* systematic position 17:33-34
Plebejus acmon larval host acceptance in Calif. 23:31-32
Plebejus icarioides foodplant specificity 15:34-42
Poecilopsis phylogeny 9:199-202
Poem on butterfly migration 17:242
Polites draco, altitudinal variation 16:239-242
Polites mystic, authorship of 20:239-242
Polski, Kucze do Oznaczania Owadow, reviewed 15:132
Polydorus aristolochiae:
biology and seasonal history 24:19-22
host plant specificity 21:127-128
larval description 23:107-108
longevity of adults 22:183-185
observations and description 22:183-185
pupa of 22:115-118
Polygonia c-aureum reared for four generations 6:43
Polygonia faunus smythi, collecting notes 5:70
Polygonia, field observations 6:73
Polygonia gracilis in Maine 2:59
Polythrix octomaculata not *procerus*, in Texas 21:278
Population analysis techniques 6:13-27; 14:227-229
Porthesia feigning death 6:42
Possible record for the occurrence of *Callophrys xami* in California 20:212
Postage stamps depicting Lepidoptera 16:145-146
Powell collection to University of Calif.(Berkeley) 15:203
Predators of butterflies 7:27, 54, 55, 168; 8:28; 12:202
Predators of Lepidoptera 3:64; 15:93-94; 16:20-22, 136
Preparing Lepidoptera for class study 6:61-63
Preservation of larvae by inflation and heat-drying 14:67-78
Preserving larvae, Peterson formula 2:14
Presidential addresses: 7:3-4; 8:1-4; 9:41-45, 150; 10:76 137-138; 11:97-99; 12:1-5, 209-210; 13:11-13, 241-244; 14:1-4; 15:57-62, 245-248; 16:247-248; 17:173-177; 19:193-195; 25:155-168
Principles of classification in Lepidoptera, reviewed 1:74
Principles of taxonomy 2:26, 38, 50, 62
Pribilof Islands Lepidoptera 4:27-30, 44
Probable second U.S. record for *Erebia discoidalis* 19:76
Problema byssus life history 19:77-81
Problema in Kansas 13:15-16
Procedure in taxonomy 3:14
Proceedings, special meeting 6:28-31
Professional and amateur research in Lepidoptera 14:1-4
Protective resemblance in Africa 13:121-150, 256
Protoparce albiplaga new for the U.S. 24:217
Psaphidini, key to the genera with descriptions of a new genus and species 20:83-91

Pseudoscorpion on moths 10:87-89
Psyche reprints 1:32
Psychidae larvae hatched from birds' excrement 10:36
Publication, "Curso de Entomologia" from Argentina 4:68
"Pumping" by *Papilio* at water 17:35-36
"Pumping" of moths at water 11:18-21; 12:170-172
Pupal color in *Papilio demodocus* related to season 25:271
Pupal colors in *Pieris rapae* 7:5-6
Pupal defenses against parasites 13:1-10
Pupal volume, correlations with wing radius & weight 7:47
Pyralid fauna of western states 13:241-244
Pyrginae, biological notes in New York 22:179-180
Pyrgus communis and *Pyrgus albescens*, a note 19:91-94
Pyrrhia exprimens life history 24:234-239
Pyrrhopyge creon lilliana new subsp. from Panama 23:127-130

Questions and Answers: 1:100; 2:10, 24, 36, 60, 76, 88, 100 110; 3:11, 24, 36, 59, 72; 4:16, 36, 60, 84; 5:16, 52, 80, 126

Ramburia antonia shivaensis new subspecies from Afghanistan 15:17-18
Ramosia fragariae in a flight trap 24:189
Range extension of *Callophrys comstocki* 21:114
Range extension to Missouri for *Monodes georgei* 18:83
Rearing:
and preserving larvae of Macrolepidoptera 19:233-236
cages 15:102-106
containers for larvae 14:159-160
European Lepidoptera in Ohio 4:13
house for Lepidoptera 5:12
larvae of cecropia moth, techniques for reducing mortality 20:119-121
larvae of *Lonomia cynira* 21:138-140
methods, leaf-mining Lepidoptera 4:3-6
notes from Nova Scotia 4:12
observations 3:25, 106
Speyeria in captivity 7:56; 25:247-256
techniques 13:182-183
Recent Literature on Lepidoptera (abstracts:
1:4-7, 16, 28-29, 33-34, 40-41, 52-55, 64-68, 74, 76-80, 82, 97; 2:8-9, 18-21, 32-33, 44-45, 56-58, 68-72, 84-85, 96-97, 106-107; 3:10, 21-22, 33-34, 55-58, 69-70, 81-82, 109-112; 4:17-22, 32-34, 53-58, 77-82; 5:13-14, 47-50, 73-78, 121-124; 6:45-52, 81-86, 112-119; 7:29-30, 59-62, 129-130, 173-179; 8:51-54, 105-118, 147-150, 175-180; 9:23-34, 83-102, 161-182, 216-218; 10:58-73, 121-134, 179-190, 223-237; 11:63-93, 179-198, 233-244;

12:57-78, 133-156, 213-237; 13:36-56, 105-120, 184-192, 247-253; 14:83-88, 161-168, 208, 244-253; 15:75-80, 133-144, 204-208, 253-261; 16:76-80, 151-160, 204-208, 251-253; 17:50-62, 118-128, 191-200, 246-253; 18:119-128, 197-200, 243-266; 19:63-68, 126-128, 245-251; 20:57-64, 124-128
Recent range extension for *Pieris beckeri* in Wyoming 19:221
Record:
charts for collectors 17:44-46
keeping 18:58-63
of *Eumaeus atala* from the Florida Keys 20:216
of *Glaucopsyche lygdamus*
Recurvaria milleri in California 8:91-93
Recurvaria speciation 11:97-99
Rediscovery of *Euptychia mitchellii* in New Jersey 20:43-44
Regional lists 3:15-16
butterflies from Cahulila, Mexico 22:227-231
butterflies of Madison County, Illinois 24:305
butterflies of Roosevelt County, New Mexico 25:213-214
Central Brazil Plateau 21:77-106, 145-168; 22:1-20
Hesperiidae from Honduras 21:243-247
new or unusual butterfly records from Florida 24:240-244
northeastern Arkansas 21:206-209; corrected 21:277-278
south Florida Lepidoptera 25:137-142
Sphingidae of Honduras 25:150-152
uncommon butterflies of northeastern Indiana 23:186-188
unusual butterflies of northern Pennsylvania 25:222
Regulations, shipping live insects 3:13
Relaxing box for pinned and papered specimens 24:81-83
Remarks on "Distributional notes on the genus *Mestra* in North America" 25:146
Remarks on "The use of net-traps at Palawan, Phil." 25:293
Reply to F. Martin Brown 4:75-76
Reply to Prof. Nabokov 4:76
Reprints available 5:45
Research requests 3:66, 74, 114; 4:9, 48, 69, 110; 5:10, 80, 126
Reversed wings in dying Lepidoptera 6:44
Review of:
Acanthopteroctetes with desc. of new species 23:137-147
Arcas with descriptions of new species 25:87-108
collections of Lepidoptera by airplane 19:129-137
Glaucopsyche in California 23:149-154
Limenitis lorquini complex 20:172-176
Reviews of publications: 4:42, 50, 107-108; 5:45, 60-62; 10:56-57, 175-178; 11:57-62, 171-178; 13:33-35,245-246
Revised Constitution and By-laws 9:35-38
Revised synonymic catalogue with taxonomic notes on some

Nearctic Lycaenidae 24:26-38
Revision of the *Limenitis weidemeyerii* complex 21:213-234
Revision of *Speyeria nokomis* 24:44-52
Revision, Synonymic list of Nearctic Melitaeinae 23:115-125
Rhopalocera in Oregon and Washington 18:47-50
Rhopalocera of Afghanistan, further additions 18:78
Rhyacionia subtropica new sp. from Caribbean 14:231-236
Riker mounts 4:47
Riley, N. D. (Personalia) 5:10
Rileyiana new name for *Thecophora* 17:169-170
Rimsky-Korsakoff, M. N. (Personalia) 5:40
Rindge & Smith, Revision of *Annaphila* 6:111
Rindge, The type material in the J.B. Smith & C.D. Hulst collections of Lepidoptera in the AMNH 9:160
Roepke, Walter Karl Johann (1882-1961), obituary 14:241-242

Salmon's Fluid, new slide mounting medium 7:170-171
San Francisco's vanishing butterflies 10:113-115
Saturnia walterorum new name 12:17, 160
life history and description 12:17-25
Saturniidae first instar larvae, classification 14:89-111
Saturniidae foodplants in Arizona 16:61-62
Satyridae, new genera for Brazil 25:12-19
Satyrinae chromosomes 14:127-147
Satyrium acadica in Pennsylvania 16:199
Satyrium behrii in Nevada 15:56
Satyrium behrii in Oregon 16:44, 199-200; 17:245
Satyrium caryaevorus distribution 16:195-196
Satyrium kingi in Virginia 16:197-198
Satyrium saepium, unidirectional mass movement 21:204
Scarce literature for sale 2:95
Schinia cupes deserticola life history 25:109-114
Schinia felicitata life history 21:22-26
Schinia ligeae life history 25:274-277
Schinia niveicosta life history 20:29-33
Schinia olivaceae larval foodplant & dist. notes 19:105-106
Schinia separata life history 25:177-180
Schinia walsingham life history 25:181-185
Schoenobius maximellus life history in Texas 17:234-236
Schwarz, MOTYLI. 3. 9:158
Scopula ancellata immature stages 24:54-55
Season (Annual) Summaries: 1:49, 50, 72, 87-96; 2:89; 3:27 48, 73, 84-102; 4:45-46, 85-107; 5:68, 81-110, 120; 6: 40, 88, 89-91; 7:66-118
Seasonal dimorphism in *Pieris napi oleracea* 14:78-80
Seitz' Macrolepidoptera of the World 9:104
Seppänen, Suomen Suurperhostoukien Ravintokasvit 9:159
Seppänen, The Foodplants of the Macrolepidoptera larvae of

Finland 9:159
"Sex-ratio" in *Pieris* hybrids 20:189-196
Sexual dimorphism, induction in *Lycaena phlaeas* 17:105-106
Shapiro collection to Cornell 25:82-83
Shipping live insects, regulations 3:13
Siamese twins of *Parthenos* 14:155-156
Sierra San Pedro Martir Lepidoptera 13:229-235
Siphoning water by *Venusia cambrica* 6:43
Society activities, 1947 2:1
Society of Systematic Zoology 4:6
Solar eclipse, butterfly collecting at time of 18:109-110
Some properties of cuticular materials of *Pieris rapae* 20:217-225
Sound production in pupae of Lycaenidae 20:129-155
Sounds produced by *Neptis hylas* 22:254
South American Congress of Zoology 14:81-82
South Carolina, notes on two rare Lepidoptera 25:143
Southwestern research station of American Museum 9:76
Special Meeting, Amsterdam, 1951 5:10, 42
Special notices 2:40, 48, 61, 76
Species descriptions, components 5:46; 6:37-40
Specimens damaged by carpenter ants 25:83
Specimens, handling of 1:26, 44
Speyeria:
aphrodite attracted to UV light 25:150
asymmetry 13:14-15
atlantis breeding notes 17:143-147
atlantis subspeciation 5:31-35
segregation in Black Hills 17:129-143
callippe and *Artemisia* a possible foodplant 19:186-187
collecting in California and Oregon 17:111-116
cybele at light 20:44
diana collecting notes 3:74; 4:72
saga of a larva 13:93-95
egleis secreta collecting 2:91-92; notes 10:102-106
egleis tehachapina field observations 17:243-244
nokomis apacheana search for type locality 23:62-68
nokomis revision 25:44-52
zerene overlapping subspecies 16:81-97
Sphingidae:
at mercury vapor lamp in Michigan 15:191-194
collecting in Jamaica and Haiti 9:136
collecting in Venezuela 3:78
foodplants in Wisconsin 15:64
in Florida 13:78
name changes 9:135-136
of Honduras 25:150-152
rarities in Pennsylvania 17:102

subspeciation in West Indies 5:20-23
Sphingids over water 1:34, 46, 82
Sphinx:
franckii in Pennsylvania 17:102
larval stages 11:15-17
kalmiae foodplant in Wisconsin 17:36
sequoiae engelhardti early stages 21:27-31
Spindasis takanonsis in Japan 9:13-14
Spreading device for Lepidoptera 4:47
Spreading technique 2:86; minute moths 19:115-116
Stallingsi jacki new species from Mexico 17:84-86
Stallingsia new Genus 12:87-88
Statistics for taxonomists 5:4-6, 43-45, 64-66, 112-120; 6: 64-66, 67; 11:1-3
Status of *Heliopetes sublinea* 21:205-206
Stenoptycha Zeller, 1863, name replaced 11:178
Stinson, Walter C. (1898-1958), obituary 15:121-123
Stoll, names proposed by 12:197-198
Storage box 3:104
Storing and shipping papered Lepidoptera 7:27; 13:25-26
Story of a "mixed up" *Thorybes pylades* 18:169-170
Striated band of Thorax of Arctiidae 11:147-150
Structures employed by *Actias luna* in effecting emergence from the cocoon 21:249-252
Strymon:
borealis a redescription 24:83-86
falacer habits 9:105-117
liparops host plant 9:11; observations in Ontario 15:108
melinus foodplants 13:170
titus titus mutant 6:107
Stuttgärter Beiträge zur Naturkunde 12:40
Subeana (Tortricidae) new subgenus 16:176-177
Subspeciation: equation 13:173; general outline 5:17-20
symposium at 1950 meeting 5:17-35
Subspecies:
concept 10:35-36, 207-208; 11:155-157
naming in Lepidoptera 8:100
proposal for restriction of term 9:1-3
Suggestion regarding hindwing diversity among *Catocala* 23: 261-264
Swanepoel, Butterflies of South Africa 9:81
Swarm of butterflies in Colorado 22:172
Symbiotic relationship of an Ecuadorian skipper and Maxillaria orchids 25:20-22
Symposium on phylogeny at Amsterdam meeting 5:10
Synchloe callidice hazara new subspecies from Afghanistan 15:6
Synonymic list and dist., *Philotes* in N. A. 23:49-62

Synonymic list of the nearctic Melitaeinae with taxonomic notes, revised 23:115-125
Synonymy and systematic position of some Texas Lycaenidae 20:65-70
Synonymy of *Leucophebia lineata brunnea* 21:8
Synopsis of nearctic Adelid moths with desc. of n. sp. 23:211
Syssphinx bicolor in Ontario 18:89-90
Syssphinx heiligbrodti larval foodplants in Texas 18:42
Systema Nature, 1758; N.A. butterflies described in 10th edition of 22:77-88
Systematic:
  Catalogue of *Speyeria* with designations of types and fixations of Type Localities 2:5
  categories for Lepidoptera 15:57-62
  definition 3:14
  position of *Papilio antimachus*, note on 24:224-225
  review of the Megathymidae S1:1-58

Tamaulipas, Mexico, spring collecting 15:197-199
Tannenschmetterlinge der Slowakei, reviewed 14:201; 15:74
Taxonomic
  categories, on the "lower" 2:3
  list of philatelic Lepidoptera 22:241-252
  notes on the Nearctic *Holomelina* with partial key to the species 19:69-76
Taxonomists' glossary of genitalia in insects, rev. 12:125
Taxonomy:
  distrib. and biology of the genus *Cercyonis* 23:165-175
  of Lepidoptera 13:24; symposium 11:1-12
  procedure in, symbols 2:78, 90; 3:14
Taygetis, draft key to the species 6:97-98
Techniques:
  bait trap for Lepidoptera 23:97-101
  carbon tetrachloride danger 5:6
  device for spreading Lepidoptera 4:47
  for hand-pairing *Limenitis* butterflies 23:109-112
  for investigations of population structure 25:22-29
  for Lepidoptera collectors 11:53, 54, 168, 225-229
  for reducing mortality when rearing larvae of cecropia moth 20:119-121
  incubation apparatus for larvae and pupae 5:35
  inducing captive Rhopalocera to oviposit 2:74
  killing butterflies 3:26, 49
  notes 4:47, 70
  papering Lepidoptera in glassine envelopes 25:65-68
  rearing leaf-mining Lepidoptera 4:36
  simple method for preparing male hesperiid genitalia 25:295

venation made visible 4:70
Termination of saturniids' diapause 20:47-48
Tests of Lepidoptera 11:8-9
Texas and New Mexico collecting 13:89-93, 99-100
Texola new genus 12:161
Theclini collecting eggs in winter 13:175-181
Theclini, revised catalogue of Oriental 3:79-80
Thecophora Lederer, renamed 17:169-170
The ontogeny of insects 17:38
Thorybes bathyllus first Minnesota records 20:34
Thorybes pylades, "mixed up" 18:169-170
Three letters from J.A.B.D. DeBoisduval to W.H. Edwards 19:197-211
Three new species of Hesperiidae from Mexico 21:115-119
Thymelicus lineola:
colony at Edmundston, New Brunswick 20:28
first Connecticut records 19:195-196
form "pallida' in New Jersey 23:249-250
in Illinois 22:21
in Indiana 22:20
in New Jersey 12:174
in North America 10:151-152
in Virginia 23:76
new Canadian records 20:243-244
Thymelicus lineola kushana new subsp., Afghanistan 15:18
Thysania zenobia in Connecticut 4:13
Tineidae, Sect. 3, Tineinae of USSR, reviewed 15:130-132
Toll, Sergiusz (1893-1961), obituary 17:47-49
Tortricid abdominal organ 10:153-156
Transplanted colony of *Mitoura gryneus* in Kansas 21:120
Trap nets for Rhopalocera 8:26
Treasurer's report for 1952 7:132
Trogonoptera trojana of Palawan-Philippines, notes on 21:49
True identity of *Melitaea pola* and *Melitaea callina* 19:197
Turgorator for rearing insects 8:121-123; 11:225-226
Turnerina, new genus 12:84-85
Two new subspecies of *Megathymus yuccae* from Texas 19:83-90
Two rare Lepidoptera in South Carolina 25:143
Two species of Hesperiidae previously unrecorded from the U. S. 18:214-216
Two weeks of butterfly hunting in central Labrador 19:242
Tympanic organs as phylogenetic characters 6:7-12
Tympanic structures and classification of the Lepid. 17:1-6
Type:
localities of some Neotropical Lycaenidae taken by Gervase Mathew and described by W.C. Hewitson 21:181
locality of *Cercyonis stephensi* revisited 19:161-164
locality of *Speyeria nokomis apacheana*, search for 23:62

nomenclature 2:26
specimen figures available 9:140-141
specimens in Paris Museum 5:9
Types of genera 3:14
Types, photography of 3:41, 42
Typhedanus crameri, new name for *orion* 13:66

Uitlandsche kapellen, dates of publication 12:195-198
Uncommon moths in central and southern Ontario 21:33-39
Unidirectional mass movement by *Satyrium saepium* 21:204
Unpublished field notes of H.G. Lacey, Lepidoptera 25:29-44
Unusual butterfly records from central California 19:171-176
Unusual species of Lepidoptera from southern Florida 25:294
Urbanus dorantes and *Pieris protodice* in southern Florida 24:244-247
Urbanus proteus head measurements and weights of larvae 24: 47-51
Use of color for species-recoginition in *Heliconius besckei* 24:25
Use of the suffix -ellus, -ella, -ellum in species-group names 24:38-41
Utetheisa ornatrix complex, chromosome numbers 17:231-233
UV light, *Speyeria aphrodite* attracted to 25:150

Vancouver Island Rhopalocera 10:169-171
Vanessa atalanta taken at night 13:18
Vanessa cardui:
experimental aberration 22:146
in Mississippi and Tennessee 25:147-148
migration 16:229-233; 24:157-175
migration in Montana and Wyoming 21:39-40
migration in Ohio 22:118-119
migration in Utah 8:17-22
migration southward, late summer and fall 20:123-124
Vanessa kershawi of Australia and New Zeland 25:114
Vanessa tameamea biology and host preferences 22:111-114
Van Someren collection to AMNH 25:143
Variant females of *Colias cesonia* 22:253
Variant of *Colias philodice* in New Hampshire 25:108
Variation in *Junonia coenia* in Mississippi 21:59-71
Variation in related species of *Catocala* 22:231-236
Variations in the markings of *Pieris rapae* induced during pupal stage 24:125-135
Venation aberration in *Papilio glaucus* 18:157-158
Venation made visible 4:70
Vidius perigenes first U.S. record 18:214-216
Viette, Pierre E. L. (Personalia) 5:40
Voting results 9:40; 10:80

Wallace, A. R., Quotation 3:80
Wallengrenia otho nomenclature corrected 9:218
War damage and losses to Lepidopterology in:
  Europe 2:80
  France and Belgium 2:80
  Germany 2:61
  Japan 2:95
  Malaya and Java 2:79
  Poland 3:4
Washington, Yakima County, butterflies of 18:217-218
Water holes attracting Arizona butterflies 7:146-147
Weights of fresh and dried butterflies 6:73
West Indian Sphingidae subspeciation 5:20-23
What's your collection worth 24:51-54
Wheeler expeditions to Southwest (1869-1876) 9:124-126, 185
Wheeler survey, Itineraries of naturalists 20:71-82
White Mountains (New Hampshire) collecting 10:53-54
Wild silk moths of the U.S., reviewed 16:58
Williams, Evelyn Gilstrap (Personalia) 6:78
Wing radius estimation method 14:63-66
Wing venation preparations 6:61-63
Winter butterflies in California 2:42
Winter collecting in California 2:34
Wyoming butterfly collecting 7:49-52; notes 4:76
  Rhopalocera of SE part 10:91-101

Xanthocles and *Heliconius hecale* in Venezuela 23:104-105
X-irradiation on larvae of *Papilio p. asterius* 22:173-177

Yellow forms of *Coenonympha tullia* in Oregon 21:1-7
Yucatecan forests, devastation of 20:201-205
Yukon and Alaska, collecting trip in 19:57-62

Zegris fausti lucullus new subsp. from Afghanistan 15:6
Zellerina, nomen novum 11:178
Zeuzera and *Adopaea* in Kentucky 9:142
Zoological Nomenclature 18:100, 164; 19:176, 224; 20:56, 211; 21:76, 260; 22:195-196; 23:125, 242; 23:242; 24:76

Scientific Names

Names are arranged alphabetically by species followed by the genus except for subspecies where the latter is omitted. *New taxa* are shown in **BOLD FACE** type, including Family, Genus, Species and Subspecies which are new.

Excluded from the species index are names appearing in: Annual (Season) Summaries; the *News* (1959-72); revisions to *Memoir No. 1*; and List of the butterfly fauna of Pocas de Caldas, Minas Gerais, BRAZIL (S3, 1969). Reference citations include volume number of the *News/Journal* followed by the page number; the prefix "S" indicates Supplement series.

Illustrations are indicated by symbols in parenthesis following the name, i.e.: I= illustration; IA= adult(s); IC= components (head, antennae, wing, etc.); IF= larval foodplant; IG= genitalia; IH= habitat; II= immatures. Larval foodplants recorded in the text are given immediately below the species, indented four spaces.

ADDENDA, see last page of index to scientific names.

---

aaroni, Poanes 4:44
aaroni howardi 14:28; 24:243, 244
absinthiata, Eupithecia 2:94
abbottii, Sphecodina 15:121; 23:100, 194
abdelkader, Satyrus 10:215
abdelkader lambessanus 10:221
abdelkader taghzefti 10:221
abdominalis gerhardti, Nordmannia 10:117
abdul, Cogia 21:161
abencerragus, Philotes 8:8; 10:222
abesa, Euryphene (IA) 13:137
abia, Limenitis 21:98
abortivaria, Dyspteris 11:18; 12:171
ABRAHAMI, *Carystoides (IA,IG) S2:34*
abrasaria congregata, Xanthorhoe 5:54
abretia, Yphthimoides 21:91
absolon, Euryphene (IA) 13:137
acaciae, Strymon 18:110
acaciae, Strymonidia 20:135; 24:196
acadica, Strymon 9:77; 10:55, 107; 12:44; 14:20, 215; 16: 199, 200, 244; 20:135; 22:129, 132; 23:187; 24:305; 25:82
acadica coolinensis, Satyrium 22:163

acaste, Callophrys 21:152
acastus, Chlosyne 16:139; 22:162
acastus, Melitaea 2:92
accius, Lerema 4:25; 9:193; 17:21; 19:133, 137; 20:241; 21:207, 277; 25:34
Paspalum Panicum sp. 4:25
Stenotaphrum secundatum 13:221-227
accius accius 17:157
accusatrix, Symmachia 21:196
aceris, Neptis 7:38; 14:196
aceris oda 8:99
Lespedeza, Robinia, Rhynchosia, Milletia, Wistaria, Glycine (LEGUMINOSAE)
aceris passerculus 8:99
(same as for oda)
achaemenes, Charaxes 9:129; 13:124
acheloia, Byblia 25:146
Tragia brevipes
Dalechampia hildebrandti
achemon, Pholus 11:43
acheronta, Coea 3:40; 5:56; 17:37; 21:195
achillaena paulista, Morpho 21:89; 22:150
achine, Lopinga 9:204
achine, Pararge 11:164; 14:204
achine achinoides 8:99
Gramineae & Cyperaceae
acis, Strymon 14:20
acis bartrami 9:208
acmon, Plebejus 7:146; 13:74, 238; 16:69; 17: 205, 244; 18:225; 19:112, 232; 20:134, 161; 22:96, 97, 98, 103
Eriogonum nudum 16:35
Lotus scoparius, Trifolium obtusiflorum, Eriogonum latifolium, E. fasciculatum v. foliolosum 23:31-32
acmon acmon 12:31; 13:234
acmon cottlei 17:218
acmon lupini 16:35; 17:114
Eriogonum ? 16:35
acmon lutzi 5:63; 10:98; 22:165; 25:214
Eriogonum effusum ? 22:165
acontius, Catonephele 21:100
acraea, Estigmene 2:13; 13:26; 22:204
acrisione acrisione, Roswellia 14:26
acrisione deflavata 15:26
acronyctoides, Panthea 5:54
actaea, Satyrus 9:145

actaea simillima 10:221
actaeon, Adopaea 10:222
acteon, Thymelicus 9:145, 204
actorion actorion, Bia 24:18
actoris cuyabaensis, Cremna 21:87; 21:146
acuta, Taygetis 6:98
acuta paracuta, Curetis 7:37; 8:97; 14:138
  Wistaria, Milletia, Pueraria, Sophora 8:97
  Robinia pseudoacacia 18:107
acutilinea, Schinia 25:177
adenensis doris, Powellia 10:215, 222
adenostomatis, Satyrium 14:20; 17:111; 20:135; 22:96, 97, 99, 100, 103
adenostomatis, Strymon 1:58; 20:160
adelina, Chlosyne 13:165
adepta, Pyrgus 21:246
adippe, Fabriciana 9:145; 11:163; 14:193, 204
adjuncta, Mamestra 23:5, 7
admixta, Quasisalebria 24:254
adonis, Semiothisa 13:219
adrastus, Hyarotis 16:144
adultera, Pseudaletia 9:57
aecas, Aecas 21:165
aegeria, Pararge 10:221; 11:164; 13:4; 14:139; 18:41; 22:128 10:215
Aegiale 23:244, 246
aegita, Aides 20:228
aegeus, Papilio (IA,II) 16:161-173; 25:58
aegeus aegeus 16:163
  Apium graveolens; Aristolochia elegans; Capparis nobilis; Choisya ternata; Cinnamomum camphora; Citrus aurantium, aurantiifolia, limon, paradisi, reticulata, sinensi; Daucus carota; Flindersia bennettiana, collina, xanthoxyla, australis; Geijera salicifolia; Microcitrus australasica, australis; Owenia venosa; Petroselinum crispum; Zanthoxylum brachyacanthum
aelia, Hipparchia 9:145
aemula, Epizeuxis 23:5
aemulus, Epamera 10:39
aeneas, Papilio 4:41
aenescens, Prodoxus 20:54
  Yucca whipplei
aenus, Amblyscirtes 13:91, 92; 13:100
aenus aenus 25:34
aenus linda 21:207, 277
AEOLUS, *Lycaena* (IA) 15:17

aepeona, Thecla 21:153
aequaliaris, Caripeta 13:220
aerata, Dynamine 21:99
aerea, Plusia 13:26
aeruginosella, Adela 23:212, 216, 218
aesculapius, Amblyscirtes 17:157; 20:241
aethalia metana, Limenitis 21:99
aethalodaria, Selidosema 17:28
aetherie, Melitaea 10:215, 221
aethiops, Erebia 14:140
aethra, Melinaea 13:84
aetna, Wallengrenia 14:29
affinis, Acrocercops (IG) 25:198, 199
affinis, Callophrys 10:115; 14:21; 18:228
affinis, Euptychoides 21:91; 22:164
affinis, Mysoria (IA) S2:3-4
afflictus, Phlegethontius (IA,IG) 11:107
afflictus bahamensis 5:22; 11:112
afranius, Erynnis 7:146; 16:36
   Aquilegia ? 16:36
agacles, Dynamine 21:99; 22:152
agamemnon, Graphium 9:74; 21:50
agamemnon admiralis (IA) 25:75, 76
agamemnon argynnus, Papilio 25:77
agamemnon, Zetides 11:229
aganisa, Didonis 12:172
agarithe, Phoebis 6:44; 12:130; 15:72, 199;
agarithe agarithe 8:25; 17:225, 226; 22:34
agarithe maxima 9:207; 18:15; 21:190; 24:119;
   25:37
agathina, Doxocopa 21:87, 103
agathon caphusa, Aporia 11:168; 16:144
AGATHYMINI *12:93*
AGATHYMUS *12:82; 23:243-248*
agave pallida, Eurema 21:154
agavus, Battus (Parides) 22:154, 156
agavus, Papilio 4:41
agesilas agesilas, Lasaia 21:148
agesilaus, Papilio 4:66; 12:171
agestis, Aricia 9:204; 11:165; 14:137; 22:62;
   25:159, 160
agestis agestis 5:28
agestis artaxerxes 5:28
aglaia goda, Delias 9:76
aglaope, Papilio 4:41
aglaura, Nessaea 21:194
aglea, Danaus 11:167, 229; 16:144
agnesata, Eupithecia 13:218

agnosia agnosia, Ithomia 21:95; 22:151, 156
agricola, Agarista 19:165
agricola, Ochlodes 20:160; 22:94, 97, 99, 100, 102
agricola agricola 12:32
agricolana, Eana 16:179
agrifoliella, Caloptilia (IA,IG) 25:196, 199, 201, 206
Quercus agrifolia
agrippina, Thysania 13:83
ahola, Thecla 18:23
aidea, Anaea 14:190; 15:72; 22:230
aidea morrisoni 15:216; 19:113; 25:39
aidea floridalis 6:33; 25:137
ajax, Papilio 1:75; 4:11, 12, 63, 110; 5:67; 15:211
Heracleum sp.; *Ligusticum scothicum* 4:11-12
alexandra, Colias 18:220
Astragalus serotinus
alabamae, Atrytone 6:76
alabamae, Megathymus 14:27
alalia, Actinote S3:35
alana, Heliopetes 21:164, 246
alardus alardus, Astraptes 21:160
alardus latia 21:244
alaskae, Epipsiliamorpha 4:27, 28
alaskalis, Diasemia 4:29
alatella, Myelopsis 24:254
albania, Enantia 21:189
albanus, Papilio 12:194; 21:271
albastrigulella, Pleurota 13:232
albata sedecia, Strymon 21:183
albellus, Crambus 5:29
alberta, Œneis 15:223; 22:159; 23:155
albertalis, Loxostege 13:242
albescens, Pyrgus 19:91-93; 24:212
albicans, Agrodiaetus 14:137
albicans, Lysandra 10:218
albicans berber, Lysandra 10:222
albiceralis, Loxostege 5:29
albicosta, Cirphis 9:57
albifera, Hydrelia 5:54
albifrons, Symmerista 5:18
Oak
albilinea, Cirphis 9:57
albilinea, Faronta 9:57
albilinea, Metalectra 6:53

albimargo, Urbanus S2:13
albimargo takuta 21:88, 160
albina, Appias 3:39; 13:22
albina agatha 24:304
albinotata, Taygetis 6:97
albipicta, Cardepia 10:201
albiplaga, Phlegethontius 3:78
albiplaga, Protoparce 24:217
albiplaga aegides, Theagenes 21:246
albiplagiatella occidentalis, Pima 24:254
albipuncta, Cargolia 18:111, 114
albipunctata, Eupithecia 13:217
albistria albistria, Aguna 21:159
albistrigella, Ethmia S6:35
Phacelia ramosissima
albitogata, Ethmia S6:22
Amsinckia lunaris, A. spectabilis
albocaerulea sauteri, Celastrina 8:98
Prunus (Rosaceae); *Dicalix* (Symplocaceae);
Ilex (Aquifoliaceae); *Viburnum* (Caprifoliacea)
albociliatus albociliatus, Achalares 21:244
albofasciata, Saturnia 15:251
albogalleriella, Opostega 17:237-242
albovenosa, Arsilonche 2:94
albovittata, Neodezia 5:54
albovittata tenuifasciata 13:217
albugo albugo, Lymnas 21:148
albula, Eurema 18:15; 21:154, 190; 22:154
alcaeae, Carcharodus 14:51
alcaeae australis 10:222
alcaeus, Eudamus 15:72
alcaeus alcaeus, Codatractus 21:244
alcestis, Satyrium 14:21
alcestis, Speyeria 12:44
alcetas, Everes 14:136
alcestis alcestis, Phaeostrymon 18:145-148
Sapindus drummondii
alcidamus, Pap. Eq. Achivi. 12:196
alcinous, Byasa 11:9
alcinous bradanus 18:105
Aristolochia liukiuensis
alcinous loochooanus 18:105
Aristolochia liukiuensis
alcinous miyakoensis 18:105
alcinous, Menelaides 8:96
Aristolochia, Hocquartia
alcinous, Papilio 4:60; 5:52
Aristolochia sp.

alcionea, Pap. Helicon. 12:196
alciphearia, Selenia 13:220
alciphron, Chrysophanus 10:218
alciphron herakleana, Chrysophanus 10:222
alciphron, Heodes 11:165
alcmena, Eunica 12:172
alcmena, Euptychia 12:172
alcon, Maculinea 11:165
alcyone, Satyrus 10:215
alcyone maroccana 10:221
alea, Strymon (IA) 20:65-67; 25:38
alector, Eudamus 12:194
aletas, Phanes 21:165
alexanor, Papilio 4:74; 10:117; 13:199
alexanor HAZARAJATICA, *Papilio (IA) 15:2; (IA) 18:78*
alexandra, Colias 5:25; 8:163; 10:99; 14:42; 19:232 (Legume feeding)
alexandra christina 19:61
alexandra edwardsii 17:115
alexandrae, Ornithoptera(IA,II) 25:58-64
Aristolochia schlechteri, A. tagala
alexis, Glaucopsyche 14:138; 18:110; 24:196
alfacariensis, Colias 8:4
algae, Archanara 2:94
Typha latifolia
algidata, Xanthorhoe 5:54; 12:50
algira, Ophiusa 3:76
ali, Powellia 10:222
aliaska, Papilio 2:25
alicia, Apatura 19:200
alienaria, Hyperitis 23:206
alienus, Epamera 10:39
alienus, Iolaus 20:135
aliphera, Eueides 20:203
aliphera gracilis 18:19
aliphera gracilis, Heliconius 21:192
alipherus, Heliconius (Eueides) 21:97; 22:152
allardi, Polyommatus 10:215, 218, 222
allearea, Eupithecia 3:76
alleghenienis, Crambidia 8:93
alliae, Agathymus 18:45-46; S1:53
allottei, Ornithoptera (IA,IC,IG) 24:88-105
allubita, Mnasilus 21:165
allusa, Catocala 14:240
alma, Melitaea 18:44
almana, Precis 11:229; 14:195; 18:107
Lippia nodiflora, Lindernia verbenaefolia

almeida, Aloeides 10:38
alope, Cercyonis 9:77; 10:55, 166; 17:180
alope, Minois 6:33
alope, Erinnyis 2:86; 3:78; 13:78; 25:152
alopius, Papilio 4:40
alphaeus, Capys 10:39
alphenor, Papilio 13:76
alpheus, Pholisora 13:91; 22:168
alphonsus, Monethe 21:147
alsimo, Callimormus 21:167
alternata, Abagrotis 4:26
   Rhus copallina
alternata, Amblyscirtes 4:62; 13:89
alternata, Epirrhoe 5:54; 13:219
alternata alternata, Talides 21:166
alticola, Boloria 8:65
alveus, Pyrgus 14:51
alwina, Kalkasia 14:204
alwina kaempferi, Neptis 8:99
   Prunus (Rosaceae)
amabilis, Anepia (IA,IG) 21:235-239
amadis cyrene, Xylophanes 5:22
amadis meridanus, Xylophanes 3:78
amakosa, Durbania 10:39
amanda, Lysandra 11:166
amanda, Ocnerogyia (II) 25:53-57
   Ficus (cultivated)
amandus, Agrodiaetus 14:137
amandus abdelaziz, Polyommatus 10:222
amantes, Amblypodia 16:144
amarah, Anthene 10:38; 13:145
amarynthina monogramma, Parcella 21:148
amathea roeselia, Anartia 21:98; 22:152
amathonte centralis, Morpho 25:223-233
amathusia, Brenthis 3:75
amatrix, Catocala 24:108, 111, 115
amazon, Chorinea 21:87, 147
amazoula, Alaena 10:39
ambica, Apatura 16:144
ambigua, Mellicta 14:204
ambigua niphona, Melitaea 8:99; 15:220
   Melampyrum, Siphonostegia, Veronicastrum
   (Scrophalariaceae); *Plantago* (Plantaginaceae)
amblycasis, Cirphis 9:57
amblyspila, Pompeius 21:167
ambrax egipius, Papilio 13:76
ameliae, Charaxes 13:124

amella, Safia 4:26; 18:150
Quercus borealis 4:26
americalis, Epizeuxis 23:5
americana, Acronicta 2:13
ameriicana, Acronycta ? 4:26
Dasystoma pectinata
americana, Epicnaptera 18:37
americana, Harrisina 21:34
americus, Papilio 9:8
amestris, Catocala 15:124
Amorpha
amica, Catocala 10:167; 23:5; 24:108, 110, 113, 114; 25:43
amicaria, Hyperetis 13:220
amicia eurytion, Euphydryas 21:261-270
aminias, Codatractus 21:159; 22:155
AMIRICA, *Lyela (IA) 15:16-17*
ammon bethunebakeri, Hemiargus 25:137
amoena, Cyanospiza 20:76
amor, Rathinda 20:135
amorata, Callizzia 5:54
amorphella, Walshia 3:48
Amorpha fruticosa
parasites
ampelos, Coenonympha 13:74, 20:161
ampelos mono 9:141
amphidamas, Lycaena 3:76
amphimachus, Prepona 18:22
amphinome, Ageronia 4:67
amphinome aegina, Hamadryas 21:103; 22:153
amphinome mexicana, Hamadryas 21:195
amphion marpesus, Telemiades 21:161
amphione, Dismorphia 24:299
amphirrhoe, Catoblepia 21:93
amphrysus, Troides 9:75
amurensis, Leptidea 8:96; 14:53
Vicia, Lotus, Lathyrus (Leguminosae) 8:96
amymone, Mestra 12:130; 15:72; 17:151; 18:20; 20:161; 22:228, 230; 24:203-207; 25:39,146, 214
amyntor, Ceratomia 15:193; 21:252
Elm, basswood 21:252
amyntor distractus, Callophrys 21:195
amyntula, Everes 10:98, 170; 11:48, 12:30; 13:232; 14:130; 16:70, 244; 18:226; 23:115
amystis epigona, Myscelus 21:158; 22:155
amystus, Myscelus 15:72
anacreon anacreon, Ebrietas 21:163

anadyomene, Argynnis 7:37; 14:194, 204
anadyomene midas 8:98
Viola (Violaceae)
ananta, Neptis 11:167; 16:144
anaphissa, Ithomia 15:221
anaphus, Pap. Pleb. Urbicol. 12:196
anaphus anaphus, Astraptes 22:155
anaphus annetta, Astraptes 21:244
ANAVERTA, *Pseudocopivaleria (IA,IG) 20:84-90*
Quercus chrysolepis
anaxibia anaxibia, Morpho 21:89; 22:150
ancellata, Scopula 13:217; 24:54-55
anchemolus, Pholus 3:78; 25:152
anchises, Coeliades 18:74
anchises, Papilio 4:41
anchises orbignyanus, Battus 21:87, 156
anchisiades, Papilio 4:64; 13:194
anchisiades capys 21:157; 22:154
anchisiades idaeus 21:189
ancilla, Philotes 8:8
ancillaris, Lerema (1st U.S.) 24:266
ancipitalis, Celama 3:76
andalusica, Hadena 5:28
ander, Mylon 21:164
anderida zuleika, Heliconius 21:192
andersoni, Itame 25:267
andraemon, Papilio 4:64
andraemon bonhotei 9:207
andria, Anaea 2:22; 3:17; 6:33; 7:65; 10:165; 12:130; 17:23, 151; 19:34; 21:207, 277; 23: 182; 24:44; 25:140
Croton capitatum 10:165
andria andria 20:102; 25:39, 214
androgeus, Papilio 4:64; 21:157
andromeda, Taygetis 6:98; 18:17; 21:191
andromeda, Cithearias (IA) 24:16-17
andromedae, Catocala 23:5, 262; 24:108, 110,114
andromica, Hymenitis 24:298
ANEURETUS, DECODES *(IG) 14:121-122*
angasii, Caligatus 11:215
angelica, Apatelodes 4:26; 18:89; 21:35
Fraxinus 4:26
angelicata, Lithostege 13:233
anginus anginulus, Urbanus 21:278
angolanus, Papilio 9:129
anguina, Dasylophia 2:S-viii
Robinia pseudoacacia

anguinella, Teleia 3:76
anguitea, Hesperocharis 21:155; 22:154
angularis, Yphthimoides 22:151
angulifera, Callosamia 4:36; 22:186
tulip tree
angulifera, Hyalophora 7:58; 14:103; 21:34
angulosa, Taygetis 6:97
angusi lucetta, Catocala 10:168
angusii, Datana 21:36
augustinus, Callophrys (Incisalia) 21:136
anicia, Euphydryas 2:92; 6:105; 8:3; 12:165; 18:228; 25:246
anicia alena 22:160
anicia eurytion 10:96; 14:184
anicia helvia 19:59
anicia hopfingeri 16:148
anicia magdalena 19:114
animata, Marmopteryx 13:218
ankeraria, Argiopis 3:76
anna, Diaethria 21:194
anna, Plebejus 16:34
annada, Erebia 11:167; 16:144
annexa, Agrotis 2:S-viii
Taraxacum officinale
anonaria, Epimecis 5:59
anonyma, Limenitis 14:196
ansorgei, Charaxes 13:124
antaeus, Cocytius 2:86; 9:207; 13:78
antalus, Deudorix 10:39
anteas, Actinote 21:192
antenor, Pharmacophagus 9:66
Quisqualis grandidieri (Combretaceae)
antevippe, Colotis (IA) 13:138
anthedon, Lethe 6:32
anthedon anthedon 25:256-261
antholia, Myscelia 5:56
anticaria fumata, AEthalura 13:220
anticlea, Charaxes 13:124
ANTIOCHELLA, *Lithocolletis (IA, IG)* 25:195, 203, 204, 208
Quercus agrifolia
antiochus salvinii, Heliconius (IA) 23:199-201
anticostiensis, Papilio 18:44
antilope truentus, Hypolimnas 18:108
Villebrunea sp. ?
antimachus, Daphne 15:248
antimachus, Druryia 15:248
antimachus, Papilio 24:224-225

antinympha, Catocala 12:201; 23:5, 7; 24:108, 110, 112, 113, 114
antionia SHIVAENSIS, *Ramburia* (IA) 15:17-18
antiopa, Nymphalis 2:13; 3:39, 44; 4:14; 5:26, 27, 42; 6:33, 73, 105; 7:146; 8:99; 9:78, 143; 10:55, 97, 107, 169, 171; 11:47, 164; 12:29, 44, 130; 13:73, 238; 14:195, 204, 206, 207; 15:111, 218; 16:32, 69, 244; 17:151; 18: 40, 223; 19:49, 113, 182, 183, 232; 21:207, 241-242, 277; 22:98-100, 103, 120, 128, 230; 23:85, 100, 273-274 (II); 25:140, 214
Salix, Toisusu (Salicaceae) 8:99
Salix interior 23:273-274
antiopa creta 15:219
antiopa grandis 15:218
antiopa lintnerii 18:148; 25:39
antiope, Telemiades 21:164
antiphates, Papilio 13:22
antiphila, Eana 16:179
antiphulus philippus, Papilio 13:76
antiqua, Orgyia 18:37
antiqua, Phemiades 5:55
antiqua, Pyrrhocalles 14:31
antistia equa, Synapte 21:164
antodyca, Pereute 22:154, 156
antonia, Asterocampa 6:33
antonius, Saliana 20:228
anubus, Xylophanes 3:78
apama, Callophrys 14:21; 19:112, 231
apama apama 22:172
apama homoperplexa 10:97; 22:164
Ceanothus fendleri ? 22:164
apapinaria, Stenaspilates 13:232
aphaca, Thecla 21:153; 22:154
aphirape, Boloria 5:8; 12:49-50
aphirape alticola 1:42; 8:64
aphirape triclaris 2:65; 5:8
aphnaeoides nasissii, Iolaus (IA) 13:145
aphnaeoides nasissii, Epamera 13:145
aphrodite, Speyeria 2:5; 9:78; 10:55, 102, 107, 112; 11:163; 12:38, 41, 43, 44; 14:215; 15: 23; 17:107; 22:120, 128; 25:141, 150, 196,197
aphrodite alcestis 21:137; 22:35
aphrodite byblis 19:114; 22:160
aphrodite ethne 10:95; 14:180, 238
aphrodite winni 16:243; 21:136
apicalis, Adeloneivaia 14:103

apicalis, Schizura 19:190; 21:37
apiformis, Aegeria 18:37
apiformis, Alcathae (IA) 7:164
apollo, Parnassius 9:203; 11:162; 13:199
apollonia apollonia, Lyropteryx 21:146
appalachia, Lethe 6:32; 25:256, 261
appassionata, Papaipema 8:58
  Sarracenia sp.
appendiculata, Eupithecia 13:233
appias, Anaea (Memphis) 22:153, 156
approximaria, Plagodis 13:220
aprica, Acontia 2:S-viii
  Althaea rosea
aquata, Oleria 21:94; 22:151
aquilina, Bibasis 14:51
aquilina chrysaeglia 8:95
  Kalopanax (Araliaceae)
aquilo, Agriades 23:35, 38 (IH)
aquilo, Plebeius 2:65; 5:8, 26; 7:43; 10:171
aquilo aquilo 5:9; 19:243
aquilo bryanti 7:125; 11:52
aquilo rusticus 2:92
aquilonaris, Boloria 14:193
arabus, Lerodea 18:215-216
arachne victoria, Anaea 21:105; 22:153
araliae, Papaipema 8:58
  Aralia spinosa
arana, Cargolia 18:115
arana, Caripeta 18:115
arana, Erilophodes 18:115
arana, Neodesmodes 18:115
aranda, Aloeides 10:38
araratensis, Agrodiaetus 14:137
arata, Rapala 8:97; 14:204, 206
  Wistaria (Leguminosae); *Deutzia* (Saxifragaceae); *Rosa* (Rosaceae); *Vaccinium* (Ericaceae); *Palura* (Symplocaceae); *Rhamnus* (Rhamnaceae)
araxes eupolis, Narathura 20:135
arbela arbela, Eurema 21:154; 22:154
arcas, Papilio 4:41; 12:196
arcas mylotes, Parides 18:14; 21:189
arcania, Coenonympha 9:204; 11:164; 14:139; 18:110
arcanioides, Coenonympha 10:218, 221
arcesilaus, Papilio 4:66
archea, Antirrhea 21:90
archidamus, Papilio 4:41

archippus, Limenitis 2:22; 3:17; 5:17; 6:33; 9:78; 10:55, 97, 107; 12:130; 14:190; 16: 244; 17:180; 18:48, 224; 19:183; 21:32, 207, 277; 22:120; 23:100; 24:305; 25:140, 278
willows 18:224
archippus, Pap. Dan. 12:196
archippus archippus, Limenitis 18:165
archippus floridensis 9:205, 209; 25:146, 147
archippus obsoleta 8:129; 19:113
archippus watsoni 17:23, 151; 25:39
archippus X astyanax 7:53
archippus X lorquini (IA) 24:27
arcigera, Schinia 6:53
arcius amycus, Rhetus 21:147
arcius thia 18:24; 21:196
arcticus, Papilio 4:69
arctivorens, Papaipema 8:58
Cirsium spp., Arctium lappa
arctostaphylella, Ethmia (II, IF) S6:42; 13:232
Eriodictyon californicum
arcuata, Drepana 5:54; 11:18
Arcas 25:87-108
arduinna, Melitaea 3:76
arene, Ancyloxypha 22:231
arethusa, Eumenis 9:145
arethusa aksouali, Satyrus 10:218, 221
areolata, Euptychia 9:206, 209; 14:127, 209
areolata areolata 17:23, 149; 19:139-159 (IA); 22:34
areolata septentrionalis 14:209; 19:151, 155
areolatus, Euptychia 6:76
argane, Pieris 3:75
argante, Phoebis 4:67; 21:190
argante argante 18:15; 21:154; 22:154
arge, Apantesis 25:42
argentana argentant, Eana (G) 16:179, 182
argentana argentana 16:176
argentana colossa 16:176
argentana plumbeana 16:176
ARGENTUS, *Vettius (IA, IG) S2:24*
argester, Agrodiaetus 14:138
argester, Lysandra 9:204
argiades, Cupido 18:110
argiades, Everes 3:39; 7:37; 8:144; 14:136; 20:134; 24:196
argiades seitzi 8:98
Lespedeza, Indigofera, Pisum, Pueraria, Trifolium Medicago, Phaseolus (Leguminosae);

Humulus (Moraceae)
argillacea, Alabama 15:111; 18:1, 9; 19:132, 133; 25:43
argiolus, Celastrina 7:37; 13:170; 14:136; 15:199; 16:20-22 (IA), 70; 18:226; 19:231; 21:207, 277; 22:120; 24:77, 305
argiolus, Cyaniris 6:75
argiolus, Lycaenopsis 2:25; 5:26; 6:75; 10:55, 107; 11:165; 15:111; 16:136; 24:196
argiolus cinerea, Celastrina 19:113
argiolus echo 16:35; 17:212; 22:96-99, 103
Actinomeris, Ceanothus, Vaccinium, Spiraea
argiolus ladonides 8:98
Lespedeza, Robinia, Sophora, Wistaria (Leguminosae); *Malus, Prunus* (Rosaceae)
argiolus mauretanica, Lycaenopsis 10:222
argiolus pseudargiolus, Celastrina 17:153; 19:50; 20:42
argiolus pseudargiolus, Lycaenopsis 10:99; 14:239; 16:244
Argus (suppression of) 8:174
argus, Plebeius 11:165; 14:137; 15:58; 18:41; 20:134; 22:129; 24:199
argus micrargus, Plebejus 8:98
Synurus (Compositae)
argus, Ypthima 7:37; 8:99; 14:139
Oplismenus (Gramineae) 8:99
argyrodines, Calephelis 18:25; 21:196
argyrognomon, Plebejus 11:52; 18:226; 19:61, 167; 20:134
argyrognomon, Lycaeides 5:26; 6:41; 7:54; 8:98; 9:148; 14:136
Indigofera 8:98
argyrognomon anna, Plebejus 17:114
argyrognomon aster 4:11, 12; 23:38
argyrognomon kodiak, Lycaeides 15:216
argyrognomon longinus 7:50, 51
argyrognomon longinus, Plebejus 16:235
argyrognomon scudderi, Lycaeides 7:50, 198; 10:107; 19:243
argyrognomon sublivens 4:51; 6:35
Lupinus parviflorus 6:35
argyrospila, Archips (I) 7:164; 10:154
aria, Matapa 18:162
ariadne, Ergolis 3:40; 25:146
Tragia spp. 25:146
ariane, Cercyonis 19:161
ariarathes, Graphium 21:157

ariarathes, Papilio 4:66
Aricia 25:159
aricye, Pap. Dan. 12:196
arida, Cisthene 11:148
arion, Maculinea 11:165; 14:138; 18:41; 22:129
arion orion, Nomiades 16:104
arionides, Maculinea 7:45
Plectranthus exisus, Myrmica
arionides takamukui 8:98
Isodon (Labiatae), later carnivorus on ant-grubs & eggs of *Myrmica*
aripa elodia, Leptophobia 21:189
arisanus, Teratozephyrus 3:80
arisbe, Caligo 22:151, 156
aristeus, Papilio 3:40; 4:65
aristodemus, Papilio 4:64
aristodemus ponceanus 9:207, 210; 25:126-136 (II)
Amyris elemifera, Zanthoxylum americanum
aristolochiae, Polydorus 21:127-128; 22:115-118 (II), 183-184; 23:107-108; 24:19-22
Aristolochia bracteata, A. indica, Lagenaria vulgaris, Luffa aegyptiaca, Diascorea wallichii
aristolochiae, Tros 11:229
aristolochiae antiphus, Papilio 9:75
aristor, Papilio 4:64
arizonensis, Ceratonyx 21:13, 14 (II)
Vigueria multiflora
arizonensis, Gloveria 3:106
arizonensis, Limenitis X angustifascia 8:130
armigera, Heliothis 6:53
armilla, Praefounala 21:92
armillata, Taygetis 6:98; 21:90
armoricanus, Pyrgus 11:166
armoricanus maroccanus, Hesperia 10:222
arogos, Atrytone 4:26; 7:56; 20:177-181 (IA,II)
Andropogon gerardi 20:177-181
Panicum sp. 4:26
arogos, Hesperia 12:122
arota, Lycaena 17:243; 20:134; 22:96, 97, 99, 100, 103
arota, Tharsalea 8:128
arota schellbachi, Lycaena 14:238
arota virginiensis 16:34
Ribes montigenum
arota virginiensis, Tharsalea 20:52
arpa, Hesperia 12:122
aristoteles, Sophista 21:164

arsace, Thecla 12:122
arsalte arsalte, Heliopetes 21:164, 246
arsinoe palawanica, Vindula 24:304
artaxerxes, Aricia 25:159, 160
artella, Arivaca 24:40
artemides, Spioniades 21:161
artemisia, Dynamine 21:99; 22:152
arthemis, Limenitis 5:54; 6:33, 105; 11:163; 14:179; 15:99, 217; 16:244; 19:182; 23:100, 193; 25:140, 257, 278
arthemis astyanax, Limenitis 21:207, 277; 25:81, 257, 278
arthemis astyanax X L. archippus (IA) 25:281
arthemis proserpina, Limenitis 10:55
arthemis rubrofasciata 7:125; 10:107
arthurianus, Rhetus 21:87, 147
artona, Vettius 21:166; 22:156
arvalis, Annaphila 4:71; 13:78
  blossoms/seeds of annuals 4:71
aryxna, Agathymus S1:38; 13:204; 18:46; 20:242; 23:245
  Agave palmeri 18:46
aryxna, Megathymus (I) 8:77-87; 11:29-34, 116; 14:27
  Agave palmeri 11:116
asamana, Oeneis 8:99
  Carex (Cyperaceae)
asander, Aguna 13:90
asander asander 21:159; 22:155
ascanius, Papilio 4:41
ascaphalus, Staphylus 21:245
ascolia, Peplia 21:197
ascolius, Papilio 4:65
asine, Polythrix 21:244; 22:129
asius, Papilio 4:66
asopialis, Palthis 23:192, 194
aspasia aspasia, Gonepteryx (IA) 24:214
aspasia coreensis (IA) 24:214, 215
aspasia iwateana (IA) 24:214
aspasia kansuensis (IA) 24:214-216
aspasia niphonica (IA) 24:214
assimilans, Phragmatobia 4:12; 11:148
assimilis, Oeneis 5:42
assimilis, Polia 4:12
  Myrica asplenifolia
assimilis shirakii, Hestina 18:108
  Celtis boninensis
astala, Diaethria 21:194

astarte, Boloria (IA) 11:142-146; 15:221; 18:228; 20:103
astarte distincta 19:60
asterius, Papilio 8:141; 12:196; 18:149
asterius X P. machaon (IA) 10:49
asterope, Ypthima 11:167; 16:144
astrarche, Polyommatus 16:144
astrologa, Autographa 4:71
astuta, Papaipema 8:58
  Collinsonia canadensis
astaroth, Satyrus 18:44
astyalus, Papilio 21:157
astyalus astyalus 22:154, 156
astyanax, Limenitis 1:42; 3:25, 48; 6:33; 7:146; 9:16, 77; 10:44, 55; 12:130; 13:73; 14:188; 15:23, 99, 217; 17:151; 19:229, 232; 22:120; 23:100, 194; 24:305; 25:140
  parasites 3:48
astyanax arizonensis 8:129, 130; 19:113
  Salix 8:130
astyanax astyanax 17:23; 25:39, 278
  Salix sp. 17:23
astylus, Paonias 4:25; 19:189
  Kalmia latifolia 4:25
astynome astynome, Dismorphia 21:156
astyocha, Dismorphia 21:156; 22:154, 156
astyoche, Pierella 24:17, 18
asura, Pantoporia 11:167; 16:144
ASURA, Paralasa (IA) 15:12-13
asychis, Chiomara 8:25
asychis, Pap. Pleb. Urbicol. 12:196
asychis autander, Chiomara 21:163
atahuallpai, Bolla 21:164
atahualpa, Elzunia 15:29
atala florida, Eumaeus 6:77; 9:208, 209; 15:237 (IA, II, IG); 20:216; 22:35
  Zamia integrifolia 15:237
atalanta, Vanessa 3:17, 18, 39; 4:61; 5:26, 27; 7:40, 146; 8:141; 9:77; 10:55, 97, 171; 11:164; 12:130; 13:18, 73; 14:79, 179, 195, 238; 16:32, 233, 244; 17:180, 225; 18:40, 223; 19:113, 180, 182, 232; 21:136, 207, 277; 22:35, 98, 99, 103, 120, 128, 228, 230; 23:37, 100, 182; 24:305; 25:39, 140, 214
atalanta atalanta 17:23, 151
atalanta italica 15:218
atascaderalis, Noctuelia 13:242
ataxus, Neozophyrus 3:79; 13:178, 180

Cyclobalanopsis acuta 13:178
ataxus kirishimaensis 8:97
Cyclobalanopsis (Fagaceae)
athalia, Melitaea 9:204; 18:110
athalia, Mellicta 9:145; 13:207 (IG); 14:194
athemon maeon, Dynamine 21:100; 22:152
athenion, Polites 17:37
atinas, Teriocolias 15:119
ATLANTEA 12:162
atlanticus, Lysandra 10:215, 218, 222
atlantis, Satyrus 10:215, 218, 221
atlantis colini 10:221
atlantis, Speyeria 1:58; 2:1, 5, 92; 5:31, 54; 10:102, 107; 16:243; 17:107, 109, 129-147 (IA); 19:62; 22:120, 172; 23:37 (IA); 25:141
atlantis atlantis 10:102
atlantis canadensis (IA) 23:36, 37
atlantis dodgei 13:14; 18:222
atlantis electa 14:238
atlantis greyii 5:33
atlantis hesperis 7:49; 10:95
atlantis irene 16:31
atlantis lais 6:106
atlantis nausicaa 19:114, 231
atlantis nikias 14:182, 238
atomariella, Lithocolletis 24:272, 274, 277 (IA, IG)
Populus ?
atra, Acanthopsyche 10:36
atricapillus septentrionalis, Parus 20:73
atricolor atricolor, Lepricornis 21:147
atrifasciata, Eustroma 13:218
atroliturata, Cladara 13:217
atronia, Phyciodes 21:193
attalion, Thecla 16:1
attalus seminole, Hesperia 3:17; 14:30
attilia, Antigius 3:80; 8:97; 13:178; 14:204, 205
Quercus spp. (Fagaceae)
aubyni, Charaxes 13:124
augusta, Phyciodes 21:97
augustinus, Callophrys 14:21; 16:70; 18:235; 20:41; 22:226
augustinus, Incisalia 7:33-35; 9:11; 10:89-90, 97, 107; 11:47; 14:127; 22:26, 252
Arctostaphylos uva-ursi 7:34; 10:89-90
Azalea calendulacea 10:89-90
Kalmia angustifolia ? 9:11; 7:33-35; 10:89

Vaccinium vacillans 7:33-35
augustinus iroides, Callophrys 16:33; 18:224; 22:96, 97, 99, 100, 102, 129
Ceanothus, Sedum 16:33
aulestis, Pap. Eq. Achiv. 12:196
aurantiaca, Copaeodes 7:146; 12:31; 19:1, 31; 22:231; 25:34
Cynodon dactylon 19:4, 31
aurantiaca, Holomelina (IA) 19:69-74
aurantiaca, Incita 3:74
Gilia virgata v. *dasyantha*
aurantiaca BUCHHOLZI, Holomelina (IG) 17:100-102; 18:118
aurantiaca rubricundaria 17:101
aurata, Entephria 12:50
aurata, Pyrausta 5:29
aurelia, Amathuxidia 9:74
auresiana, Argynnis 10:217, 221
auretorum, Satyrium 14:20; 17:111; 20:135; 22:96, 97, 99, 100, 103
auricyanea, Mnemonica 4:6; 10:161
Quercus sp., Castanea sp. 4:6
aurilacus, Euphydryas 13:172
aurinia, Euphydryas 10:215; 14:194; 15:58
aurinia ellisoni 10:221
AURIPURPURA, Hydroecia (IA, IG) 22:136, 137, 138, 139
aurivilii, Euryphene 13:128
aurora, Hyparpax 18:89; 21:36
aurorina, Colias 4:74
aurorina libanotica 10:117
aurorinus, Neoxophyrus 3:79; 8:97; 13:178
Quercus (Fagaceae) 8:97
Quercus crispula 13:178
aurota, Anapheis 3:39
aurota, Glycestha 10:117
ausonides, Euchloe 7:125; 10:107; 11:47, 162; 14:42; 15:113; 16:70; 18:221; 19:112; 22:96-98, 100, 102; 24:45
ausonides andrewsi 12:28
ausonides coloradensis 1:42; 2:92; 10:99
ausonides mayi 15:113; 16:245; 21:136; 22:40-41; 24:226
austerus, Gesta 21:163
australella, Zamagiria 24:254
australis, Calephelis 25:37
australis, Colias 7:58; 9:145

australis, Phanus 21:158
autolycus, Strymon 7:146
autosilaus viridis, Graphium 21:157
autumnana, Phaneta (IA, IG) 25:285-287
auxo, Camptopleura 21:163
avalona, Strymon 14:20
avanta, Ypthima 16:144
avis, Callophrys 10:215, 222; 24:196
avuncularia, Dasyfidonia 13:220
axena, Ematurgina 21:149
axine, Semioptila 12:160
azanoides, Nymphidium 21:87, 150
azia, Calycopis 21:153
azia, Strymon 14:22
azia, "Thecla" 17:37
azia, Tmolus 22:164; 25:38
azuba, Tmolus 21:152

ADDENDA TO ABOVE LIST:

aegeria, Pararge 10:215
alexiares, Papilio 4:64
anthedon borealis, Lethe 25:256-261
aspersa, Borolia 9:57
australis, Pseudaletia 9:57

----

babbi, Melitaea 13:235
bacchus, Lepidochysops 10:38
bachmanii, Libytheana 1:70; 3:17, 28; 7:54, 146; 9:148; 12:36, 123, 130; 14:190; 15:72; 16: 128; 17:152; 19:232; 21:207, 277; 22:35; 23: 105, 106; 24:305; 25:214
bachmanii larvata 6:44; 14:190; 16:249; 19:113, 223, 224; 22:35; 25:38
bachus, Papilio 4:65
badia, Catocala 24:110, 112, 114
baffinensis, Xanthorhoe 2:65
bagrada, Thecla 21:152
bairdii, Papilio 1:75; 4:63; 7:146; 8:3; 9:8; 18:65; 19:109, 232; 22:166; 23:278
Artemisia dracunculus 22:166
Foeniculum vulgare 18:65
bairdii bairdii (I) 9:51-53; 23:272
bairdii brucei 9:10
bairdii oregonia 9:10
bairdii oregonius 18:220; 22:167
Artemisia dracunculus 18:220

baiva, Poanes 14:28
baldur, Parnassius 15:210
baldus, Ypthima (IA) 18:159
ballus, Thestor 10:222
ballus, Tomares 24:196
balsamitae, Cucullia 3:76
balteata, Macroglossa 12:123
bambana, Axiocerces 10:38
banghaasi, Taygetis 6:98
banksaria, Chlorissa 13:317
baptisiae, Erynnis 4:62; 8:3; 9:78, 194; 13: 71; 14:33, 103; 19:1, 11, 31; 24:46
Baptisia tinctoria, B. laevicaulis 19:11, 31
baptisiae, Papaipema 8:58
Baptisia tinctoria, B. alba; Cacalia tuberosus; Apocynum sp.
barbinerivis, Clethra 14:206
barcastus barta, Mysoria 21:158; 22:155
barnesi, Phyciodes 15:115
barnesi, Synaxis 13:220
baroma, Dirphia 14:103
basalides, Thecla 18:23
basiplagana, Sciaphila 14:112
basiplaganus, DECODES *14:116-118*
basirei, Neocarnegia 14:103
basoches basoches, Carystoides 21:166
basutus, Thestor 10:39
batabano, Phocides 9:207
batea, Opsiphanes 22:151, 156
batesii, Phyciodes 5:126; 8:3; 11:39; 20:245-247; 21:136; 22:35; 25:139, 141
batesii, Thecla 21:151
bathyllus, Thorybes 4:26; 9:77, 194; 10:55; 12:101, 130; 14:34; 17:155; 19:1, 25, 31, 180, 196; 20:34; 21:207, 277; 22:35, 120, 179; 24: 305; 25:81
Astragalus engelmannii, Centrosema virginanum, Desmodium ciliare, D. panculatum, Lespedeza hirta 19:1-31
Strophostyles helveola, S. leiosperma 20:34
battoides, Philotes 8:8; 11:7; 17:112
battoides battoides 23:50, 54
battoides bernardino 13:231; 17:203-206 (IA,IG); 19:95-101; 23:51, 55
Eriogonum fasciculatum, E. f. foliolsum, E. f. polifolium, E. f. flavoviride
battoides centralis 5:63; 22:165; 23:50, 55
Eriogonum umbellatum

battoides glaucon 17:112; 23:50, 55
Eriogonum umbellatum
battoides intermedia 16:35; 17:113; 23:50, 54
Eriogonum nudum ?
battoides martini 23:51, 56
Eriogonum fasciculatum polifolium
battoides oregonensis 18:226; 23:50, 55
battus aufidena, Thecla 21:196
BAUERI, *Megathymus (I) 8:80-87*
Agave parryi
baueri, Agathymus 18:45-46; S1:40
baumanni, Charaxes 13:124
bavius fatma, Philotes 10:215, 217, 222
Salvia (flowerheads) 10:217
bazalus turbata, Arhopala 8:97
Lithocarpus, Kuromatea (Fagaceae)
bazalus turbata, Narathura 18:106
Lithocarpus edulis
bazochii, Polyommatus 12:120
bazochii, Strymon 14:22
bazochii, Thecla 13:90
bechina, Evonyme 21:87, 102
beckeri, Pieris 3:106; 5:26; 9:148; 12:29; 14: 46; 16:69; 18:221; 19:221-222; 22:34, 166
Stanleya pinnata, Isomeris arborea, Bassica nigra
beckeri, Plusia 4:74
beckeri pseudochloridice, Pieris 9:141
beeriana, Papaipema 8:58
Liatris pycnostachya spicata
begga, Gorgythion 14:33
begga begga 21:161
begga pyralina 13:90
beggina escalophoides 21:162
behrensarius, Pero 13:220
behri masoni, Cercyonis 22:172
behrii, Callipsyche 10:97; 20:160; 22:59
behrii, Colias 5:25; 8:163; 17:112
behrii, Minois 10:113-114
behrii, Satyrium 14:20; 15:56; 16:33, 44, 69, 199; 17:115, 245; 18:48, 224; 20:135
Lupinus ? 16:33
Purshia tridentata 18:224
belae, Didasys 11:148
belemia, Anthocaris 10:215
belemia distincta 10:220
belemia röberi 10:220

belesis, Papilio 4:66
belia, Anthocaris 4:72
belia butleri 10:220
belia turatii 10:220
bella, Adela 23:212, 216, 218, 233 (IA)
bella, Utetheisa 4:62; 6:108
bellargus, Agrodiaetus 14:137
bellargus, Lysandra 9:145, 204
bellargus, Polyommatus 10:215
bellargus punctifera 10:222
bellela, Nemophora 23:214-217, 233 (IA), 237
bellerophon, Papilio 4:65, 66
belli, Amblyscirtes 4:44; 6:77; 9:193; 13:90
belli, Megathymus 11:117; 20:166; S1:42
Agave parryi
belli eberti, Cumbre 21:166
bellona, Boloria 8:65, 66
bellona, Brenthis 9:54
bellona, Pap. Heliconius 12:196
belus, Papilio 4:41
belus varus Battus 21:189
benjaminii, Choaspes 14:51, 206
benjaminii japonica 8:95; 18:105
Meliosma oldhami (Sabiaceae)
benjamini formosana 18:105
Meliosma oldhami
beon, Calycopis 20:135; 21:153, 196; 24:196; 25:38
beon, Pap. Pleb. Rural. 12:196
beon, Strymon 14:21; 15:56
bera, Neocoenyra 12:6
berania, Marpesia 21:195
berecynthia, Catoblepia 21:88
berecynthia berecynthia, Catoblepia 21:87, 93
berenice, Danaus 1:42; 3:17; 7:146; 8:123-124; 9:67
Asclepias syriaca 8:123-124
berenice, Pap. Nymph. Phaler. 12:196
bernardino, Philotes 8:8
bernia, Marpesia 25:223, 225, 230
berryi, Atrytone 4:31, 62; 9:204
beskei, Heliconius 21:96; 22:152; 24:25
betulae, Salebria 13:8
Betula spp.
betulae, Thecla 3:79; 11:166; 14:138; 24:196
betularia, Amphidasis 6:93; 9:199
betularia, Biston 17:162, 174; 19:228
betulina, Thecla 3:79

BEULAHAE, *Megathymus (IA, IG) 11:121-25; 15:68; S1:25*
Agave schottii
bianor, Papilio 5:52; 7:37; 11:9; 13:199; 14: 203; 15:182
orange 5:52
Orixa japonica, Fagara ailanthoides, Phellodendron amurense, Poncirus trifoliata
bianor dehaanii 8:96
Orixa, Phellodendron, Fagara, Zanthoxylum, Poncris (Rutaceae)
bianor junia 18:106
Evodia glauca
bianor okinawaensis 18:106
Evodia glauca
bibulus, Clito 21:164
bibulus, Lachnocnema 10:39
bechina, Evonyme 22:153, 156
bichroma, Eresinopsis 13:138
BICOLOR, *Decodes (IG) 14:122-123*
bicolor, Lexis 5:54
bicolor, Syssphinx (Sphingicampa) 18:89-90; 21:34
bicolor, "Thecla" 21:184
bicolorago, Sunira 9:56
bifascia, Schinia 6:53
bifasciata adamas, Sostrata 21:164
bifasciatus casyapa, Lobocla 16:144
bifilata, Drepanulatrix 13:219
bilineata, Drepana 5:54
biloba, Autographa 13:26
bilunaria, Selenia 13:2
BIMACULA, *Acanthopteroctetes 23:139 (IA), 140, 141 (IC), 143 (IG), 144-146 (IC,IG)*
bimacula, Atrytone 16:243
bimacula, Euphyes 19:196; 23:186; 25:143
bimaculata, Alypiodes 19:165
bina, Heliophana 6:53
binda, Grotella (IA, IG) 20:92-94
bipuncta, Nisoniades 21:161; 22:155
bipunctata, Adalia 25:161
bipunctatus, Charaxes (IA) 13:124
birchalli, Papilio 4:65
birivia, Euxoa 3:76
birói, Oreopsyche 3:76
biruptana, Eana 16:179
biselliella, Tineola 4:28, 29
bistortata, Ectropis 8:4

bistriaris, Parallelia 23:206-209
bistriatella bistriatella, Apomyelois 17:39
bistriatella neophanes 17:39
bistrigulella, Opostega 17:237-242
bitactata, Itame 13:219
bithys, Catasticta 22:154, 156
biundata, Heterocampa 2:50
Carya sp.
blakei, Eurema 2:51
BLANCHARDI, *Grotella (IA, IG) 20:194, 254*
blanda arsaki, Eurema 18:106
Caesalpinia nuga?, Wagatea spicata
blandula, Catocala 12:201; 24:111, 114
blanka argentea, Pratapa 20:135
blomfildia, Smyrna 21:104
blomfildia datis 12:172; 18:22; 21:195
bocchoris bocchoris, Phaenochitonia 21:149
bochus, Jamides 16:144
bochus ishigakianus 18:106
Pueraria montana?, Pongamia pinnata, Oromacarpum cochinensis
bore mackinleyensis, Oeneis 19:61
bodenheimeri, Caradrina 10:201
boeticus, Cosmolyce 3:39-40; 24:196
boeticus, Lampides 4:43; 8:98; 10:38, 222; 11: 167; 14:136; 16:144; 18:106
Dolichos, Phaseolus, Vicia 8:98
Vigna marina, Phaseolus angularis, Pisum sativum, Crotalaria sessiflora 18:106
bogotana, Mestra 24:207
BOHARTIELLA, *Neurobathra (IA, IG) 25:196, 199, 200, 201, 206*
Quercus agrifolia
bohemanni, Charaxes (IA) 13:124
boisduvali, Gerydus 20:131
boisduvaliana, Eurema 2:51; 14:44; 15:72; 18: 16; 21:190
boisduvaliana, Terias 12:194
bola, Panoquina 21:87, 167
bolina, Hypolimnas 11:229; 16:143
bolina nerina 22:121
bolina philippensis 18:108
Ipomea adulis
bolorica hodja, Karanasa (IA) 15:12
bombycoides, Lapara 15:193
bonfilius bonfilius, Vacerra 21:167
bonplandii bonplandii, Elzunia 15:29
bonplandii descandollesi 15:29

boopis incana, Cercyonis 17:113
boothii, Colias 2:13, 63; 7:125; 11:12
borbonica, Pelopidas 18:74
boreale, Satyrium (IA,IG) 24:83-86
borealis, Calephelis 1:70; 3:25; 6:75; 7:119-122, 133-138
Senecio obovatus
borealis, Hesperia 5:8, 9
borealis, Lephelisca 10:159; 22:35
bosqueella, Stegasta 19:134
bourkei, Thecla 7:169
bowkeri, Stugeta 10:39
bowkeri, Tarucus 10:38
BRADUS, RADIATUS *(IA,IG) 22:4, 7-10*
brahma, Ochlodes 11:168
branchus, Papilio 4:66
brasiliensis, Vanessa 3:40
brassicae, Autographa 13:26; 19:133
brassicae, Pieris 3:39, 40, 76; 8:145; 11:162, 168; 13:4; 14:53; 16:144; 17:70; 22:211, 219, 221, 223
brassicae venata, Pieris 10:220
brassolis, Liphyra 20:131, 152
BRASUS, *Oeonus (IA, IG) 22:16-19*
brebisson brebisson, Gindanes 21:162; 22:155
bredowii, Adelpha 7:147; 15:217; 18:40, 41
bredowii, Limenitis 1:42; 13:100; 17:244; 18:48; 19:113, 232; 22:128
bredowii californica 18:228; 22:96, 98, 103
bredowii eulalia 25:39
bredowii californica, Adelpha 11:44; 12:199
Quercus garryana ? 12:199
bredowii californica, Heterochroa 3:106;23:278
brennus, Bolla 25:293
brennus, Nymula 20:131
brennus brennus 21:150
brettus, Hesperia 12:122
brevicauda, Papilio 5:26; 9:141; 23:35 (IH),41
brevicauda bretonensis 4:11, 12; 9:141
Ligusticum scothicum
brevicauda gaspeensis 5:53; 9:141
brevis, Schinia 6:53
brevistriga brevistriga (II) S6:17
Phacelia distans
brevistriga ardicola S6:20
Phacelia distans subsp. *australis*
bridgei, Papilio 25:78
brigitta, Eurema (IA) 18:160

brigitta australis 23:280
brillians, Harrisina 8:11
wild and domestic grape
briseis, Catocala 24:109, 111
briseis, Eumenis 9:145
briseis, Satyrus 10:215
briseis major 10:221
brisottii, Citheronia 14:103
britomartis, Melitaea 9:204; 15:220
brixia brixiola, Caeruleuptychia 21:87, 91
brizo, Erynnis 3:48; 4:26; 9:194; 10:55; 12: 122, 130; 21:207, 277; 22:120, 168; 25:35
Tephrosia or *Indigophera sp. 4:26*
brizo brizo 13:72
brizo burgessi 20:231; 22:168
Quercus fusiformis 20:231
brizo lacustra 22:97, 99, 100, 105
brizo somnus 14:37; 4:62
brizo somnus, Thanaos 13:72
BRODKORBI, *Epargyreus (IA, IG) S2:5*
bromias, Pellicia 13:90
bromius, Pap. Pleb. Urbicol. 12:196
bromius bromius, Papilio 13:76
bromus, Pap. Pleb. Urb. 12:196
brontes, Protoparce 2:86
brontes, Phlegethontius 5:21
Tecoma stans
brontes brontes 5:22
brontes cubensis 5:22
brontes haitiensis 5:21, 22
brontes smythe 5:21, 22
brontinus bronta, Gindanes 21:164
brookeana, Troides 9:75
Aristolochia
brookiana, Troides 4:15
browerellus, Crambus 13:241
browerellus katahdini 12:51
bruceata, Operophtera 7:127
brucei, Clostera 21:35
brucei, Papilio 1:42, 75; 4:46; 10:99; 16:36
Cymopterus terebinthinus 16:36
brucei yukonensis, Oeneis 7:124; 9:141; 19:60
brumosa, Acronicta 19:189
brunneata, Dysstroma 13:218
brutus, Charaxes 13:124
brutus natalensis 18:104
Melia azedarach (toxic)

bryoniae, Pieris 11:47; 14:53; 16:5-19 (IA), 105-127; 22:223; 25:6, 9-11, 160, 161
Biscutella laevigata, Thlaspi, Arabis halleri
bryoniae pseudobryoniae 7:125
bubastus bubastus, Thecla 15:215
bubastus butleri, Madoryx 5:22
buchholzaria, Pseudoboarmia 8:3; 13:28; 19:190
buchholzi, Agrotis 24:3
buchholzi, Geltia 13:28
buchholzi, Megathymus 12:183
buchholzi, Strymon 14:22; 20:65
buddha, Calinaga 12:173
bulenta, Hesperia 12:122
bulenta, Problema 4:44
buniae, Ascia 21:156
bunichus, Battus (Parides) 22:154, 156
burchellanus, Battus (Parides) 21:87, 156
burchellanus, Papilio 4:41
burejana, Araschnia 7:37; 14:195, 204-206
burejana strigosa 8:99
Boehmeria, Urtica (Urticaceae)
burgessi, Erynnis 7:146; 13:91
burneyi catharinae, Heliconius 23:105
burnsi, Hemileuca 2:81
Spinosum ?
buena, Staphylus 21:164
busirus, Pap. Pleb. Urbicol. 12:196
busirus heros, Achylodes 21:245
busirus rioja 21:163; 22:155
butes, Diorhina 18:24
butleri, Antigius 3:80; 8:97; 13:178; 14:206
Quercus dentata, Q. crispula
butleri, Ceramidia 17:66
butleri, Dione 15:198
büttneri, Sedina 2:94
byssus, Problema 9:191-192; 12:130; 13:15; 19:77-81 (IA, II)
Tripsacum dactyloides

caberata, Sabulodes 6:94 (IA); 7:150 (II)
cacicus, Papilio 4:65
cadmus, Pap. Eq. Achiv. 12:196
caelina, Evonyme 21:102
caenia, Precis 12:25
caerulea, Caenurgina 19:165-170; 21:122-124
caesonia, Colias 10:99
caesonia, Zerene 3:25; 6:75-76; 12:28, 130; 14:238

caespitatis, Pyrgus 14:32
caestrum, Hypopta 3:75
caffrariae, Lepidochrysops 10:38
caicus, Grammodia 2:86; 25:152
caicus caicus, Cogia 21:244
caiguanabus, Papilio 4:64
cairnesi, Œneis 9:141
caja, Arctia 5:19; 18:37
calanus, Satyrium 14:20
calanus calanus 19:49
calanus falacer 16:195, 197; 18:148; 24:2; 25: 37, 81
Carya texana, Juglans microcarpa 18:148
c-album, Polygonia 6:43; 7:38; 10:215; 11:164; 14:195, 204, 206, 207; 18:40; 22:128; 23:83
c-album hamigera 8:99
Ulmus, Celtis (Ulmaceae); *Humulus* (Moraceae); *Ribes* (Saxifragaceae)
c-album hokkaidensis 8:99
(same foodplants as preceding)
c-album imperfecta 10:221
calcarius, Erebia 14:139
calchas, Cogia 20:35, 37, 38; 21:161, 245; 22:129, 155
Mimosa berlandieri; M. pigra v. *berlindieri*
calice, Castalius 25:80
california, Coenonympha 3:106; 20:161;22:98, 99, 103; 4:70
california galactinus 13:74
californica, Coenonympha 9:148
californica, Chionobas 19:203
californica, Chlosyne 7:146; 16:61; 24:226
Viguiera deltoidea v. *parishii, Helianthus annuus*
californica, Hemileuca 2:81
Populus sp., Sali- sp.
californica, Mechanitis 15:221
californica, Nymphalis 2:13; 3:17, 106; 4:13, 46; 5:26; 6:107; 7:9, 10, 15, 51, 147; 8:94, 103; 9: 15; 10:23, 97, 169; 11:164; 13:64; 15:112; 16:32, 63, 69; 17:10-20; 18:223; 19:232; 20:124; 22:98, 99, 100, 103; 23:274; 25:140
californica, Strymon 10:97; 14:20; 16:33, 69; 17:111, 245; 18:224; 22:96-100, 103
Quercus, Quercus vaccinifolia ? 16:33
caliginosa, Acosmetia 2:94

calixta, Hermeuptychia 21:91
calleta, Eupackardia 8:5; 13:215
callias, Erebia 5:63; 6:103; 14:127, 140, 147
callias, Papilio 4:67
callias callias, Erebia 7:49
callicles, Kedestes 13:146
callidus, Erynnis 14:33
callidice, Synchloe 8:144
callidice HAZARA, *Synchloe* (IA) 15:6
callidus callidus, Erynnis 13:233
callimachus, Tomares 10:117; 24:196
callina, Melitaea (IA) 19:207-211; 21:273
calliomenae, Hemeroplanes 3:78
calliomma, Amphidecta 21:87, 90
callippe, Argynnis 11:11
callippe, Speyeria 2:5; 5:31; 17:112; 19:172; 20:8; 22:98-100, 103, 172
callippe callippe 19:171
callippe comstocki 3:106; 4:70
callippe elaine 17:114
callippe gallatini 19:186
  Artemisia trifida ?
callippe inornata 17:111
callippe juba 16:37; 19:172
callippe laurina 17:111
callippe macaria 17:243-244; 25:255
callippe meadii 10:95; 14:182
callippe nevadensis 10:102
callippe rupestris 17:113
callippe semivirida 16:69; 18:222
callipteris, Lethe 8:99
  Pleioblastus (Gramineae)
calliste, Papilio 4:67
callitrichoides, Phiprosopus 2:S8
  Smilax rotundifolia
calyce, Pieris 14:48
calyce calyce, Nymula 21:150; 22:153
calydonia, Prothoe 9:74
camarensis, Diestogyna (IA) 13:137
cambes, Thecla 21:196
cambrica, Venusia 5:54; 6:43; 11:21; 13:219
camilla, Ladoga 14:205
camilla, Limenitis 7:38; 11:163; 14:196
camilla japonica 8:99
  Lonicera (Caprifoliaceae)
camillus, Phyciodes 2:92
campestris, Atalopedes 9:195; 12:31, 130; 13: 221-227; 15:110; 16:203; 17:156; 19:180; 20:

160; 21:207, 277; 22:94, 97, 99; 24:305
Stenotaphrum secundatum 13:221-227 25:213
campestris, Hamearis 21:145; 22:153
campestris, Phyciodes 8:3; 7:125; 10:170; 11:52; 16:69; 17:112; 18:223; 20:161, 245; 22:98, 99, 103
campestris camillus 10:96; 19:114; 22:172
campestris campestris 22:35
campestris montana 16:32; 17:112; 20:161
Aster; Aster integrifolius ? 16:32
cana, Mestra 25:186
canace, Kaniska 7:38
canace, Nymphalis 14:195
canace, Vanessa 11:167; 16:144
canace ishima, Kaniska 18:107
Smilax bracteata, Smilax sebeana
canace no-jaconicum 8:99
Lilium, Tricyrtis, Streptopus, Smilax
canace siphnos 18:107
Smilax bracteata, Smilax sebeana
canadaria, Melanolophia 23:4
canda, Gorgythion 21:162
candida, Lycaena 16:104
candida virgo, Eurema 22:121; 23:280
candiope, Charaxes 18:74
candrena, Diaethria 21:101; 22:152
canescana, Sciaphila 16:177
canescana, Eana (SUBEANA) 16:177
canescens canescens, Carrhenes 21:245
canida, Pieris 11:168
canidia, Pieris 16:144
canidia juba 8:97
Cruciferae
cannae, Quinta 21:246
cantheus, Satyrus 12:120
capella, Euphydryas 1:42
capronnieri, Pholus 25:152
capucinus, Asbolus 9:206
cara, Catocala 19:52; 23:5; 24:108, 110, 115
cardamines, Anthocaris 11:162
cardamines, Euchloe 14:52; 18:94
cardamines isshikii, Anthocaris 8:96
Arabis, Uarbarea (Cruciferae)
cardui, Nymphalis 16:70
cardui, Vanessa 3:17-18, 39, 40, 48, 76; 4:13, 61, 72; 5:26-27, 69; 7:51, 55, 146; 8:17-22, 99, 162; 9:16, 78, 145; 10:97, 118, 169, 171; 11:164; 12:130; 13:4, 73; 14:195, 238; 15: 251; 16:32, 144, 229-233; 17:151; 18:40, 74,

107, 160, 223; 19:172, 180, 232; 20:123; 21: 39-40, 207, 277; 22:98-99, 103, 118-120, 128, 146; 23:37; 24:157-174, 305; 25:39, 140, 147, 148, 214 *19:113*
(aberration - IA) 22:146
Arctium, Gnaphalium (Compositae); *Glycine* (Leguminosae) 8:99
Arctium lappa 18:107
Malva parviflora, Malvaceae, Compositae, Leguminosae 24:157-174
thistle, peas, round-leaf mallow, sunflowers, cockle burr, Canada thistle, hollyhock, dandelion blossoms, beets, potatoes, alfalfa, clover, chrysanthemums, petunias, *Lantana 8:19*
cardui carduelis, Vanessa 15:218
carcinia, Episcada 22:151, 156
carinenta, Libytheana 3:40; 12:171; 13:62-63; 17:37; 18:229; 21:105; 22:153
carinenta, Pap. Nymph. Phaler. 12:196
carinenta mexicana, Libytheana 14:190; 21:195; 22:230; 25:39
cariosa, Omphalocera 7:28
pawpaw
carlia, Pteronymia 22:150-151
carlsbadensis, Agathymus 13:99-100, 204-206 (IA); 23:245; S1:35
Agave parryi 13:99-100
CARMELITA, *Cargolia (IA, IG) 18:111-115*
carnearia columbiaria, Drepanulatrix 13:219
carolina, Amblyscirtes 4:44, 62; 13:71
carolina, Eucereon 11:148
carolina, Hyalophora 14:103
carolinianus, Papilio 15:211
carolus, Heliolonche 23:26-30 (IA,II,IF):25:288
Rafinesquia neomexicana
carpathicola, Erebia 13:104
carrizoi, Mathania 15:119
cartereti, Papilio 25:74-75, 77-78 (IA)
carterii, Protambulyx 13:78
caryaevorus, Strymon 4:14; 7:58; 9:105; 12:44; 14:20; 15:97; 16:195; 23:187; 24:84 (IA), 85
caryatis, Charmona 21:87, 148
carye, Vanessa 3:40; 7:53, 146; 10:97, 169; 13: 73; 16:32, 233; 18:40, 224; 19:113; 20:161; 22:98-99, 103
Lupinus ? 16:32
caryi, Œneis (IA) 11:22
cashmiriensis, Vanessa 11:167; 16:144

casica, Achalarus 25:35
cassandrina cassandrina, Elzunia 15:29
cassandrina microguttata 15:29
cassiae lucullus, Opsiphanes 21:92; 22:151
cassiensis, Parnassius 4:15
cassina fabricii, Opsiphanes 12:172; 21:192
cassioides, Erebia 14:139
cassius, Leptotes 15:216; 21:151; 22:153
cassius, Pap. Pleb. Phalor. 12:196
cassius striatus, Leptotes 13:225-227
Phaseolus vulgaris
cassius theonus 9:80, 207
cassotis, Hypoleria 21:191
casta, Crambidia 24:227
casta, Fumea 18:37
casta, Melitaea 10:117-118
castalia, Kricogonia 22:227, 229; 25:124, 125
castalia lyside 18:229
castanea, Phragmatoecia 2:94
Phragmites sp.
castigata, Eupithecia 5:54; 13:217
castolatus, Nisoniades 21:245
castolus, Nisoniades 21:161
castor, Charaxes 9:129; 13:124
catalina, Lycaena 25:240
catalpae, Ceratomia 19:189
cataphracta, Papaipema 8:58
Arctium sp.
catemacoensis, Astraptes (IA,IG) 21:116-119
catilina, Hemiargus 15:216
catillus, Pap. Pleb. Urbicol. 12:196
catillus, Chioides 14:38; 15:72
catillus albius 21:244
catillus albofasciata, Chioides 21:244
catillus albofasciatus 19:1, 29, 31, 32; 20: 35, 39, 40; 25:36
Phaseolus atropurpureus, Rhynchosia minima, Tephrosia lindheimeri
catillus catillus 21:159
catoria, Eressa 8:138
CATUA, *Myrinia (IA,IG) 22:1-5*
catullus, Pholisora 9:77, 194; 10:55, 100; 12: 31, 130; 16:69; 17:21, 155; 18:227; 19:1, 4, 31, 51, 196; 21:207, 277; 22:94, 97, 99, 102, 120, 168, 180, 231; 23:278; 24:305; 25:35,213
Malva rotundifolia 18:227
Amaranthus caudatus, A. spinosus, A. retroflexus, Chenopodium album, C. ambrosioides,

C. berlandieri 19:1-51
caunus, Polythrix 21:160
c-aureum, Polygonia 6:43; 7:38; 8:99; 14:195
Humulus (Moraceae) 8:99
cauter, "Thecla" 21:152
cavernosa, Hyssia 3:75
ceacigena, Perisomena 3:75
cecropia, Platysamia 2:1; 9:142; 10:166; 13:75, 213; 14:94, 103; 20:47, 119; 25:42, 68
cecrops, Calycopis 14:130; 16:174; 17:152; 21:207, 277
cecrops, Strymon 1:9; 3:48; 9:207, 12:130; 14:21; 16:199; 17:22, 108; 19:180; 20:102
parasites 3:48
cedreatis, Charaxes (IA) 13:124
celadon, Graphium 15:211
celadon, Papilio 4:66
celebris licus, Urbanus 15:215
celeno, Jamides 15:228; 20:134
celeris, Heliosea 6:53
celia, Amblyscirtes 13:90, 221-227; 25:34
Stenotaphrum secundaum 13:221-227
celia belli 21:207, 277
celia, Taygetis 6:98; 21:90; 22:151
cellus, Autochton 6:75; 9:193; 21:207, 277; 25:36
cellus, Eudamus 12:122
cellus, Rhabdoides 6:75
celmis, Yphthimoides 21:91; 22:151
celmus, Thecla 18:23
celsiae, Cucullia 3:75
celtis, Apatura 12:122; 13:212
celtis, Asterocampa 3:17, 76; 4:13, 14: 6:33; 7:146; 9:78; 10:55; 12:130; 14:190; 19:180; 21:137, 207, 277; 23:192, 193; 24:305; 25:140
celtis alicia 12:36; 17:152; 22:35
Celtis laevigata 12:36
celtis antonia 19:113
celtis celtis 22:35
celtis, Libythea 3:75; 7:37; 14:196, 204
celtis amamiana 18:107
Celtis boninensis
celtis celtoides 4:43; 8:98
Celtis (Ulmaceae)
centaureae, Pyrgus 5:8, 19
centaureae freija, Pyrgus 5:9; 7:50; 10:107; 18:228; 19:242

centaureae loki 10:100
centaureae wyandot 24:45
centaurus, Narathura 20:135
centralis, Philotes 8:8
centuriella, Scoparia 5:29
ceos, Staphylos 13:232
cephalus, Papilio 4:65
cepheus, Acraea 13:146
cephise, Cephise 24:69
cephise hydarnes 21:160
cerasana, Pandemis 13:6, 8
cerasivorana, Archips 10:154
ceratomioides, Xylophanes 3:78
ceraunus antibubastus, Hemiragus 9:208; 22:35
ceraunus gyas 19:112
ceraunus zachaeina 18:154-155; 21:151, 196; 22: 153; 25:38
Acacia angustissima v. *hirta, Rhyncosia minima* 18:154-155
cercocarpana, Epinotia 25:119-123 (IA,IC,IG)
Cercocarpus ledifolius, Arbutus menziesii, Arctostaphylos sp.
CERCYEUPTYCHIA 25:13-19
cerea, Emesis 21:87, 149
ceres, Pap. Helicon 12:196
ceres, Euphaedra (IA) 13:128
ceres, Lycorella 9:207; 24:298
ceres atergatis, Lycorea 21:191; 22:108
ceres ceres 21:93; 22:151
ceres cleobaea 5:56
ceriales, Quadrus 21:162
cerina, Papaipema 8:58
Lilium superbum, Silene stellata, Podophyllum peltatum, Hystrix patula
cerisyi, Smerinthus 15:193
cerisyi saliceti 5:22
cernes, Hesperia 12:122
cerogama, Catocala 12:201; 24:108, 111, 115
cerussata, Papaipema 8:58
Vernonia noveboracensis
cervinaria, Synaxis 13:220
cerymica, Zophopetes 18:74
cesonia, Pap. Dan. Cand. 12:196
cesonia, Colias 6:75, 76; 13:224-227; 15:100, 198; 17:154; 18:149; 19:34 (IA), 110, 111, 232; 21:207, 277; 22:253 (IA); 23:182, 187, 251-253 (IA); 25:37
cesonia, Zerene 14:44; 15:72

cespitana, Celyphoides 19:134
cestri, Strymon 14:22
cestri, Thecla 13:90
cestus, Atrytonopsis 14:28
cethura, Anthocaris 12:194; 21:272
cethura cethura 12:28
chabrias, Papilio 4:41
chalcedona, Euphydryas 8:3; 10:35; 12:33, 165; 22:98-100, 103
chalcedona, Melitaea 11:11; 20:8
chalcedona colon, Euphydryas 18:223
Penstemon subserratus, Symphoricarpos albus
chalcedona dwinelli 17:113
chalcedona hermosa 17:104
chalcedona klotsi 15:220; 17:104
chalcedona macglashanii 16:36
chalcedona quino 12:29
chalcedona sierra (IA) 9:174, 175; 15:220; 16:36; 17:112
chalestra chalestra, Xeniades 21:167
chalco, Urbanus 22:155
chamberlaini, Eurema 18:31
chamissonia, Papilio 4:41
CHAMULI, *Euphyes* (IA,IG) S2:43
CHAPA, *Cymaenes* (IA,IG) 22:10, 13, 14
chapada, Panoquina 21:87, 167
chapada, Pellicia 21:161
chara, Melitaea 9:148
chariclea, Boloria 5:8, 26; 7:124; 19:59
chariclea boisduvalii 5:8
chariclea butleri 19:58
charithonia, Papilio 12:119
charithonia, Heliconius 3:25
charitonius, Heliconius 8:102; 9:205; 13:30, 62; 15:71; 21:248; 22:34; 25:186
(abarrant, IA) 21:248
charitonius churchi 5:56
charitonius simulator 9:80
charitonius tuckeri 25:135
charitonius vasquezae 18:19; 21:193; 22:159; 25:40
charlonia, Anthocaris 4:72
charlonia, Euchloe 10:215, 220
charlonia levaillanti 10:220
charlotta, Argynnis 7:37
charlotta, Fabriciana 9:145
charlotta, Mesoacidalia 9:204; 10:217; 11:163; 14:194, 206

charlotta basalis, Argynnis 8:98
Viola (Violaceae)
charlotta fortuna 8:98
Viola (Violaceae)
charltonius voigti, Parnassius (IA) 15:3
charon, Amphidasis 18:111
charonda, Sasakia 7:38; 8:99; 14:196
Celtis (Ulmaceae) 8:99
charops, Pereute 21:189
charybdis, Ethmia (II) S6:31
Amsinckia tessellata, A. intermedia
chela, Cymaenes 21:165
chersis, Oechydrus 21:164
chersis, Sphinx 15:193
chersis mexicanus 5:22
chiansiades, Papilio 4:64
CHIAPA, *Aethilla (IA, IG) S2:13*
CHIAPAENSIS, *Mimia (IA, IG) S2:15*
childrenae, Papilio 4:41
childreni, Argynnis 11:167
chilica, Thecla 21:153
CHINATIENSIS, *Agathymus (IA, IG) 18:172-185; S1:45*
Agave lecheguilla
chione, Nymphidium 21:150
chiridota, Lacosoma 18:89
chiridotum, Lacosoma 21:34
chiron, Eumedonia 9:204; 14:137
chiron, Marpesia 5:56; 9:207; 15:199; 17:37; 18:22; 21:99, 195; 22:152; 25:39
chiron, Megalura 3:40
chiron, Xylophanes 9:136
chiron chiron 5:22
chiron cubanus 5:22
chiron lucianus 5:22
chiron neccus 2:86; 3:78; 5:22; 25:152
chisosensis, Agathymus 13:91; S1:37
CHISOSENSIS, *Apodemia (IA,IG) 18:75-77*
chisosensis, Megathymus 8:77
chitone, Speyeria 5:33
chloe rhea, Hamadryas 21:87, 102
chloeropis, Euryphene 13:128
chlorargyra, Dyops 16:54
chlorinde, Anteos 21:190
chloropha, Caenurgia 17:226
chonida, Thecla 21:183
christina, Colias 2:13; 7:125
Hedysarum sp.

chromus, Hasora 17:110; 18:162 (IA)
chromus inermis 18:105
Pongamia pinnata
chrysalus, Hypaurotis 7:147; 8:28; 13:164
Quercus gambelii 13:164
chrysidipennis, Ramosa 24:189
chrysippe, Euselasia 21:196
chrysippus, Danaus 10:221; 11:167, 229; 13:146; 16:131-134 (IA); 18:74, 107, 163; 25:80
chrysogone, Taygetis 6:98
chrysomelas, Colias 19:39
chryson, Phytometra 2:94
Eupatorium cannabinum, Salvia glutinosa
chrysophrys, Mnasitheus 21:246
chrysotheme, Colias 2:25; 5:25; 15:213
chryxus, Œneis 6:103; 10:94; 11:22-26 (IA); 14:127; 16:68; 18:227; 20:161; 23:155-160
chryxus caryi 11:47, 49; 19:57
chryxus ivallada 17:112; 19:174
chryxus stanislaus 17:112; 19:173
chryxus strigulosa 2:34; 9:141; 12:39
chryxus VALERATA *(IA, IG) 11:23-26*
chydaea, Conga 21:166, 247
ciliata, Schinia 6:53
cilissa, Lemonias 21:197
cillene, Pap. Helicon. 12:196
cindra, Urbanus 21:159
cinerea, Pleroma 2:42
cinereus, Prodoxus 20:54
Yucca whipplei
cingulata, Herse 2:86; 3:78; 13:78; 10:112; 19:189; 25:152
cingulis, Phaenochitonia 21:87, 149
cinnamomea, Miltomiges 22:156
cinnamomea, Phoegoptera 21:273
cinnara, Borbo 18:105
grasses (Gramineae)
cinxia, Melitaea 9:145; 13:207 (IG); 14:194
cinxia amardaea 10:118
cinxia atlantis 10:221
circe, Aulocera 9:145
circe, Brintesia 9:204
circe, Kanetisa 14:140
cipris, Pap. Dan. Candidi 12:196
cirsii, Pyrgus 9:145
cissusa, Thecla 21:153
cithaeron, Charaxes 13:124
Cithearis 24:15, 16

citrata, Dysstroma 13:218
citrinellus, Oxylos 6:53
cladonia, Phaeoura 18:111
clarus, Pap. Pleb. Urbicolae 12:196
clarus, Epargyreus 6:75; 7:52, 146; 9:77; 10: 55, 100; 12:130; 13:100, 232; 14:216; 15:98, 108; 17:22; 18:226; 19:133-137, 180, 196; 21: 277; 22:120, 207; 23:192, 194; 24:305
Robinia pseudoacacia 17:22
clarus clarus 17:155; 19:1; 20:102; 25:36, 81
Robinia pseudo-acacia, Wisteria sinensis 19:1
clarus, Proteides 3:48; 6:75; 9:194
parasites 3:48
clathrata, Chiasma 18:110
claudia, Pap. Nymph. Phalerat. 12:196
claudia, Euptoieta 7:51, 52, 147; 9:148; 10:23, 55, 95, 107; 12:130; 13:99, 100; 17:23; 19: 104, 114, 232; 21:207, 277; 22:128, 228, 229; 23:182, 188; 24:305; 25:40, 138, 141, 214
claudia claudia 17:150
claudia godmani, Agrias 21:87, 104
claudina guatemalena, Phyciodes 18:19; 21:193
clavicornis clavicornis, Epargyreus 21:159
clavicula, Vehilius 22:156
clavus, Damas 20:227
clavus, Mellana 21:167
claxon, Aguna 25:293 (1st US record)
claytoni, Lycaena 16:95
cleanthes cleanthes, Thracides 21:167
clearista, Athesis 25:149
clearista clearista 14:26
clearista colombiensis 15:26
clearista vitrala 15:26
clelia, Precis 18:74, 25:80
clemataria, Abbotanna 23:206, 207
CLENCHI, *Virga (IA, IG)* 24:122-124
cleobaea, Heliconia 12:120
cleobaea zorcaon, Eueides 18:19
cleobaea zorcaon, Heliconius 21:192
cleomes, Pieris 12:122
cleopatra, Gonepteryx 9:145
cleopatra mauretanica 10:220
cleotas, Papilio 4:65
cleothera, Danais 12:120
clercana, Eana 16:179
clinias, Eracon 21:164
clio, Phyciodes 21:193

clito, Clito 21:164
cloanthus, Zetides 16:144
clodius, Parnassius 4:16; 10:171; 13:74; 15:
210; 18:188, 220; 22:34
clodius baldur 16:30; 17:10
Sedum 16:30
clodius sol 17:113
clonius, Pap. Pleb. Urbicol. 12:196
clonius, Spathilepia 13:90; 21:160, 244; 22:155
clorinde, Anteos 13:62; 14:44; 15:72, 198; 21:
155; 22:154
clorinde, Colias 12:120
clorinde nivifera, Anteos 14:156; 22:34; 25:37
clymena, Pap. Nymph. Phalerat. 12:196
clymena, Diaethria 9:209
clymena janeira 21:101; 22:152
clymene, Haploa 21:38; 24:271
clymene, Pararge 3:75
clytemnestra forbesi, Anaea (Hypna) 21:104, 88
clytemnestra hubneri 22:153, 156
clytie, Strymon 14:22
clyton, Asterocampa 4:13; 6:33; 9:77; 12:122,
130; 13:212; 19:180; 21:137, 207, 278; 22:
120; 24:305; 25:140; 23:192, 193
clyton clyton 17:152
clyton flora 4:44; 6:33; 9:206; 12:36; 22:35
Celtis laevigata 12:36
clyton louisa 6:33
cnidia, Colias 12:120
cnejus, Euchrysops 18:106
Vigna marina, Phaseolus angularis, Crotalaria
coagulata, Eupithecia 13:218
cocala viola, Limenitis (Adelpha) 21:98
coccinata, Catocala 10:168; 12:201; 24:109, 111,
114
cocyta, Pap. Nymph. Phaler. 12:196
codrus auratus, Graphium (IA) 25:74, 75
coelebs, Catocala 4:12; 12:202
Myrica gale
coelodactyla, Zellerina 11:178
coelus, Aguna 25:293
coelus, Papilio 4:41
coenia, Junonia 6:38; 8:1; 10:44; 14:186; 19:
133, 137; 21:207, 277; 22:98, 99, 103, 120,
162; 24:305; 25:138, 140
coenia coenia 17:23; 21:59-70 (IA); 22:35; 25:
39, 127
coenia nigrosuffusa 19:113

coenobita, Panthea 3:75
coenobita aino, Neptis 8:99
Spiraea (Rosaceae)
coenobita insularum 8:99
Spiraea (Rosaceae)
coenosa, Laelia 2:94; 3:75
Phragmites sp.
coenus albidula, Dynamine 21:100
COENYROPSIS *12:6*
cofaqui, Megathymus (I) 8:153-154, 160; 9:204; 12:186; S1:22
Yucca filamentosa, Yucca aloifolia
cogina, Plebejus 21:151
cognataria, Glena 19:190
cognatus, Favonius 14:206
cognatus jozanus 13:180
colchis, Hamearis 21:87, 145
Colias 25:164
collina, Pseudolucia 21:184
colon, Euphydryas 16:70
colon stinga, Telicota 18:105
Miscanthus sinensis, Microstegium sp.
coloradalis, Orenaia 13:241
coloradensis, Itame 13:219
coloradensis arizonae, Megathymus S1:19
coloradensis browni S1:15
coloradensis coloradensis S1:13
coloradensis elidaensis 25:213; S1:14
coloradensis kendalli 25:34; S1-22
coloradensis louiseae S1:22
coloradensis martini S1:17
coloradensis maudae S1:18
coloradensis navajo S1:14
coloradensis reinthali S1:16
coloradensis reubeni S1:19
coloradensis stallingsi S1:16
coloradensis wilsonorum S1:21
coloradensis winkensis S1:20
colorado, Hesperia 14:30, 37
colombati, Satyrus 10:215, 221
colossus, Astraptes 20:227
colossus rhoda 21:160
colotes transvaaliae, Spialia 13:146
colubra colubra, Ancyluris 21:87, 147
columbaria distigma, Onophas 22:156
columbia, Glaucopsyche 13:74
columbia, Hesperia 13:239; 22:94, 97, 99, 100, 102, 130

columbia, Hyalophora 13:213; 25:68-73
columbia, Samia (Platysamia) 8:2
columbiata pernigrata, Hydriomena 13:218
columbina, Argynnis 12:120
columbus, Papilio 4:67
columella, Strymon 9:206; 12:30; 14:20; 15:215; 17:226
columella istapa 22:228, 230; 25:38
combinata, Polites 14:30
comes, Thecla 3:79
comes, Triphaena 25:161
comis, Lacinipolia 8:101
cultivated clover (dry foliage)
COMITANA, *Niconiades* (IA, IG) S2:26
comma, Hesperia 5:26; 9:145; 10:170; 11:52, 166; 18:188; 22:170
comma, Polygonia 6:73; 9:78; 10:55; 12:130; 15:99; 19:48, 180; 21:207, 277; 22:35; 23:100, 193; 24:305; 25:40, 141
comma, Vanessa 19:205
comma borealis, Hesperia 19:243; 23:41 (IA)
comma colorado 10:100
comma florinda 8:95
Carex (Cyperaceae)
comma hulbirti, Hesperia 18:228
comma manitoba 18:228
comma ochracea 14:30
comma ruricola 14:30
comma yosemite 14:30
comma comma, Melinaea 15:32
comma isocomma 15:32
comma simulator 15:32
commixtalis, Loxostege 13:242
commoides, Leucania 9:56
communis, Pyrgus 4:13; 7:52, 146; 9:194; 10:100; 12:130; 13:99, 100; 14:37; 16:35, 70, 203; 17:10-20, 112; 18:227; 19:91-93, 105; 21:207, 277; 22:94, 97, 98, 102, 179, 180; 24:305; 25:213
sphaeralcea lindheimeri 19:105
communis albescens 12:31
communis communis 17:21, 155; 19:1, 10, 32; 20:102; 25:35
Althaea rosea, Callirhoe leiocarpa, Sida filicaulis, Sida rhombifolia, Sida sp., Sphaeralcea angustifolia, Sphaeralcea cuspidata, Sphaeralcea lindheimeri

communis orcynoides 21:164
comnena, Perisama (IA, IG) 24:10-13
complana, Eilema 8:94
complana, Eutychide 14:31; 21:246
compressipalpis, Plusiodonta 13:26
comstocki, Agathymus 20:181; 23:246, 247, 248; S1:52
comstocki, Callophrys 21:114; 24:228
comstocki, Megathymus 11:115; 12:26
comus, Euryphene (IA) 13:128
comyntas, Polyommatus 12:120
comyntas, Everes 9:77; 10:55, 107, 112; 12:130; 13:99, 100; 14:130, 215; 16:136; 18:24; 19: 180, 232; 20:134; 21:135, 207, 277; 22:96, 97, 100, 103, 120, 122; 24:196, 305
Lespedeza texana (Leguminosae) *22:122*
comyntas amyntula 16:34
Astragalus whitneyii
comyntas comyntas 17:23, 153; 20:102; 21:196; 22:35, 129; 25:38
comyntas herrii 19:113
comyntas texanus 22:129
comyntas valeriae 22:165
Astragalus flexuosus ?
concinna, Schizura (II) 7:148
concinusella, Battaristis 19:134
concisa, Epizeuxis 5:54
concolorella, Eriphia 15:81
concolorella, Ithome (IA, IG) 15:84-88
Prosopis chilensis
concolorella, Periploca 19:134
concumbens, Catocala 12:201; 23:5, 263; 24:108, 110, 113, 115
condigna, Anadelosemia 24:254
confinis, Eressa 8:135
confinis intensa 8:139
confinis malaccensis 8:136
confinis SZECHUENIENSIS, *Eressa 8:136*
confusa, Lethe 11:167
CONFUSA, *Panoquina (IA, IG) 22:16, 19, 20*
confusis, Thorybes 4:62; 21:207, 277; 24:305
congolensis, Euryphene (IA) 13:128
conjugaria, Epimecis 5:59
conjuncta, Pelopidas (IA) 18:162
connubialis, Catocala 14:177
connubialis pulverulenta (IA) f. BROWERI *14: 177-178*

conradti, Eresina (IA) 13:138
consors, Catocala 10:168; 25:42
conspicua, Actinote 22:151, 156
conspicua, Atrytone 9:40, 78; 10:55; 12:42, 44; 13:16; 14:215
conspicua, Euphyes 23:186
conspicua buchholzi, Atrytone 13:28
conspicua conspicua, Euphyes 22:130
conspicua buchholzi 14:28
consimilis pedia, Neptis 22:121
consulis, Melitaea 10:118
consulis mea 10:117
CONTA, *Cynea (IA, IG) 22:16-19*
contorta, Melipotis 17:226
contracta, Datana 18:89
contrastata, Anthene 13:145
contubernalis contubernalis, Quadrus 21:245
conversa, Catocala 3:76
convolvuli, Herse 3:76
coön, Atrophaneura 25:164
coön delianus, Papilio 9:75
Apama tomentosa
coquillettella, Ethmia (II, IF) S6:8
Phacelia distans ?
corallina, Phlegethontius 3:78
coras, Papilio Urbicola 12:196
coras, Polites 19:196; 21:207, 277; 22:170; 24:305
corbulo, Pyrdalus 21:164
cordiaria, Eilicrinia 3:76
cordigera, Anarta 18:236
core, Euploea 3:39; 11:229; 16:144
core corinna 22:121
coresia, Nymphalis 12:120
coresus, Automeris 14:103
coreta, Euploea 3:39
coridon, Agrodiaetus 14:138
coridon, Lysandra 9:145; 20:134
cornelia, Dismorphia 15:199
cornelius, Hesperia 12:120
cornifrons, Ceratonyx 21:13
coronado, Dichogaster (II) 11:99-102; 18:164
Quercus (lifeoak)
coronis, Speyeria 2:5; 5:31; 10:102; 12:203; 15:221; 17:112; 22:98-100, 103, 104
coronis halcyone 10:95; 14:180
coronis semiramis 4:70
coronis simaetha 16:69; 18:222

coronis snyderi 16:31
corrivalaria, Acidalia 2:94
palustrine plants
corrosus, Antigonus 21:245
corticea corticea, Corticea 21:246
corusca, Oxynthes 21:167
coruscans, Neozephyrus 3:79
cos, Euxoa 3:75
cosa, Atlides 22:153
costa, Neperigea 24:3
costata, Holomelina 19:75
costata pallipennis 19:75
costiguttata, Perizoma 13:219
costimacula, Pellicia 13:90; 14:33
costimacula costimacula 21:161
cottiana, Eana 16:178
cottoni, Euryphene (IA) 13:137
cotytto, Pteronymia 18:15-17; 21:191
COVADONGA, *Pheraeus (IA, IG) S2:28*
crabroniformis, Sphecia 3:76
crambusa,"Thecla" 21:153; 22:153
cramera calida, Plebeius 10:222
crameri, Euchloe 14:52
crameri, Erynnis 2:86; 25:152
crameri, Typhedanus 13:66
crassus, Battus 21:156
crassus, Papilio 4:41
crataegi, Aporia 7:37; 9:145, 203; 11:162; 14: 52; 18:41; 18:110; 22:126
crataegi adherbal 8:97
Chaenomeles, Crataegus, Malus, Pyrus, Prunus (Rocaceae)
crataegi mauretanica 10:220
crataegi, Catocala 12:201; 23:5; 24:108, 111
crataegi, Polyommatus 12:122; 13:212
crathis crathis, Olyras 15:27
crathis montagui 15:27
crathis staudingeri 15:27
crathis sticheli 15:27
crathis weeksi 15:27
cratios, Anaea (Memphis) 21:105
crawshayi niloticus, Argiolaus 13:145
crawshayi niloticus, Iolaus (IA) 13:145
crebus, Papilio 9:75
crenana, Epinotia (IA,IC,IG) 25:115, 119-122
Salix sp.
crenda, Chiomara 21:163

crenulata, Orthodes 2:S8
Taraxacum officinale
creola, Lethe 12:46; 17:149; 24:143-151 (IA); 25:145, 146, 256, 259
Arundinaria tecta ?
creon creon, Pyrrhopyge 23:128
creon LILLIANA *(IA,IG) 23:127-130*
creona, Anapheis 3:39
creona infida, Glycesthes 18:74
creona severina 18:74
crepuscularia, Boarmia 6:93
crepuscularia, Ectropis 4:26; 8:4
Lupinus sp.
cresphontes, Pap. Eq. Achiv. 12:196
cresphontes, Papilio 3:17, 25; 4:13, 14, 63, 64; 9:143, 206; 10:55, 164; 12:130; 13:74, 195, 199; 15:71; 17:37; 18:134-135; 19:109; 21: 207, 208, 277; 22:34; 24:305, 239; 25:213
Xanthoxylon sp. 4:14
Citrus, Ptelea trifoliata, Ruta graveolens, Zanthoxylum fagara, Z. Clava-herculis, Z. Clava-herculis v. *fruticosum 18:134-135*
cresphontes forsythae, Papilio 6:76
cresphontes cresphontes 17:22, 153; 25:36, 129-130, 133-136
Zanthoxylum fagara, Z. americanum
cressida cressida, Cressida 22:121
cressida cressida, Eurycus 16:100
Aristolochia pubera, A. elegans (toxic to)
cretaceata, Eupithecia 13:218
cretea, Opstega 17:237-242; 20:232-234 (IA)
creteus siges, Astraptes 21:160
creton, Adelpha 24:294-296 (IA, IG)
Quercus ?
creusa, Dasyophthalma 21:92
creusa, Euchloe 6:105; 7:146; 16:139; 18:221; 19:112; 22:166
creusa hyantis 16:30
Cruciferae
creusa lotta 3:48
crocala tepici, Phlegethontius 5:22
crocale, Chlosyne 13:165
crocale, Catopsilia 3:39; 6:69; 13:22; 18:106, 162
Cassia siamea, Cassia fistula 18:106
crocataria, Xanthotype 13:75
crocea, Colias 9:145; 11:163
crocea, Heliochroma 22:189, 190

croceus, Colias 3:39; 7:139; 8:76; 10:117, 221; 11:12; 13:24; 14:52; 15:43, 45; 22:127
Medicago sativa 15:45
cronion, Sostrata 21:162; 22:155
crotonis, Xylophanes 3:78
crotopus, Pap. Pleb. Rural. 12:196
cribraria, Orthostixis 3:76
crison, Ridens S2:12
crispus, Mictris 21:164
critola, Strymon 14:22
critola critola, Hypostrymon 21:183
cruralis, Zanclognatha 23:206, 207
crysalus, Thecla 10:209
crysalus, Hypaurotis 15:227; 16:200; 19:55-56, 112, 232; 20:135, 160; 22:129, 132
crysalis citima 22:172
cubensis, Phlegethontius 5:21
Bignoniaceae ?
cuculla, Lophopteryx 3:76
cunea, Hyphantria 7:55; 25:42
cuneana, Amorbia 10:154
cuneifer, Troides 9:75
cuparina, Metacharis 21:87, 147
cupentus, Tmolus 21:152
cupes, Schinia 25:109, 110
cupes deserticola 25:109-114 (IA, IF, II)
Oenothera clavaeformis
cuprescens, Dyops (IA,II,IG) 16:47-54
cupressana, Laspeyresia 23:278
cupressi, Isoparce 9:211; 13:78
cupreus, Lycaena 16:34; 17:112; 18:49
Calyptridium umbellatum 16:34
cupreus snowi 22:172
cupreus snowi X Chlosyne acastus 16:235
curassavica, Wallengrenia 14:29
curiosa, Noctasota 11:215
curitiba, Dirphia 14:103
curius, Lamproptera 9:74; 21:50
curvata, Semiothisa 13:219
curvifascia, Notocrypta 8:95
Alpinia (Zingiberaceae)
curvifascia curvifascia 18:105
Alpinia speciosa, A. intermedia
curvifascia yaeyamana 18:105
Alpinia speciosa, A. intermedia
curvilinea, Epimecis 5:59
cuspidea, Euclidina 18:169
cutorina, Papilio 4:41

cutteri, Euryphene (IA) 13:128
cuvieri, Libythina 21:87, 102
cyananthe, Myscelia 12:194
cyanargyria, Dyops 16:54
cyanecula, Myscelia 12:194
cyanescana, Eana 16:178
cybele, Speyeria 1:58; 2:5; 5:53; 9:77; 10:55, 107; 11:163; 12:41, 130; 14:215; 15:23; 16: 128, 243; 20:44; 21:207; 22:120, 128; 24: 305; 25:81, 82, 141, 196, 197
cybele carpenterii 22:160; 25:47
cybele charlottii 10:95, 102; 14:180; 22:172
cybele cybele 22:35
cybele leto 16:31, 69; 17:113; 18:222
  Viola
cybele novascotiae 9:141
cybele pugetensis 17:116
cydno galanthus, Heliconius 21:192
cydrara, Tmolus 21:152
cyllabarus, Narope 21:87, 88, 92
cyllarus, Narope 22:151, 156
cyllarus aeruginosa, Glaucopsyche 10:117
cyllastros, Narope 21:92
cymbalariae, Omia 3:75
cymela, Euptychia 6:32; 10:55, 107; 12:43; 14: 209; 15:93, 99; 16:243; 21:207, 277; 22:120, 127, 132; 23:100; 24:305
cymela, Megisto 12:130
cymela, Pap. Pleb. Rural. 12:196
cymela cymela, Euptychia 17:23, 149; 18:137-138; 19:47; 22:34; 25:41
  Stenotaphrum secundatum 18:137-138
cymela viola 22:34
cynea, Cynea 21:246
cynira, Lonomia (II) 21:138-140
  Robinia pseudo-acacia, Fagus sylvaticus, Carpinus betulus, Quercus ilex
cynorta, Papilio 13:76
cynthia, Samia 5:16; 14:103
  Ailanthus sp., *Lauraceae*, castor bean
cyparissa aurata, Euphaedra (IA) 13:128
cypria, Arcas (IA,IC,IG) 25:89, 92, 94, 96, 99, 100, 104
cypris, Phoebis 23:253
cypris mayae, Argynnis 6:38
cyrriana, Thecla 21:183
cytherea herennia, Limenitis 21:99; 22:152

cytherea marcia 18:21; 21:195
cytheris, Adelpha 4:67

dabanensis, Erebia 11:50
dacela, Caenides 13:146
dactylina, Acronicta 24:3
daedalus, Papilio 13:23
daedulus, Hamanumida 9:129
daeta, Hypothyris 21:95; 22:151
dahli, Diarsia 3:75
daira, Eurema 2:51; 3:25; 15:72, 198; 18:16; 22:127; 23:182
daira, Pieris 12:120
daira biskrensis, Teracolus 10:220, 221
Capparis
daira daira, Eurema 6:75; 9:208; 10:204-206; 17:154; 21:190; 24:240, 241
daira lydia 15:114
daira nouna, Teracolus 10:220, 221
Capparis
daira palmira, Eurema 6:77; 9:210; 24:240, 241
daira thruppi, Colotis (IA) 13:138
daisetsuzana, Oeneis 8:99
Cyperaceae
dalman, Hesperia 12:120
dalman, Thespieus 21:167
dama, Agroiaetus 14:137; 16:144
damaris, Eurema 2:51
damaris, Terias 12:194
damastus, Polyommatus 12:120
dammersi, Lathosea 2:42
Ericameria palmeri
dammersi, Philotes 8:8
damoetas, Chlosyne 14:186; 22:162
damoetas malcolmi 17:112
damoetus, Microtia 22:172
damon, Agrodiaetus 14:137
damon, Mitoura 1:58, 70; 2:22; 3:48
damon, Pap. Pleb. Rural. 12:196
danae, Pap. Nymph. Phalerat. 12:196
danava, Limenitis 16:144
danieli, Boarmia 2:94
danna, Taractrocera 11:168; 16:144
daphne, Brenthis 9:145; 14:205
daphne daphne 14:193
daphne rabdia 8:98; 14:193
Sanguisorba (Rosaceae)
daphnis, Meleageria 14:137

daphnis, Pap. Nymph. Phalerat. 12:196
daplidice, Pieris 4:72; 14:53
daplidice, Pontia 3:39; 4:43; 10:117, 215; 11:162
daplidice nitida, Pontia 10:220
dardania, Cargolia 18:115
dardania, Hasodima 18:115
dardanus, Papilio 4:41; 7:4; 10:47; 13:216; 17:176; 24:72; 25:164, 273
dardanus cenea (IA) 10:50
dardanus polytrophus 7:4; 13:76
dardus misellivestis, Eurybia 21:146; 22:153
darius darius, Dynastor 21:92
darwiniata, Nemoria 13:217
daunus, Papilio 1:42; 4:64; 12:101
daunus, Pap. Pleb. Urbicol. 12:196
daura, Oeneis 3:25
dawsoni, Agathymus 23:246-248; S1:52
debora, Eumaeus 14:130; 20:135
DECODES 14:113-114
decolorata, Everes 14:136
decora, Pyrgus 12:49
defectaria, Anacamptodes 2:58
  Prunus serotina
defensaria, Xanthorhoe 13:219
definita, Anthene 10:38
definita, Melitaea 13:99
degeerella, Nemophora 23:217
  Anemone remorosa
deidamia, Pararge 14:203, 206, 207
deione, Melitaea 9:204
deione, Mellicta 9:145
delaware, Atrytone 21:207, 277
delaware, Hesperia 17:103
delaware lagus, Atrytone 22:170; 25:34
delessertii, Papilio 9:74
deleware, Atrytone 20:177
delia, Eurema 10:204-205
delia, Pap. Dan. Cand. 12:196
deliana, Eressa 8:137
DELPHIA, *Arcas* (IA, IG) 25:90-94, 101-104
delphica, Chresmarcha 10:77
delphinii, Euchlea 4:26
  Cornus florida
delphis niveus, Eulepis 24:303
dejecta, Catocala 24:109, 110, 114
demeter, Lycorea 12:194
demodocus, Papilio 9:129; 13:76; 18:74, 104, 170; 25:271-274

demoleus, Papilio 3:39; 11:229; 18:104, 163; 22:187
demoleus demoleus (IA) 16:131-134
demoleus sthenelus 16:100
Psoralea tenax; Citrus (toxic to)
demolion, Papilio 21:50
demophile calydonia, Itabalia 21:189
demophon, Prepona 12:172
demophon antimache 21:104
demophon extincta 21:104; 22:153
demylus demylus, Patricia 15:27
demylus gemellus 15:27; 25:149
dennisi, Speyeria 5:34
denticularia, Synchlora 22:37
denticulata, Semiothisa 13:219
dephis, Charaxes 9:76
depicta, Annaphila 3:74
Nemophila menziesii
dercyllidas, Patricia 25:149
dercyllidas dercyllidas 15:27
dercyllidas hazelea 15:27
derivana, Eana 16:178
dermaptera, Myrina 10:39
dero, Dircenna 21:94; 22:151
deserticola, Melitaea 10:221
desfontanei, Euphydryas 10:215
desfontainei gibrati, Euphydryas 10:221
designata emendata, Xanthorhoe 5:54; 13:218
destinata, Lygris 25:264
Salix, Populus
destinata bowmani 13:218
detexta, Epimecis 5:59
detracta, Mamestra 23:7
deva, Eurema 21:154; 22:154
deva, Pratapa 20:135
devia, Eupsilias 2:42
devilliers, Papilio 4:41
devilliersi, Battus 15:210
dia, Boloria 18:110
dia, Clossiana 9:145
diabloensis, Agathymus S1:36
diamina, Melitaea 9:204; 14:194
diana, Lethe 7:37; 8:99; 14:139
Pleioblastus (Gramineae) 8:99
diana, Pap. Nymph. Phaler. 12:196
diana, Speyeria 1:42; 2:5; 3:74; 4:72; 5:70; 6:33; 13:93;15:23;16:174;17:148,150;22:35; 23:278
Viola cornuta 13:93

dianassa, Eueides 13:84
diaphanella, Lithocolletis 25:204
dichroa, Sephisa 11:167; 16:144
dichrous, Theagenes 21:163
dicksoni, Thestor 10:39
dicoma, Phyciodes 21:97
didyma, Melitaea 9:145, 204; 10:215; 14:194
didyma interposita 10:221
didyma mauretanica 10:221
didyma occasus 10:221
diffinis, Hemaris 4:12; 10:166; 18:89; 19:189; 25:41
Symphoricarpos orbiculatum 10:166
diffissa tropicalis, Phlegethontius 3:78
diffundaria, Epimecis 5:59
dilucida, Eutresis 15:26
diluta, Cynea 21:166
dimidiata, Pellicia 21:245
dimidiata zamia 21:161
dina dina, Eurema 2:51
dina helios 16:246
dina westwoodi 2:51; 21:190; 22:34
diocles, Adelpha (IA, IG) 24:295, 296
diocles, Deudorix 10:39
diodorus, Battus (Parides) 21:87, 156; 22:154
diogenia, Emesis 21:149; 22:153
dion, Atrytone 9:192; 12:41, 43; 15:98
dion, Euphyes 23:186; 24:243, 244
dion alabamae, Atrytone 6:76
dione, Hypanartia 21:194
diores, Theope 18:25
dioszoghyi, Psodos 3:76
dioxippus, Papilio 4:67
dipoea, Dodona 11:167; 16:144
diraspes, Dalla 21:164
dirce, Colobura 21:104; 22:153
dirce, Gynaecia 12:172; 18:22; 21:195
dirce wolcotti, Colobura 5:56
disa, Erebia 5:42; 7:124; 11:47; 14:140; 18:228; 19:57
disa, Œneis 11:48
disa mancina, Erebia 16:243
disa mancinus 19:94
disa subarctica 9:141
disaffecta, Yphthimoides 21:91
discoidalis, Erebia 5:42; 11:47; 15:95; 19:57, 76, 125; 22:252; 24:43
discoidalis discoidalis 21:136

discoidalis, Œneis 11:48
discoloralis, Renia 23:206
discospilata, Eufidonia 8:3
discostrigella, Ethmia S6:46
  Cercocarpus ledifolius
disippe, Nymphalis 12:120
dispar, Lymantria 7:7-8; 9:146; 15:240; 17:176, 18:37
dispar, Thersamonia 11:165
dispar rutilus, Lycaena 8:128
dispuncta, Semiothisa 13:219
dissimulatrix, Actrix 24:255
dissitana, Archips 10:154
dissoluta, Archanara 2:94
disstria, Malacosoma 12:50; 18:150; 25:43
distincta, Boloria 11:142-146 (IA); 15:221; 19:62
distincta, Caenurgina 19:165
distorta, Noctasota 11:215
diva, Annaphila 3:74
  Montia perfoliata
diversicolor, Oligia 8:27
diversilineata, Lygris 25:263
  Vitis, Parthenocissus, Epilobium
diversilobiella, Gracilaria 23:278
diversus diversus, Vettius 22:156
divina barine, Sinia 8:98
  Sophora (Leguminosae)
divisata, Caripeta 5:54; 13:220
dodana, Eucosma 4:29
dodecella, Exoteleia 21:18, 20, 21
  Pinus sylvestris, Pinus mugho
dodgei, Hesperia 8:5; 9:198
dodgei, Speyeria 5:34
dodonea, Amblypodia 16:144
doeri, Zellerina 11:178
dohertyi, Virachola 13:145
dolicaon, Graphium 21:157
dolicaon, Papilio 4:67
dolichos, Prodenia 2:S8
  Ricinus communis 2:S8
dolon dolon, Taractrocera 22:121
dolorosa, Euchrysops 10:38
dolus, Agrodiaetus 14:138
domicella, Heliopetes 25:35
domicella willi 21:164; 22:155
domingonis, Erinnyis 3:78; 9:136
domingonis pallescens 5:22; 9:135

dominicana, Eana 16:179
dominula, Panaxia 17:175; 25:165
domitia, Appias 13:22
domitianus, Caria 21:53
domitianus domitianus 21:54, 55
domitianus vejento (IG) 21:54, 55
donahuei, Euphyes (IA, IG) 21:116, 118, 119
donysa, Adelpha 24:292-296 (IA, IG)
Quercus ?
donysa dariensis, Amplypterus 5:22
donzelii, Aricia 3:76
dorantes, Pap. Pleb. Urbic. 12:196
corantes, Urbanus 22:35; 25:36, 294
Desmodium tortuosum 25:294
dorantes cramptoni 5:55
dorantes dorantes 21:159, 244; 22:155; 24:241, 242, 245, 246
dorantes rauterbergi 12:204
dorcas, Cystineura 9:80
dorcas, Lycaena 10:23, 107; 11:165; 12:44; 14: 215; 16:244; 17:24-26; 19:61
Potentilla
dorcas, Mestra 24:207
dorcas dospassosi, Lycaena 1:58
dorcas florus 14:238; 22:172
dorcas michiganensis 5:120
doria, Timochreon 21:164
dorilas, Syrmatia 21:147
doripe, Colias 12:120
doris transiens, Heliconius 18:18; 21:193
dorothea, Euptychia 14:238; 16:203; 18:140; 19: 114, 232; 22:159, 172
Cynodon dactylon ?
dorothea, Neonympha 7:146
dorothea, Speyeria 5:33
dorus, Coenonympha 9:145; 10:218
dorus, Zestusa 7:146
dorylas, Lysandra 9:145; 10:218
doryphora, Chamaelymnas 21:147
doryssus albicuspis, Urbanus 21:160
doryssus doryssus 21:88, 160, 244
doson, Graphium 9:74; 11:9; 13:199
doson albidum 8:96
Michelia, Magnolia (Magnoliaceae)
doson perillus 18:105
Michelia formosana ?
dospassosi, Lycaena 16:95
DOSPASSOSI, *Papilio 23:255, 256 (IA)*

dotata, Dyops (IG) 16:54
doudoroffi, Callophrys 14:21
draco, Polites 4:37; 10:100; 16:239-242 (IA)
drexelii, Datana 21:36
dromulus, Erebia 14:140
drucei, Megathymus (IA) 8:77-87
drucei, Papilio 4:41
drucilla neumoegenii, Appias 22:34
drupiferarum, Sphinx 15:193
drury, Hesperia 12:120
drusilla, Appias 14:46; 15:212; 18:16; 19:110, 111 (IA)
drusilla, Pap. Dan. Cand. 12:196
drusilla, Phyciodes 21:193
drusilla boydi, Appias 5:55
drusilla drusilla 21:155; 22:126, 154
drusilla neumoegenii 9:206
drusius, Thorybes 9:194; 13:91
dryas, Minois 14:140
dryas, Satyrus 7:37
dryas, Vanessa 19:205
dryas bipunctatus, Satyrus 8:99
 Agrotis, Poa (Gramineae)
drymo, Brachyglenis 21:147
drymo, Ithomia 21:94; 24:298
drymo drymo 22:151, 156
dryope, Eurytela 25:146
 Dalechampia hildebrandti, Ricinus communis
dryope, Satyrium 14:20; 17:243; 22:96-103
duaria, Metarrhanthis 23:206
ducalis, Arcas (IA,IC,IG) 25:88-104
ducens, Feltia 5:69
duena, Moeris S2:32
dukesii, Atrytone 11:42; 18:191; 19:148; 23: 186
 Carex lacustris (oviposited on)
Dulcedo 24:15
dulcinea, Elbella 20:227
duma, Neozephyrus 3:79
dumeril, Drephalys 21:158
dumetellus, Crambus 5:29
dumetorum, Callophrys 10:115; 12:30; 14:21; 18:225; 22:96-102; 23:280; S5:3-41 (IA,II)
 Lotus scoparius, Eriogonum latifolium latifolium, E. heracleoides, E. compositum, E. elatum, E. fasciculatum, Trifolium obtusiflorum S5:3-41 (IF)
dumetorum X augustinus iroides 16:235

dumetorum perplexa 16:34; 21:113

Eriogonum ? 16:34, Lotus scoparius

dumi, Lemonia 18:37

dumonti, Eana 16:179

duovata, Papaipema 8:59

Solidago sempervirens, Solidago gigantea leiophilla

duplex, Neocoenyra 12:6

duplicata, Papaipema 8:59

Collinsonia canadensis

duponchel, Cocytius 2:86; 3:78; 9:135; 25:152

duponcheli, Leptidea 9:145; 14:53

durbani, Lachnocnema 10:39; 18:74

durhya, Euthalia 9:74

durga, Dodona 11:167; 16:144

dux, Hübneria 9:66

dyari, Gabriola 13:220

dydimaon, Atlides 21:152

dymas, Dymasia 18:140-141; 25:40

Siphonoglossa pilosella 18:140-141

dymas, Melitaea 7:147; 15:220

dymas chara 19:114

dymas, Microtia 12:97

dymas chara 12:97

dymas dymas 12:98

dymas IMPERIALIS *12:98*

dymas, TEXOLA *(IC, IG) 12:161-163*

dyonis, Dynamine 18:21

dysaules, Lerodea (1st US record) *18:214-215*

earis, Graphium 21:157

earis, Papilio 4:66

eberti, Euphaedra 13:128

ebriola, Eurema 2:52

echemon, Papilio 4:41

echerioides, Papilio 24:72

echerius, Abisara 18:162

echina echina, Aethilla 21:164

echion, Strymon 14:22

echion, Tmolus 20:135

echion echiolus 18:24

echo, Taygetis 6:97; 21:90

ecksteini, Amicta 3:76

ectrapelaria, Somatolophia 13:220

edata, Lerodea 13:90

EDAX, *Ithome (IG) 15:88-89*

Byra ebenus ?, ebony blossoms

edessa, Heterossais 21:93
editha, Lycaena 16:34; 18:49, 228
Potentilla drummondii ? 16:34
editha montana 22:164
editha, Euphydryas 8:3; 10:96; 12:33, 165; 13: 171; 14:227; 16:70; 17:113; 18:223; 22:161, 169, 200; 23:279; 24:139
editha aurilacus 12:165
editha bayensis 12:33-36; 22:98-103
Plantago erecta, Plantago lanceolata 12:33
editha colonia 13:172
editha lawrencei 17:114
editha monoensis 17:112
eithha nubigena (IA) 19:172, 174; 17:112
editha REMINGTONI *12:168-170; 13:171; 17:114*
editha taylori 10:170; 13:172
edithana, Euphydryas 13:172
educatalis, Lomanaltes 5:54
edwardsata, Neoterpes 13:232; 18:32-34 (II)
Dendromecon rigida, Romneya coulteri 18:32-34
edwardsata, Phengommataea 13:220
edwardsi, Hemihyalea 19:82
scrub oak
edwardsi edwardsi, Trichoclea 9:57
edwardsi, Lymire 11:148; 25:138
edwardsii, Oarisma 13:91; 16:203; 22:169
edwardsii, Satyrium 14:20; 16:195, 197, 199; 21:135, 207, 277; 22:120; 23:187; 24:305
edwardsii, Speyeria 2:5, 92; 5:31; 10:95, 102
edwardsii, Strymon 1:58; 9:77; 10:159
egaensis, Mechanitis 18:15
egaensis doryssus 18:16
egea, Polygonia 3:75, 76; 14:195
egeon, Dodona 16:144
egeremet, Hesperia 20:241
egeremet, Wallengrenia 19:196
egina agricola, Melinaea 15:31
egina egina 15:31
egina paraiya 15:31
eglanterina, Pseudohazis 23:278
egleis, Speyeria 2:5; 5:31; 7:49; 17:112, 115, 243; 18:222
egleis secreta 2:91; 9:146; 10:95, 102-106; 14:182
eichhorni, Euploea 16:101
Plumeria acutifolia (toxic to)
elada, Eresia 12:161
elada, TEXOLA *(IC, IG) 12:161-163*

elada, Texola 13:226-227; 18:140-141; 25:32
Siphonoglossa pilosella 13:226-227
elada ulrica 25:40, 214
elana elana, Synale 21:167
elara, Aeria 21:87, 94
elathea, Pap. Dan. Candidi 21:196
elathea, Eurema 2:52; 5:55; 9:80
elathea elathea 21:154; 22:154
ELDORADA, *Adela* 23:212, 215, 216, 221-224, 234 (IA), 237
electa, Speyeria 5:32
electo, Colias 11:168; 16:143-144
electra, Anaea 18:21
electra, Hemileuca 2:81
Polygonaceae
electra, Yphthimoides 21:88, 91
elegans, Packardia 21:33
eleuchea, Marpesia 25:185, 186
eleuchea BAHAMENSIS 25:188-189 (IA)
eleuchea DOSPASSOSI 25:190-191 (IA)
eleuchea eleuchea 25:187-188 (IA)
eleuchea pellenis 25:191-192 (IA)
elevatus roraima, Heliconius 23:105
elevatus tumatumari 23:105
eligius, Pap. Pleb. Urb. 12:196
elingua, Manruta 6:53
elis, Colias 5:25
ella, Spindasis 9:14
Mundulea (Leguminosae)
ellida, Rekoa 21:152
ello, Erinnyis 2:86; 3:78; 13:66, 78; 25:152
elmorei, Keiferia 24:7, 8
elocata, Catocala 18:37
elpenor, Deilephila 14:73 (II); 18:37
eltola, Polytrema 16:144
eluina, Diaethria 21:87, 101; 22:152, 156
elva, Microtia 12:97; 15:72; 17:37; 21:193
elvina, Leucidia 21:155; 22:154, 156
elvina tephrias, Eurybia 21:146; 22:153, 156
emarginana, Epinotia 25:115-118, 120-122 (IA, IC, IG)
Quercus agrifolia, Q. kelloggii, Q. lobata, Q. dumosa, Q. wislizenii
emasculata, Anacamptodes 13:220
emigratella, Amorbia 10:154
emmedonia, Nephelodes 6:108
emyra, Hypoleria 21:87, 93
enaemargyrea, Chrasmarcha 10:77

encedon, Acraea 13:146; 18:74; 25:80
endeis, Chlosyne 25:40
endymion, Strymon 14:22
endymion cyphara, Electrostrymon 21:184
enispe, Epargyreus 21:158
enoptes, Philotes 7:146; 8:8; 11:7; 16:20, 35
  Eriogonum nudum 16:35
enoptes ancilla 10:98; 22:165, 172; 23:51, 56
  Eriogonum umbellatum ?
enoptes BAYENSIS *(IA, IG) 17:208-223*
  Eriogonum latifolium auriculatum, E. latifolium nudum
enoptes bayensis 19:95-101; 22:93, 96-105;23:51
  Eriogonum latifolium auriculatum & nudum
enoptes columbiae 18:226; 23:51, 56
enoptes dammersi 15:91; 23:52, 57
  Eriogonum elongatum, E. wrightii trachygonum
enoptes enoptes 23:51, 56
  Eriogonum latifolium nudum, E. l. saxicola
enoptes smithi 17:206-208; 19:95-101; 23:51, 57
  Eriogonum parvifolium 17:206-208
enoptes TILDENI *(IA, IG) 17:212-223; 19:95-101; 22:105; 23:52, 57*
  Eriogonum latifolium?, E. auriculatum?, E. nudum, E. latifolium indictum, E. l. saxicola
enthea, Araragi 3:80; 13:178; 14:206
  Juglandaceae
entheum, Araragi 8:97
  Juglans (Juglandaceae)
enys, Polites 14:29
eos, Amblyscirtes 13:90-92, 99; 25:34, 213
eous, Paryphthimoides 21:91
epalia, Mesene 21:149
epaminondas, Aranda 8:99
  Gramineae
epaphus, Metamorpha 21:194
epenetus, Papilio 4:64
ephelidaria, Neoterpes 18:34
  Argemone platyceras v.*hispida*
ephestion, Pap. Eq. Troës. 12:196
ephora, Nisoniades 21:245
epicalus, Telemiades 21:164
epicaste melangon, Staphylus 21:161
epictetus, Anthoptus 21:164, 246; 22:155
epidaus, Papilio 4:66; 12:171; 17:35
epidaus epidaus, Graphium 17:37; 21:189
epidaus tepicus, Papilio 12:128

epijasius, Charaxes 9:129
epijessa calicene, Chalodeta 21:148; 22:153, 156
epilais, Syntomeida 22:37
epimachia epimachia, Myscelus 21:158
epimensis, Psychomorpha 21:38
epimethea epimethea, Augiades 21:158
epinephele, Neonympha 12:194
Epinotia 25:115
epiolina, Dirphia 14:103
epione, Catocala 10:167; 23:100; 24:108, 110, 114
epiphron, Erebia 14:140
epipsodea, Erebia 2:37; 6:103, 105; 10:94; 14: 127, 140; 18:222; 19:57, 94; 20:161; 25:22, 23, 27
epipsodea freemani 15:223
epipsodea hopfingeri 16:148
epipsodea remingtoni 6:105; 7:124; 19:57
epipsodea sineocellata 15:223
epithore, Boloria 5:26; 8:66; 14:158; 16:31; 17:171; 18:188, 222
Viola 16:31
epithore caelestis 22:160
epithore CHERMOCKI *(IA)* 20:109-117
epithore epithore (IA) 20:103-117
epithore sierra (IA) 20:110, 111
epitus epitus, Aides 21:167
epius, Spalgis 7:45
epixanthe, Lycaena 5:25, 120; 8:128; 9:78; 10: 107, 120; 15:124; 23:187; 25:222
epixanthe, Polyommatus 12:122
epixanthe amicetus, Lycaena 4:12; 16:244
epixanthe michiganensis 16:244
epixanthe phaedrus (IA) 23:38, 40
eponina, Acraea 18:74
epulus, Hamearis 21:145
equitalis, Callimorpha 17:224
eracon, Papilio 4:41
eranites, Phyciodes 21:194
erate, Colias 11:12; 13:24; 14:52, 205; 15: 212; 16:143-144
erate poliographus 7:139; 8:96, 166; 18:106
Glycine, Hedysarum, Indigofera, Lupinus, Lotus, Medicago, Pisum, Robinia, Trifolium, (Leguminosae)
erato, Heliconius 17:176
erato hydara 24:25
erato phyllis 22:152

erechtea, Drasteria 15:111
eremitoides, Sphinx 25:41
eremitus, Sphinx 15:193
eresimus, Pap. Dan. Festiv. 12:196
eresimus, Danaus 9:210; 14:132; 18:16; 25:294
eresimus montezuma 8:25; 21:191
ergane, Pieris 14:53; 25:9
ERIANTHALIS, *Chilo (IA, IG) 17:31, 32*
Erianthus sp.
ericetorum, Heliopetes 7:146; 12:31; 13:232; 16:69, 70; 18:49, 50, 227; 22:94, 97-102, 105, 168
Iliamna rivularis , Bush mallow
eridania, Prodenia 4:26
Petasites ?
erigone probata, Yphthimoides 21:91
erinnyn, Erebia 19:57, 59
erinome, Actinote 4:50
erippus, Danaus 3:40; 21:93; 22:151
erithalion, Papilio 4:41
erlaces, Papilio 4:41
erna, Amblyscirtes 13:91
erodyle, Chlosyne 18:19; 21:193
eros, Agrodiaetus 14:137
erostratus, Papilio 4:64
erosus, Antigonus 21:162, 245; 22:155
erransella, Perimede 15:81
erronella, Unadilla 24:255
erruca, Pseudoscada 21:94; 22:151, 156
erschoffiana, Zellerina 11:178
erubescens, Taygetis 21:90
eryngii, Papaipema 8:59
Eryngium aquaticum, E. yucciflolium
eryphon, Callophrys 14:21; 16:33, 70; 18:224; 20:41, 42; 21:135, 137; 22:164
Pinus contorta?, yellow-pine
eryphon, Incisalia 10:98, 170; 11:47; 13:70; 20:160; 24:45, 258, 260
Pinus contorta 13:70
erythemaria, Deilinia 13:219
erythrostictus, Lerodea 21:167
erythus, Bungalotis 21:160
eryx okinawana, Deudorix 18:106
Gardenia jasminoides
ESCALANTEI, *Agathymus (IA,IG) 20:167-168; S1:50*
ESCALANTEI, *Astraptes (IA,IG) 21:115-117, 118*
ESCALANTEI, *Carystoides (IA,IG) S2:33*
escheri, Agriades 20:131

escheri, Agrodiaetus 14:137
escheri, Lysandra 9:145, 204
escheri ahmar, Polyommatus 10:222
esculi mauretanica, Strymon 10:222
esma, Urbanus 20:227; 21:159
esmeraldus, Urbanus 21:159
esperi, Saliana 21:247
esta, Urbanus 21:159; 22:155
ESTELLEAE, *Megathymus (IA,IG) 11:119-121; 14: 37, 58; 15:67; 20:181; S1:49*
Agave lecheguilla
etesipe, Charaxes 13:124
ethalion, Charaxes 13:124
ethelinda ethelinda, Nymula 21:150
etheocles, Charaxes 13:124
etheocles evansi 13:124
ethillus, Heliconius 23:105
ethillus narceus 21:96; 22:152
ethlius, Calpodes 3:17, 40; 4:43; 9:193; 12: 32; 18:229, 230; 19:1, 2, 31; 21:167, 207, 208, 247, 277
Canna indica 19:2
ethlius, Pap. Pleb. Urbicol. 12:196
ethra, Melinaea 15:30
ethusa, Myscelia 6:33; 15:199
eubule, Papilio 12:121
eubule, Phoebis 2:22; 3:17, 18; 4:67; 7:52; 12:130; 14:156; 21:196
eubule, Phoebis
Cassia corymbosia 2:22
eucalypti, Antheraea 25:238, 239
euchenor euchenor, Papilio 13:76
Euchlaena ssp. 23:206
euchytma, Dircenna 18:17; 21:191
euclea, Hypothyris 24:298
eudamippus weismanni, Polyura 18:108
Celtis boninensis, Rhamnella franguloides
eudocia acosma, Theope 21:88, 150
eufala, Lerodea 9:195; 12:130, 203; 13:221-227; 17:157; 19:133, 137; 21:167, 207, 277; 22: 94, 97, 99, 100, 102, 156; 23:278; 24:46; 25:34, 137, 213
Cynodon dactylon 12:203
Stenotaphrum secundatum 13:221-227
eugenes, Dodona 16:144
eugenes laertides, Prepona 21:87, 104
eugenia, Eurema 2:52
eulalia, Adelpha 13:91

eumelus ninyas, Entheus 21:160
eumeus agathon, Nothome 21:147
eumeus hemicosmeta 21:87
eunice, Eresia 13:84
eunice esora, Phyciodes 21:97
eunomia, Boloria 5:25, 26; 11:48, 49; 18:233; 19:184, 185; 25:149
eunomia alticola 5:63; 7:49
eunomia caelestis 5:26; 14:184
eunomia dawsoni 5:26; 16:243; 18:233-237 (IA); 21:136; 24:43
eunomia denali 5:26; 7:124; 19:61
eunomia laddi 5:26; 10:96; 25:84
eunomia ossianus 14:193
eunomia triclaris 5:26; 19:242
eunus, Pseudocopaeodes 9:148
eupatorii, Papaipema 8:59
Eupatorium purpureum
eupheme, Zegris 10:215
eupheme nr *meridionalis 10:220*
euphemus, Maculinea 7:45
Sanguisorba, Myrmica
euphemus kazamoto 8:98
Sanguisorba (Rosaceae), later carnivorous on ant-grubs & eggs of *Myrmica* (Formicidea)
euphemus shiriyensis 8:98
same hosts as preceding
eupheno, Epinephele 10:215
eupheno, Euchloe 10:220
euphrosyne, Argynnis 15:221
euphrosyne, Boloria 14:193; 15:221
euphrosyne andersoni 15:220
euphrosyne, Clossiana 11:163
euribates, Pap. Pleb. Urb. 12:196
eurilochus sulanus, Caligo 21:192
europa, Lethe 11:167
europa pavida 18:108
Bambusa sp.
eurota, Evonyme 21:102
euryalus, Hyalophora 13:213; 14:103; 18:36, 38
euryalus, Samia 8:2
euryalus kasloensis 8:2
euryanassa, Ceratinia 13:84
euryanassa, Placidula 21:94; 22:151, 156
eurycles, Hesperia 12:120
eurycles, Protambulyx 3:78
eurydice, Colias 22:94, 97, 98, 100, 102
eurydice, Lethe 6:32; 9:77; 10:55, 107; 12:41,

42, 44; 14:215; 15:99, 222; 16:243; 22:34, 120; 23:100; 25:256, 261
eurydice, Satyrodes 15:222
eurydice, Zerene 12:28; 13:232
eurydice eurydice, Lethe 17:107; 24:1
eurydice applachia 23:188; 24:1
eurydice fumosus 15:97
euryleon, Papilio 4:66
eurymedia, Aeria 24:299
eurymedon, Papilio 2:42, 92; 4:64; 6:101; 9:139; 10:99, 169; 12:28, 101, 103-113; 13:74, 232; 16:29, 45, 69; 18:220; 19:109; 22:94, 97-102
Ceanothus fendleri, C. sanguineus, C. prosteratus, Rhamnus californica, Jasminum, Prunus persica, P. emarginata, Crataegus rivularis, C. brevispina, Ribes inerme, Spiraea discolor, Amelanchier alnifolia, Pyrus malus, Alnus rubra 12:103-113; Ceanothus velutinus, C. prostratus, Prunus emarginata 16:29; Prunus demissa 18:220
eurynome, Argynnis 2:5
eurynome, Speyeria 10:102
eurypilus, Plebejus 10:117
eurypyle confusa, Anaea 21:195
eurytheme, Colias 2:25, 60; 3:17; 5:25; 6:42, 108; 7:139-146; 8:3, 7, 163; 9:78, 148, 194; 10:25, 55, 95, 99, 107, 143, 166; 11:11, 12, 163, 207-214 (IA); 12:28, 130; 13:234; 14:42, 79, 215, 238; 15:44, 110, 198, 213; 16:30, 36, 70, 129, 130, 245; 17:7, 10, 76, 112, 180, 225, 226; 18:221; 19:110, 133, 137, 182, 183, 232; 20:207, 215, 216 (IA , melanic); 21:207, 277; 22:94, 97, 98, 102, 120, 127, 158 (IA, aberrant); 23:182, 279; 24:77, 117-119, 305; 25:165
Vicia villosa 11:207-214
clover and vetch 18:221
eurytheme eurytheme 17:22, 154; 20:102; 22:34; 25:36, 214
Sesbania exaltata 17:22
eurytheme X interior 10:9-14
eurytheme philodice 15:200
eurytheme keewaydin 15:213
eurythris, Satyrus 12:120
eurytulus, Callicista 21:153
eurytulus, Strymon 15:215
eurytus, Megisto 9:77
eusemoides imitans, Euphaedra (IA) 13:128

euterpinus, Papilio 4:65; 5:16
euxanthe, Colias 15:119
evadnes, Panoquina 21:247
evagaria, Itame 25:267
evagoras, Ialmenus (II) 20:133, 134
evanidas, Ebrictas 21:245
evansi, Antigonus 13:91
evansi, Agathymus 13:204; 18:45, 46; 20:166; S1:41
    Agave parryi v. *huachucensis* 18:45-46
evansi, Megathymus (IA) 8:77-87
evansi, Panoquina 20:227
evansi, Systasea 22:230
evarete coenia, Junonia 5:56; 6:74; 14:186; 21:98, 194; 22:35, 152; 25:138
evarete, Pap. Nymp. Gemm. 12:196
evenina, Colotis 13:137
evenina casta, Colotis (IA) 13:138
evenor evenor, Zopyrion 21:162; 22:155
evenus, Urbanus 21:159
eversmanni SHIVA, *Pararge* (IA) 15:9
eversmanni, Parnassius 11:9; 13:199; 19:58, 60
eversmanni daisetsuzana 8:96
    Dicentra (Papaveraceae)
eversmanni thor 7:126
eversmanni wosnesenskii 15:210
exadeus, Pap. Ple. Urbicol. 12:196
exadeus, Epargyreus 21:244
exadeus exadeus 21:158
exauspicata, Itame 13:219
excaecata, Paonias 15:193; 19:189; 21:253
    basswood, white birch
excelsaria, Stenoporpia 13:220
exclamationis, Badamia 3:40
excurvata, Semiothisa 13:219
exilis, Brephidium 7:147; 9:148; 13:223-227; 19:112; 20:134, 161; 23:96-98, 103; 23:182; 24:3, 5, 6; 25:214
    Chenopodium album 13:223-227
exilis exilis 22:228, 230
exilis isophthalma 24:4-6
exilis yucateca (IA) 24:3-6
explanata, Lygris 25:262-264
    Vaccinium
exprimens, Pyrrhia (IA,II) 24:234-239
    Polygonium, Desmodium, Juglans, Populus, Salix, Rosa, Prunus, Althea, Betula, Delphinium, Geranium

expurgata franclemonti, Hydriomena 13:218
extincta, extincta, Leucania 9:56
extincta flabilis 9:56
extrema, Tapinostola 2:94
extricalis, Phlyctaenia 5:29
exulis assimilis, Apamea 17:176
exulis, Crymodes 4:29
exulis borea 4:52
exulans, Zygaena 3:76

fabrius, Charaxes 13:19, 22
facuna, Strymon 14:22; 20:65, 69 (IG)
facuna, Thecla 13:90
fagata, Operophtera 3:75
fagella, Diurnea 13:9
fagi, Hipparchia 9:145, 204; 14:140
fagi, Stauropus 6:5
fagina, Graptolitha 15:121
falacer, Polyommatus 12:120
falacer, Strymon 1:58; 4:14; 7:58; 9:105-117; 10:55; 12:44, 130; 14:20, 216; 15:97, 108; 16:199, 243; 21:207, 277; 24:84, 85 (IA,IG); 25:296
falacer godarti 13:92
falacer ab. *heathi* 2:6
falcataria, Drepanulatrix 13:219
fallax, Chilopsephalus 3:76
fallax, Jemadia 21:158
falloui, Anthocaris 10:215, 220
farinalis, Pyralis 19:133, 135
farinosa, Gonepteryx 10:117
fasciata, Erebia 2:64; 5:42; 19:57
fasciata ab. *suffusa* 9:141
fasciata, FULVIA (IC,IG) 12:162-163
fasciata, Melitaea 12:162
fasciatus, Pholus 11:43; 13:78; 19:189
fatima, Anartia 14:188; 15:72; 22:128
fatima fatima 18:20
fatima, Emesis 21:149
faunalis, Callicista 21:153; 22:154
faunus, Ochlodes 9:145
faunus, Vanessa 19:205
faunus, Polygonia 2:59; 10:107, 169; 11:47, 164; 16:244; 18:48, 223; 19:113; 23:192, 193; 25:139, 141
faunus faunus 6:73
faunus rusticus 19:172; 22:98

faunus smythi 5:70; 6:73
fausta, Argumenia 3:76
fausta, Colotis 10:117
fausti LUCULLUS, *Zegris* (IA) 15:6
favonius, Euristrymon 22:35
favonius, Satyrium 13:71; 14:21, 127
februa, Ageronia 12:172
februa, Hamadryas 16:128
februa gudula 15:217; 18:20; 21:195
feisthameli, Graphium 13:199
feisthameli, Notocrypta 16:144
felicitata, Schinia (IA,IG,II) 21:22-26; 24: 282; 25:110
Oenothera biennis, Oenothera deltoides
FELUS, *Vidius* (IA, IG) 22:4, 10, 11
fenestrata, Hectasia 1:48
FERAX, *Ithome* 15:89-90
ferentina ferentina, Hamadryas 21:103; 22:153
fernaldalis, Chilo 17:31
ferox, Hamadryas 21:195
feronia farinulenta 18:21; 21:195
feronia obumbrata 21:103; 22:153
ferrea, Ahlbergia 8:97
Viburnum (Caprifoliaceae), *Rhododendron* (Ericaceae), *Malus* (Rosaceae)
ferrugata, Xanthorhoe 5:54
ferruginosa, Holomelina (IA) 19:69-74
ferula, Satyrus 9:145
fervidana, Archips 10:154
fessonia, Adelpha 13:90
fettigii, Coenonympha 10:215, 218, 221
ficedula, Myrina 10:39
ficus, Pachylia 2:86; 5:21; 13:66; 25:152
fidellisima vagrans, Composia 9:207, 209; 25:138
fidia, Eumenis 9:145
fidia albovenosa, Satyrus 10:221
fidia guildi, Satyrus 10:221
fiduciaria, Chondrosoma 3:76
FIELDI, *Agathymus* (IA,IG) 14:59-62; S1:50
Agave tequilana
fieldi, Colias 3:40; 11:12; 13:24
figurata, Apantesis 13:182
filamentaria, Nematocampa 23:206, 207
filifera, Casandria 17:226
filipjevi, Eana (SUBEANA) 16:177
finitima, Syntomoides 8:136

finta, Vidius 21:167
fischeri, Tongeia 8:98
Sedum (Crassulaceae)
fissimacula fissimacula, Morvina 21:161; 22:155
flaminia, Euryphene (IA) 13:128
flammea, Meliana 2:94
Phragmites communis
flammeusella, Adela (IA) 23:212, 215, 216, 224-226, 235-237
Orthocarpus ? 23:235
flava, Adopoea 11:166
flava, Potanthus 8:95
Setaria, Miscanthus, Oplismenus (Gramineae)
flaveola, Colias 7:139
flavibrunneata, Lygris 25:264
Salix, Populus
flavidalis, Polygrammodes 5:29
flavilla flavilla, Nica 21:102
flavilla canthara, Pseudonica 18:20; 21:194
flebilis, Catocala 24:111, 115
fletcheri, Brephos 13:78
Montia perfoliata
floccifera orientalis, Carcharodus 10:118
florella, Catopsilia 3:39; 18:74
florella gnoma, Catopsilia 6:69
florenceae, Agathymus 13:91; S1:35, 36
Agave parryi-like
FLORESI, *Carystoides (IA, IG) S2:35*
florestan, Phlegethontius 3:78; 25:152
(see also *floristan*)
florida, Rhodophora 6:53
florida, Schinia 21:22; 24:282-286; 25:110
Oenothera biennis *(IA,IF,II)*
floridalis, Anaea 9:207; 13:73
floridana, Mestra 24:207
floridella, Peoria 24:40
floridensis, Callopistria 10:112
floristan cabnal, Phlegethontius 5:22
floristan ishkal, Phlegethontius 5:22
florus, Lycaena 17:24, 26
Rumex crispus
fodinalis, Pyrausta 13:242
foeminaria, Drepanulatrix 13:219
folus, Udaspes 16:144; 18:105
Alpinia speciosa
forestan, Coeliades 13:146
forestans, Coeliades 9:129
forficaria combinata, Prochoerodes 13:220

formosa, Anteros 21:196
formosa, Chrysanympha 15:121
formosa, Cucullia 3:76
formosa, Dysstroma 13:218
formosaria, Aspilates 2:93-94
Eupatorium cannabinum
formosus formosus, Anteros 21:149
fornax, Hamadryas 21:103, 195
fortunata, Dismorphia 18:15, 17; 21:189
fotis, Thecla 18:44
fotis, Callophrys (Incisalia) 14:21; 16:70; 18:224; 20:134; 22:226
Sedum 18:224
fotis BAYENSIS *23:95-96*
fotis doudoroffi 19:171
fotis fotis 24:228
fotis schryveri 24:256-266 (IA, IF, IH, II)
Sedum lanceolatum
foulquieri, Pyrgus 9:145
fovea, Rileyiana 17:170
FOXI, *Callithomia (IA, IG) 24:23, 24*
fragariae, Ramosa 24:189
fragariana, Tortricodes 14:112, 113
fragarianus, DECODES *(IG) 14:115-116*
fragile, Malacosoma 4:15
fragilis, Acronicta 5:54
fragilis, Holomelina 19:74
FRANCESIUS, *Quadrus (IA, IG) S2:21*
francisca, Mycalesis 14:139
francisca perdiccas 7:37; 8:100
Oplismenus (Gramineae)
franckii, Prothoe 9:74
frankii, Sphinx 11:15-17; 17:102
ash
FRANCLEMONTI, *Wxycnemis (IA, IG) 22:137-142*
frater, Raphia 4:62
fraternaria, Epimecis 5:60
fraxini, Catocala 1:34; 15:235
frederici, Catocala 25:42
fredericus salvina, Ouleus 21:245
freemani, Agathymus S1:40
Agave deserti
FREEMANI, *Vinius (IA,IG) 24:120-123*
freija, Boloria 5:25, 63; 7:50, 124; 8:65; 11:47, 49; 14:193; 16:243; 20:103; 22:252; 23:113; 24:43; 25:140, 141
freija, Pyrgus 5:26
freija asahidakeana, Boloria 8:98

Sieversia (Rosaceae)
freija browni 8:65; 10:96
freija freija 21:136
freya, Argynnis 12:120
fridericus CANDANGUS, *Ouleus (IA) 22:3, 5*
frigida, Papaipema 8:59
Smilax herbacea, Thalictrum revoltum
frigidana, Aphania 4:29
frigidana, Nycteola 5:54
willow
frigidaria, Scopula 5:54
frigga, Boloria 5:63; 11:37; 14:193; 16:243; 18:233; 24:44
frigga gibsoni 5:26; 7:124; 11:50; 19:58, 59
frigga saga 5:26; 6:105; 11:37, 38; 19:242
frigga sagata 1:42; 5:26; 7:50; 8:65, 66; 10: 96; 20:114-116 (IA)
frisia, Phyciodes 9:207
frisia frisia 22:35
frisia tulcis 21:193
fritillarius, Pyrgus 9:145
frontalis, Carpodacus 20:73
frugiperda, Laphygma 18:4; 19:132, 133
frugiperda, Spodoptera 25:42
frustrana, Rhyacionia 14:235
frustulum, Celiptera 2:S8
Robinia pseudoacacia
fugax, Rhodinia 14:103
fujisanus, Favonius 3:80; 8:97; 13:178
Fagus (Fagaceae) 8:97
fulgens, Lycomorpha 11:150
fuliginosa, Satyrium 7:51; 10:97; 14:20; 16:33
fuliginosum, Satyrium 18:48, 224
fulginosum semiluna 22:164
fulgerator, Astraptes 13:90
fulgerator fulgerator 21:160; 22:155
fulgerator azul (IG) 21:116-118
fulginia, Taygetis 6:98
fulgurata, Charaxes 13:124
fulgurata, Neptidopsis 25:146
Dalechampia hildebrandti
fulvaria, Itame 5:54
FULVIA *12:162*
fulvia, Melitaea 7:146; 19:232; 22:161, 237-240 (II)
Castilleja integra
fulvicollis, Cisseps 13:221-227; 17:65-80
Poa pratensis, Lolium multiflorum, Lolium

perenne, Festuca sp., Festuca rubra, Stenotaphrym secundatum

fulvicollis, Scepsis 19:133

fulvicosta, Haploa 21:38

fumata, Eupithecia 13:218

fumiferana, Archips 12:51

fumiferana, Choristoneura 7:57; 14:80; 18:116

fumistrota, Epimecis 5:60

fumosus, Lethe 15:97

funebris, Pyrausta 5:29

funeralis, Desmia 11:19; 12:171; 23:192, 194

funeralis, Erynnis 12:130; 13:225-227; 16:203; 18:152

Vicia texana, Indigofera leptosepala 13:225

funerellus, Phobus 13:233

furcata, Papaipema 8:59

Fraxinus pennsylvanicus, F. americana, F. nigra

furcata, Hydriomena 5:28; 13:218

furcilla centralis, Panthea 19:189

fusca, Niphanda 6:43; 7:45; 15:246

fusca shijima 7:45; 8:97

Greenidea kuwanai (excretion), dwarf-oak aphis, later fed orally by host-ants, *Camponotus herculeanus japonicus*

fuscata, Scopula 13:217

fuscipedella, Ethmia 4:26

Pentstemon sp.

fuscipunctella, Acedes 23:192, 195

fuscoochrella, Gracilaria 25:202

fuscus, Papilio (IA, II) 16:161-173

fuscus X aegeus (IA, II) 16:161-173

fuscus capaneus 16:163

Citrus aurantium, C. sinensis, C. reticulata, C. limon, C. aurantiifolia, Zanthoxylum brachyacanthum, Microcitrus australasica

futilalis, Pyrausta 4:26

Apocynum androsaemifolium

futilalis inconcinnalis, Pyrausta 13:233

gabatha, Thecla 18:23

gabbii, Melitaea 3:106, 4:70; 7:146

gaika, Zizeeria 18:74

galathea, Melanargia 9:145, 204; 11:164; 14:139; 18:41, 110; 22:128

galathea meade-waldoi 10:221

galbina, Agapema 19:42

galbina anona 16:61
Lycium
galena, Thysanota 21:87, 150
galii intermedia, Celerio 5:19; 11:35, 36; 21:254
Epilobium angustifolium 11:35,36
gallienus, Papilio 13:76
gallaesolidaginis, Gnorimoschema 3:48; 4:4
parasites 3:48
gambiae, Diestogyna (IA) 13:137
gamma, Plusia 24:172
ganesa, Amblypodia 11:167; 16:144
ganesa loomisi 8:97
Cyclobalanopsis (Fagaceae)
gannascus, Amplypterus 2:86; 3:78
gannascus cubanus 5:22
gannascus jamaicensis 5:22
gargamelle medusa, Gloveria 3:106
garamus, Papilio 4:65
garita, Oarisma 10:100, 14:37
garleppi, Papilio 4:64
garuda, Euthalia (IA) 18:160
gaudealis, Chlosyne 21:193
gaurae, Rhodophora 6:53
gausapata, Lithophane (IA,IG) 22:42-45
gausape, Euphaedra (IA) 13:128
GAYLEAE, *Megathymus (IA, IG) 17:81-85; S1:25*
Agave falcata
geisa geisa, Morys 21:166
geminata, Eupithecia 5:54
geminata, Packardia 12:132
geminatus, Smerinthus 19:189
geminipuncta, Archanara 2:94
Phragmites communis
gemma, Megisto 12:130
gemma, Euptychia 2:22; 6:32; 21:207, 277
gemma freemani 18:17, 139, 140
Cynodon dactylon
gemma gemma 17:23, 150; 19:47
gemmatilis, Anticarsia 10:112; 18:8
genitrix, Actinote 22:152, 156
genoveva, Pap. Nymph. Phaler. 12:197
genoveva, Precis 18:20
geneoveva gela, Ogyris 20:134
genutia, Pap. Nymph. Phaler. 12:197
genutia, Anchocharis 7:39
genutia, Anthocaris 9:133; 10:55, 163; 11:162; 12:130

genutia genutia 19:50
genutia midea 5:70
genutia, Danaus 11:229; 16:144
genutia, Salatura 18:107
    Tylophora tanakae
gentianana, Endothenia 8:26
geometrina geometrina, Trina 21:161; 22:155
georgei, Monodes 18:83
georgiana, Archips 10:154
georgiana, Dasychira 9:79
georgica, Hyperaeschra 10:203, 204
    Quercus rubra
georgii, Plemyria 13:218
georgii, Eupithecia 13:218
germainii, Tetraphlebia 25:17
germen yurakani, Xylophanes 3:78
germen yurakano, Xylophanes 5:23
gerularia, Racheospila 22:37
gesta gesta, Gesta 21:163
gesta invisus 19:1, 18, 31; 25:35
    Indigofera suffruticosa, I. lindheimeriana
gibsonata, Eupithecia 5:54
    Arbor vitae?
gigantea, Archanara 2:94
gigantea, Colias 2:13; 5:25; 6:104; 11:52; 16:243
    Salix sp 2:13; willow 5:25
GILBERTI, *Agathymus (IA, IG) 18:176-185; 20:184, 185; S1:46*
    Agave lecheguilla 18:176-185
GILBERTI, *Astraptes (IA, IG) S2:8*
gilippus, Danaus 9:210; 14:130; 15:222; 19:232; 22:159, 228, 229; 23:182
gilippus, Pap. Dan. festiv. 12:197
gilippus berenice, Danaus 9:67, 68, 205, 207; 17:150; 22:34, 127; 23:196-198; 25:127, 294
    Asclepias tuberosa rolfsii, A. humistrata, A. curassavica, Sarcostemma clausa
gilippus gilippus 21:93; 22:151
gilippus strigosus 9:67, 68, 148; 10:95; 14,36; 19:114; 21:191; 22:127; 25:41, 214
gillettii, Euphydryas 12:165; 13:171
gisca, Cymaenes 21:165; 22:155
glacialis, Erebia 14:140
glacialis, Parnassius 11:9; 14:203
glacialis glacialis 8:96
    Corydalis (Papaveraceae)

glandon, Agriades 14:130, 137; 18:226
glandon podarce 16:35
Vaccinium nivictum ?
glandon rustica 22:172
glandon, Plebejus 20:134
glandon rustica, Plebejus 10:95, 98
glaphyra modesta, Anatole 21:150; 22:153
glareosa, Amathes 17:175
glauce, Dynamine 18:21; 21:194
glaucina, Euptychia 18:17
glaucinalis, Paracolax 18:110
glaucippe, Hebomoia 3:39; 13:22; 14:52
glaucippe cincia 18:106
Crataeva religiosa
glaucippe liukiuensis 8:96; 18:106
Crataeva religiosa (Capparidaceae)
glaucon, Philotes 8:8
glauconome, Ageronia 15:72
glauconome, Hamadryas 14:188
glauconome iranica, Pontia 10:117
glaucope boenninghauseni, Pseudocercyonis 25:13, 16-18 (IA, IC, IG)
glaucus, Papilio 2:37; 3:25, 48; 4:64, 66; 6:5, 6, 101, 8:28, 131; 9:46, 77, 143; 10:55, 164; 11:48; 12:36, 103-113, 130; 13:194; 15:23, 211; 17:35, 180, 18:235; 19:57, 212-214 (IA); 21:207, 277; 22:120, 199, 200; 23:191, 192; 24:305; 25:150 5:16,17
Liriodendron sp. 3:25
Fraxinus sp., F. americana, F. pennsylvanica v. *lanceolata, Liriodendron tulipifera, Betula sp., B. papyrifera, B. lutea, Populus sp., P. tremuloides, P. tacamahaccam, Acer sp., Malus pumila, Sorbus americana, Salix sp., Alnus sp., Prunus sp.,* Magnolia 12:103-113
glaucus arcticus 11:49
glaucus australis 8:131; 9:206; 22:34
glaucus glaucus 22:34
glaucus canadensis 5:42, 53; 6:104; 7:126; 10: 107; 11:47, 49; 12:50, 124; 13:74; 14:157 IA-melanic; *16:98 (IA), 245; 18:157, 158 (IA); 19:57; 21:136; 22:252; 24:45; 25:257*
glaucus glaucus 13:74; 17:22, 153; 25:36, 257
glaucus X eurymedon (IA, II) 11:201-205
Liriodendron
glorifica, Limenitis 7:38; 8:99; 14:196
Lonicera (Caprifoliaceae) 8:99
GLORIOSA, *Chlosyne (IA, IG) 13:165-169*

gloveri, Hyalophora 13:213; 20:47; 25:68-73
gloveri, Samia 8:2
gnatheme, Eurema 2:51
gnathene, Eurema 2:51
goiana, Hypoleria 21:87, 93
gola, Oriens 11:168
goliath, Ornithoptera 25:60
goniogramma, Diestogyna (IA) 13:137
gonostigma, Orgyia 18:37
goodsoni, Callophrys (IA,IG) 20:66-70
gordius, Sphinx 12:64, 193; 21:252
  Alnus incana, Myrica gale, Comptonia sp., Vaccinium sp. apple, ash, wax-myrtle
gordoni, Victoria 25:169
gorge, Erebia 14:140
gorgon, Lycaena 3:106; 8:128; 13:235; 17:212; 20:160; 22:96, 97, 99, 100, 103
gorgone, Erebia 14:140
gorgone, Melitaea 21:137
gorgone, Phyciodes 2:22; 22:35; 25:139, 141
gorgone carlota, Chlosyne 18:141, 142; 21:207, 278; 25:40
  Helianthus annuus 18:141
gorgone carlota, Melitaea 10:96
gorgone gorgone, Chlosyne 17:150; 19:48
gorta, Vehilius 21:167
goschkevitschii, Neope 7:37; 8:100; 14:139,204
  Pleioblastus (Gramineae) 8:100
gossypiella, Pectinophora 19:130, 134, 136; 22:197, 201, 202
gotama, Mycalesis 7:37; 14:139
gotama fulginia 8:100
  Oryza, Oplismenus (Gramineae)
gotama madjicosa 18:108
  Miscanthus sinensis, Oplismenus compositus
gothica, Papilio 24:229-233
gozmanyana, Bactra 17:48
gracilalis, Palpita 13:242
gracilineata, Lygris 25:263
  Vitis, Parthenocissus, Epilobium
gracilis, Catocala 12:201; 22:234-236; 23:5, 8, 9, 262; 24:108, 110, 114
gracilis, Polygonia 1:58; 2:59; 7:125; 11:50; 18:228; 20:161; 25:140
graeca, Boloria 14:193
graefi, Eupithecia 13:218
graefiaria, Merochlora 13:234
graminis, Bombyx 3:51

granadensis polybaptus, Morpho 25:223-233
grandis, Cogia 21:87, 161
grandis, Lasiocampa 18:37
grandis assecla, Pythonides 21:162
granitata, Semiothisa 5:54; 8:3; 20:44
granitella, Pima 24:254
graphica, Drasteria 19:190
grata, Eupithecia 5:54
gratulata, Mesoleuca 13:75, 219
grayi, Lomatium 16:140
GRAZIELLA, *Eucosma* (IA,IG) 22:137, 143-145
gremius, Suastus 16:144; 18:75
grisea grisea, Acronycta 10:88
griseata, Enypia 13:220
grisella, Bertelia 24:254
griseobasalis, Phyciodes 21:194
groenlandica, Arenaria 10:54
grotei, Hemileuca 2:81; 25:47
  Quercus 2:81
grotei, Lycomorpha 11:150
grotei grotei, Cautethia 24:268, 269 (IG)
gruneri armeniaca, Anthocaris 10:117
grunus, Habrodais 17:111, 244; 20:135
grunus lorquini 22:93, 96, 97, 99, 100, 102
grynea, Catocala 23:5, 100; 24:108, 110-114
gryneus, Callophrys 14:21; 16:195; 20:135; 21:137; 24:196
  Juniperus virginiana 21:137
gryneus castalia 25:38
gryneus gryneus 17:152; 19:49
  Juniperus virginiana 17:152
gryneus, Mitoura 4:48; 5:18; 7:58; 9:4, 5, 77, 78; 10:55, 164; 11:7; 21:120, 207, 208, 277
  Chamaecyparis thyoides, Juniperus virginiana 9:5, red cedar 5:18
gryneus castalis 13:90, 91
gryneus var. *octoscripta* 12:208
guadeloupe, Charis 18:44
guatemalena, Actinote 18:18
guatemalena eupolema 12:172
gueneata, Ceratodalia 13:218
gueneata, Eupithecia 3:75
gundlachia, Eurema 1:42; 2:51; 15:198
gundlachianus, Papilio 4:40
guttata, Parnara 4:43; 7:37; 8:95; 14:51; 18:105
  Oryza, Imperata, Setaria (Gramineae) 8:95

Miscanthus sinensis 18:105
guttivitta, Heterocampa 4:26
Pinchneya pubens
gyas, Hemiargus 7:147; 9:148
gyas gyas 12:30; 13:232, 234
gynaea zama, Charmona 21:87, 148, 196

habilis, Catocala 23:263; 24:108, 110, 115
hadros, Incisalia 13:89
haesitata, Triphosa 13:217
Haetera 24:15, 16
HAFERNIKI, *Piruna* 24:247, 248 (IA, IG)
hahneli, Papilio 4:41
halesus, Pap. Pleb. Rural. 12:197
halesus, Atlides 7:146; 10:159; 12:30, 200, 13: 26, 100; 16:139, 203; 17:215; 19:112; 20:134; 22:163
halesus corcorani 25:38
halesus estesi 22:96-98, 102
halesus halesus 17:152; 22:35
halicarniae, Lapara 13:78; 19:189
halimede, Eurybia 21:196
hamada, Taraka 4:13; 7:45, 46; 8:97; 15:247
predatory on bamboo woolly aphis, *Oregma japonica* 4:13, 7:46; 8:97
hamadensis, Agrodiaetus 14:137
hamza, Adopaea 10:215, 222
hamza nova, Thymelicus 10:118
hannibal mayeri, Phlegethontius 5:22
hanno, Hemiargus 15:216
hansii, Satyrus 10:215, 221
harfordi, Colias 5:25; 13:232, 234
harmodius, Papilio 4:66
harmonia, Tithorea 24:298, 299
harmonia caissara 15:29
harmonia cuparina 15:28
harmonia deltana 15:28
harmonia egaensis 15:29
harmonia furia 15:28
harmonia furina 15:28
harmonia harmonia 15:28
harmonia helicaon 15:28
harmonia hermias 15:28
harmonia hippothous 15:28
harmonia irene 15:28
harmonia manabiana 15:28

harmonia martina 15:29
harmonia megara 15:28
harmonia neitha 15:29
harmonia pseudethra 15:29; 21:95; 22:151
harmonia pseudonyma 15:29
harmonia salvadoris 15:28
harpalus, Hesperia 9:198; 16:36; 17:112; 18:227; 20:160
  grass 16:36
harpalis dodgei 13:239
harpalus dodgei 22:94, 97, 99, 100, 102, 104
harpalus idaho 10:100
harpalus oregonia 17:113; 18:188, 227
harpalus tildeni 22:105
harpax, Axiocerses 6:43
harrisii, Chlosyne 14:184; 23:268
  Aster umbellatus 23:268
harrisii liggetti 25:222
harrisii, Halysidota 2:36; 8:3
harrisii, Melitaea 9:78; 10:55; 11:39, 163; 15:92; 16:244; 20:241; 22:120; 25:139, 141
harrisii albimontana 4:11
harrisi, Megathymus 11:54; 12:175-183, 186; S1:23
  Yucca filamentosa
harrisi, Papaipema 8:59
  Angelica atropurpurea, A. lanatum, Heracleum maximum
hassan evansi, Cogia 21:164
hastata, Eulype 7:123
hastata gothicata 5:54; 13:219
hatuey, Dyops 16:54
havilae, Spaelotis 13:233
hawkeri, Harmilla (IA) 13:128
hayashii, Favonius 8:97; 13:178, 180
  Quercus dentata 13:178
haydeni, Coenonympha 11:164; 15:222
haydenii, Coenonympha 6:103
hayhurstii, Pholisora 9:194; 12:130; 15:98
HAZELAE, *Megathymus* (IA,IG) 11:127-29; 12:81,84
  Agave 11:127-129
hazelae, Turnerina S1:28, 29
  Agave
hebesana, Endothenia 13:234
hebraeus cimelium, Thecla 21:152
hecabe, Eurema 6:43; 7:37; 14:52
hecabe, Terias 13:22; 16:143

hecabe mandarina, Eurema 8:96
Lespedeza, Caesalpinia, Acacia, Albizzia, AEachynomene, etc. (Leguminosae)
hecabe phoebus 22:121; 23:280
hecabe hobsoni 18:106
Leucaena glauca, Cassia occidentalis, C. fistula, C. siamea, C. surattensis, Acacia farnesiana, Lespedeza cuneata
hecabe subdecorata 18:106
larval foodplants same as preceding
hecale, Heliconius 23:104, 105
hecale, Neozephyrus 3:79
hecebolus, Panoquina 21:167
hecla, Colias 2:13; 5:25, 42; 7:125; 11:12; 14: 52; 19:57
hecla glacialis 11:48
hecla palamedes 15:212
hecobe, Eurema (IA) 18:160
hector, Papilio 3:39
hectorides, Papilio 22:154, 156
hedae, Papilio 4:41
hedemanni, Cyllopsis 21:191
hegemone HOURI, *Boloria (IA) 15:8*
hegesia, Pap. Nymph. Phaler. 12:197
hegesia, Euptoieta 5:56; 14:180; 18:19
hegesia hegesia 22:152
hegesia hoffmanni 8:25; 19:114 (IA); 21:193; 25:40
hegon, Amblyscirtes 9:195; 10:107, 159; 14:27, 239
hegon, Hesperia 20:241
heiligbrodti, Sysphinx 14:94, 103
heiligbrodti heiligbrodti 18:42
Prosopis glandulosa, Acacia farnesiana, Acacia rigidula. Larrea divaricata ?
heiligbrodti hubbardi 18:42
mesquite
helena, Argynnis 19:202
helena, Troides 9:75
helena helena, Boloria 8:65
helena hephaestus, Papilio 13:76
helenus, Papilio 3:62; 5:52; 11:9; 13:199; 15: 180 (IA,II); 17:76, 163-169 (IA)
Phellodendron amurense, Fagara ailanthoides, Poncirus trifoliata, Xanthoxylum piperitum, Citrus junos, Citrus unshiu, C. natsudaidai
helenus nicconicolens 8:96; 13:76; 14:5; 15:175; 18:106

Citrus, Fagara 8:96; *Citrus depressa* 18:106
helios, Graphium 21:157
helios, Papilio 4:66; 5:16
Vitex montevidensis 5:16
helios ARIANA, Hypermnestra (IA) 15:2
Heliothis 25:264-266
hellanichus, Papilio 4:65
hellmanni, Tapinostola 2:94
Calamagrostis epigeios
helloides, Lycaena 2:24; 4:14; 5:120; 6:105; 10:98; 11:165; 12:41, 44; 14:128; 16:69, 244; 17:24-26; 18:225; 20:134, 161; 22:96-98, 103; 23:187, 278
Polygonum careyi 4:14; *Polygonium* 17:24-26
helloides dorcas 7:125
helloides florus 2:92
helloides helloides 12:30
helmsii, Dodonidia 23:201, 202
helva, Mellana 21:247
helvina incanescens, Pierella 22:128
hemes memba, Milanion 21:162
hemichrysos, Lycaena 16:104
hemon, Mithras 21:151; 22:153
henrici, Callophrys 14:21; 20:133 (II), 135; 22:129; 24:196
henrici solatus 18:148, 149; 25:38
Diospyros texana 18:148-149
henrici, Incisalia 2:22; 7:34; 10:163, 164 (IH); 12:130; 21:207, 208, 278; 22:26; 24:44
henrici turneri 13:89; 17:153; 19:50
henshawi, Euptychia 19:114
herberti, Euphaedra 13:128
herennius herennius, Pythonides 21:88, 162
hermas, Phyciodes 21:98
hermes, Euptychia 14:127; 18:17; 19:48; 21:191; 22:127
hermes, Lycaena 13:238
Rhamnus crocea
hermes hermes, Hermeuptychia 21:91; 22:151
hermes sosybia, Euptychia 6:32, 76; 12:172; 13:72; 15:99; 21:207
hermes sosybius, Euptychia 17:23, 149; 18:138, 139; 20:102; 21:277; 25:41
Axonopus compressus, Stenotaphrum secundatum
hero, Coenonympha 11:164
hero neoperseis 8:100
Cyperaceae

herodotus, Callophrys 17:37; 21:195
herophile parsonsi, Calisto 4:72
herse, Nymphales Gemmati 12:197
hersiliata cervinifascia, Dysstroma 5:54
herzegovinae, Eana 16:178
hesione, Euptychia 18:17; 21:191
hesperia, Chlorochlamys 13:232
hesperiaris, Ægiale 11:114; 12:81, 89 (IC), 186; 23:247 (II); S1:27, 28
Agave americana S1:28
hesperis, Speyeria 5:32
hesseli, Callophrys 14:21; 20:135
hesseli, Mitoura 4:48; 5:18; 7:58; 9:4; 11:7; 17:27
white cedar, *Chamaecyparis thyoides*
heteronea, Lycaena 10:98; 14:128; 16:34, 69; 18:225
Eriogonum nudum 16:34
heteronea heteronea 22:172
heteropterus, Gesta 21:163
hewitsoni, Ogyris 20:134
hewitsoni brevipennis, Jemadia 21:157
hezia hezia, Callithomia 21:191
hianna, Atrytonopsis 4:44; 9:195; 10:23, 55, 101; 19:215-221
hianna turneri 14:28
hiarbas, Eurytela 25:146
Dalechampia hildebrandti, Ricinus communis
hibisci quinquefasciata, Orthosia (II) 20:213-215
Alnus rhombifolia, Quercus, Salix, Prunus, Populus, Betula
hicetaon, Mnasicles 21:165
hierta, Precis 11:167
hierta hierta (IA) 18:160
hildebrandtii, Charaxes 13:124
himeros, Papilio 4:64
himeros baia 21:88, 157
hippalus, Cogia 13:91; 15:91
hippalus outis 19:1, 23, 31; 25:35
Acacia angustissima v. *hirta*
hipparchus, Papilio 4:66
hippia, Aporia 14:52, 204
hippia japonica 8:97
Berberis (Berberidaceae)
hippoclus, Symbrenthia 11:167; 16:144
hippocrates, Papilio 13:151, 199; 14:5-18 (IA,II)
hippodrome, Chlosyne 13:165; 21:193
hippothoe, Lycaena 14:136

hippothoe, Palaeochrysophanus 9:204; 11:165
hircia, Periphoba (IA, II) 21:198-204
Fagus sylvatica, Quercus ilex, Crataegus oxyacantha, Prunus spp., Robinia pseudoacacia, Malus sp., Carpinus betulus
hirlanda phanasia, Hesperocharis 21:87, 155
hirundo, Leucostrophus 18:74
hisamatsusanus, Neozephyrus 3:79; 8:97; 13:178
Fagaceae 8:97
hispana, Agrodiaetus 14:138
hispania hispania, Erebia 14:140
hispania rondoui 14:140
hobomok, Poanes 9:195; 10:55, 107; 12:46, 130; 14:28; 16:243; 18:235; 19:196; 20:241; 22:120
hobomok ridingsi 16:245
hoffmani, Satyrus 18:44
hoffmanni, Agathymus 15:67; S1:38
Agave americana, A. parryi-like
hoffmanni, Megathymus 11:116
Agave scabra ?
hoffmanni, Melitaea 9:148
hoffmanni hoffmanni, Chlosyne 16:31
Chrysopsis breweri
hoffmanni MANCHADA *(IA) 13:208-211; 18:223; 21:71-73*
Aster conspicuus
hoffmanni segregata 13:211
hollandi, Speyeria 5:34
homerus, Papilio 4:65; 5:16
homothoas, Papilio 4:64
hopfferi, Agrodiaetus 14:137
hopkinsana, Epinotia 5:29
hoplitaria, Ceratonyx 21:13
horatius, Erynnis 7:146; 9:77, 194; 10:55; 12: 130; 14:38; 17:22, 156; 19:1, 12, 31, 180; 21:207, 277
Quercus hemisphaerica, Q. nigra, Q. texana, Q. virginiana, Q. shumardii, Q. gambelii, Q. fusiformis, Q. laurifolia, Q. marilandica, Q. phellos, Q. stellata
hormos, Hypsoropha 2:S8
Diospyros virginiana
horsfieldi, Allotinus 15:228; 20:134
horsfieldi, Kallima 3:40
hortaria, Epimecis 5:59, 60
hortense, Heliconius 21:193
hortuellus, Crambus 5:29
horus, Hesperia 9:192

HOSPITA, *Eupragia (IA, IG) 23:23-25*
hospita pseudognetus, Jemadia 21:244
howardi, Daritis (II) 15:172-174
Brickellia ?, Eupatorium rugosum (in lab.)
huanaco, Piercolias 15:119
HUASTECA, *Tirynthia (IA, IG) S2:44*
hubneri, Epiphile 21:101; 22:152, 156
huegelii, Lycaenopsis 11:167; 16:144
hugon, Polyommatus 12:120
hulsti hulsti, Drepanulatrix 13:232
humboldtii, Perisama 24:12
humboldtii albomaculata, Elzunia 15:29
humboldtii humboldtii 15:29
humboldtii judsoni 15:29
humerosana, Amorbia 10:154
humidicola, Cirphis 9:57
hungariae, Eana (SUBEANA) *16:177*
hungaricellum, Ateliotum 3:76
huntingtoni, Hemiargus 18:117
hutchinsonii, Aphnaeus 10:38
hya monostigma, Mesene 21:87, 149
hyale, Colias 3:39; 7:37, 58; 8:4; 11:12, 163; 13:24; 14:52; 18:41; 22:127
hyale hyalis 16:104
hyale poliographus 10:143
hyalinus, Pierella 24:298
hyalinus lyalinus, Pierella 24:17
hyalophora, Onenses 15:72
hybrida, Erebia 11:167
hydarnes, Cephise 24:68, 69
hydaspes, Callicore 21:100
hydaspe, Speyeria 2:5, 92; 5:31; 7:56, 111, 112; 19:172
hydaspe purpurascens 7:49; 17:113
hydaspe sakuntala 10:95, 102; 14:82; 16:69; 18:222
hygenius occulta, Euselasia 22:153, 156
hylaeus, Dolba 10:166
pawpaw
hylaeus floridensis 5:23
hylaeus prini 19:189
hylas, Agrodiaetus 14:137
hylas, Neptis 11:167; 16:144; 22:253
hylas, Polygonia 10:96; 14:238; 19:113
hylas luculenta, Neptis 18:108
Pueraria montana, Glycine max
hylas virescens, Cephonodes 18:74
hylax, Zizula 18:74

hymen, Dapidodigma 13:145
hymen, Iolaus (IA) 13:145
hymetus taygetus, Ogyris 20:134
hymetus taygetus, Thysonotis 22:121
Hypaurotis 24:14
hypensor, Pyrrhogyra 18:20; 21:194
hyperanthus, Aphantopus 9:204; 11:164; 14:141; 19:159; 22:128
hyperbius, Argynnis 8:99; 11:167
Viola (Violaceae) 8:99
hyperbius hyperbius, Argyreus 14:194; 18:107
Viola pseudo-japonica
hyperborea, Diarsia 3:75
hyperborea, Rhyacia 2:94
Vacinium myrtillus
hypereia banosana, Eutresis 15:27
hypereia hypereia 15:27
hypereia hyspa 15:27
hypereia imitatrix 15:27
hypereia theope 15:26
hyperia, Pap. Nymph. Phaler. 12:197
hyperia, Chlosyne (IG) 13:165
hyperia, Biblis 14:188; 22:228, 230; 25:186
hyperia aganisa 18:21; 21:194
hyperia hyperia 21:103; 22:153
hyperici, Thecla 12:122
hypermnestra, Zerynthia 3:76; 13:199; 21:241; 23:85
hypermestra apicalis, Mestra 21:102; 22:153
hypermnestra undularis, Elymnias 18:74
hypophlaeas, Lycaena 2:25, 37; 6:38; 8:125-128; 9:77; 14:128
Rumex acetosella 2:25
hypozonius, Aguna 21:159
hyppason, Papilio 4:64
hypsenor, Pyrrhogyra 12:172
hyrcana, Polyommatus 10:117
hyselis, Symbrenthia 16:144

iarbas auriferax, Lymnas 21:148
iasis albomaculata, Epamera 13:145
iasis albomaculata, Iolaus (IA) 13:145
ibara, Coreana 3:79; 8:97; 13:178
Fraxinus longicuspis 13:178
iburiensis, Lycaeides 8:98
Leguminosae

inornata benjamini, Coenonympha (IA) 10:86
icarioides, Icaricia 14:128
icarioides, Plebeius 10:171; 13:232
icarioides, Plebejus (Icaricia) 15:34-42, 216; 16:34, 69, 17:10-20, 111; 18:226; 19:165-170; 20:54, 134; 21:122, 123; 24:196, 198 (II)
Lupinus albifrons, L. andersonii, L. arbustus silvicola, L. arbustus calcaratus, L. argenteus v. *tenellus, L. caudatus, L. chamissonis, L. densiflorus* v. *gloreosus, L. fulcratus, L. glacialis, L. holosericeus, L. latifolius, L. leptostachyus, L. meionanthus, L. micranthus, L. sericeus, L. variicolor. Trifolium ?*
icarioides ardea, Icaricea 20:52
icarioides ardea, Plebejus 17:112
icarioides blackmorei, Plebeius 18:186
icarioides buchholzi 13:28
icarioides lycea 10:98; 20:161
icarioides lycea, Icaricia 22:172
icarioides X melissa, Plebejus 16:235
icarioides missionensis, Plebeius 10:115
Lupine (chamissonis)
icarioides montis, Plebejus 23:115
icarioides pardalis 22:96, 97, 99, 100, 103, 104
icarioides pheres, Plebeius 10:115
icarus, Polyommatus 9:145, 204; 10:215; 11:165; 14:137; 18:41; 22:129
icarus persica 10:117
icarus rosina 10:222
icelus, Erynnis 10:55, 100, 107; 11:47; 14:38; 16:69, 144, 245; 18:227; 22:120
ictinus, Ialmenus 20:134
ida neapolitana, Epinephele 10:221
idae idae, Melinaea 15:31
idae vespertina 15:31
idaeoides, Papilio 12:115-117; 13:21
idaeus, Papilio 15:71
IDAHOENSIS, *Eana (Ablabia) 16:176, 182 (IG), 191*
idalia, Speyeria 1:69; 2:5; 9:77; 10:95; 12:38; 13:15; 14:215; 20:177; 21:137; 22:120, 128, 159; 23:188; 25:141, 196, 197
idas, Pap. Pleb. Urbicol. 12:197
idas, Lycaeides 9:145; 11:165; 14:137
idas, Plebejus 18:110
ideoides, Liptena (IA) 13:138
iduata, Xanthorhoe 5:54
iduna, Euphydryas 14:194
ignobilis, Fundella 24:254

ikelemba, Euryphene (IA) 13:137
ilaire, Appias 15:211; 17:37
ilia, Apatura 7:37; 14:196, 204; 18:110
ilia substituta 8:99
Salix, Populus (Salicaceae)
ilia, Catocala 10:167; 12:201; 17:159-162 (IA); 22:233, 235, 236; 23:5, 100; 24:108, 111-114
ilicis, Strymon 9:204; 18:110
ilicis, Strymonidia 11:166; 24:96
ilithyia, Byblia 25:146
Tragia brevipes, Dalechampia hildebrandti
illioneus illioneus, Caligo 21:93; 22:151
ilus, Papilio 4:66
imitata, Melanolophia 13:220
imitata, Melitaea 18:44
impecuniosa, Papaipema 8:59
Aster punicius, A. umbellata, A. lanatum
imperfecta, Leucania 9:57
imperialis, Arcas 21:151; 25:87-107 (IA,IC,IG)
imperialis, Eacles 8:29; 11:43; 13:75; 14:91, 103; 25:42, 234; gynandromorph 18:31; 20:118 (IA)
imperialis didyma 25:138
imperialis, Papilio 25:87
imperialis, Teinopalpus 15:248
impigritella, Glyphipteryx 19:134
impressale, Homoeosoma 24:255
improba, Boloria 2:65; 5:26; 11:50; 19:60
improba improbula 5:26
improba youngi 5:26
impulsa, Septis 5:54
ina, Methionopsis 21:165
inachus, Æromachus 8:95; 14:206
Gramineae
inachus, Kallima 11:167; 16:144
inachus eucerca 18:108
Strobilanthes glandulifera, S. tashiroi
inca, Argynnis 15:118
inca, Vehilius 21:165
inca inca, Ancyluris 18:24
incalis, Zellerina 11:178
incanana, Eana 16:178
incarnata, Lerina 21:273
incisa, Phlegethontius 25:152
inclusa, Clostera 21:35
inclusa, Melalopha 18:89
incognitana, Eana 16:178

inconspicua, Hulstina (II) 17:28-30
Adenostoma fasciculatum
inconspicua, Taygetis 6:98
incudella, Homosassa 24:40
indecisa, Agathymus 18:47; S1:54
Furcraea cabuya, Agave hennigan-type
indecisa, Megathymus 11:116
indica, Diaphania 17:227
indica, Vanessa 7:38; 8:99; 11:167; 14:195; 16:144
Boehmeria, Urtica (Urticaceae); *Humulus* (Moraceae); *Ulmus* (Ulmaceae)
indica indica, Vanessa 18:107
Boehmeria nivea
indra, Papilio 1:42; 4:63; 9:8, 49, 149; 10:99; 16:70; 17:112; 18:220; 19:39; 23:279
Lomatium grayi 18:220
indra fordi 8:68; 20:52; 22:49
indra indra 9:53; 16:29; 22:47, 49
Cymopterus terebinthinus 16:29
indra KAIBABENSIS *9:49-54 (IA);21:41-48 (II)*
Pteryxia petraea, Tauschia arguta 22:49
indra martini 22:46-52 (IA, II)
Lomatium parryi
indra minori 9:49-54; 16:137; 18:65-73; 20:52; 22:49, 167
Tauschia arguta?, Lomatium eastwoodae, L. grayi?
indra pergamus 9:53; 18:65; 20:52; 22:49
Tauschia arguta 18:65
ines, Melanargia 10:221
ines colossea 10:221
ines jahandiezi 10:218, 221
inexpecta, Lühdorfia 8:96
Asiasarum (Aristolochiaceae)
infans oregonensis, Brephos 13:217
infecta, Anicla 2:S8; 4:26
Zea mays, Gentian sp.
inferior, Nasusina 13:232
infernalis, Styx 11:106
infida, Aphania (IA,IG) 11:27-28
infumatana, Archips 10:154
infuscata, Eana 16:178
innocentia, Yphthimoides 21:91
innotata, Apatela 23:5, 7
innubens, Catocala 10:167; 24:109, 111
ino, Brenthis 7:37; 11:163
ino ino 14:193

ino tigroides 8:98; 14:193
Filipendula (Rosaceae)
ino, Caria 21:55
ino ino 21:56
ino melicerta 21:53, 56 (IG); 24:13, 14
inops, Pachyneuria 21:161
inornata, Automeris 13:85
inornata, Hydrelia 5:54
inornata, Taygetis 6:98
inornata, Coenonympha 6:105; 10:170; 11:164; 24:202
inornata benjamini 11:2
inornata insulana 10:170
inornata macisaaci 23:36
inornata PHANTASMA *(IA) 10:81-84*
inornata benjamini, Coenypmha 9:141
inornata columbiana 9:141
inornata insulana 9:141
inornata nipisiquit 9:141
inquaesita, Papaipema 8:59
Onoclea sensibilis
inquietana, Olethreutes 2:65
inquinatalis, Udea 5:29
insignis, Eremberga 24:255
insignis, Nastra 21:165
insignis insignis, Olyras 15:27
insignis praestans 15:27
insignis translucens 15:27
insueta insueta, Leucania 9:56
INSULARIELLA, *Acrocercops 25:196, 198, 199, 206 (IA, IG)*
Quercus agrifolia
integrifascia, Autochton 21:160
interior, Colias 5:19, 42, 53; 10:9, 107; 12:50; 16:245; 18:227; 23:267
Vaccinium canadense 10:9
interior interior 22:34; 25:108
interior laurentina 4:11, 12; 23:39, 40 (IA)
intermedia, Holomelina 19:74
intermedia, Philotes 8:8
intermedialis, Herculia 13:67
intermediata, Xanthorhoë 23:4
intermistana, Olethreutes 12:50
intermixta, Diestogyna 13:137 (IA)
interpunctella, Plodia 2:65
interrogationis, Polygonia 3:17, 48; 5:69; 6:73; 9:78; 10:55; 12:130; 17:23, 151; 19:52, 180; 21:207, 277; 22:120; 23:100, 193; 24:305; 25:

39, 140, 213
intersecta ilona, Elbella 21:158
intestinata, Horisme 13:218
introferens, Chorizagrotis 13:26
inusitatella, Lithocolletis 25:207, 209
invirae remoliatus, Opsiphanes 21:92; 22:151
io, Automeris 10:166; 13:26; 14:92; 17:44; 18: 27-31; 25:234-238
gynandromorph (IA) 18:27-31
mosaic gynandromorphs (IA) 25:234-238
io lilith 25:138
io, Inachus 9:204; 11:164; 14:204
io, Nymphalis 3:39; 7:38; 14:195; 18:40; 22:128
io geisha 8:99
Humulus (Moraceae), *Urtica* (Urticaceae)
iolas, Iolana 24:196
iole, Nathalis 7:52; 147; 10:100; 12:29, 130; 13:99, 100, 226, 227, 232, 234; 14:238; 15: 198; 16:203; 17:22, 155; 19:112, 232; 21:137, 207, 277; 22:229; 25:37, 127, 214
Thelesperma trifidum 13:226-227
iole alayoi 4:72
iphicla, Limenitis 18:21; 21:195
iphicla iphicla 21:99
iphidamas, Parides 18:14; 21:189
iphidamus, Papilio 4:41
iphigenia, Agrodiaetus 14:137
iphigenia, Boloria 8:98
Viola (Violaceae)
iphis, Coenonympha 14:139
iphita, Precis 3:40; 11:167; 16:144
iphitas, Papilio 4:67
iphthime, Ageronia 4:67
iphthime, Hamadryas 21:195
iphthime gervasia 21:103; 22:153
ipomoea, Syntomeida 9:206, 209;
ipomoeae, Syntomeida 11:148
iranica, Erebia 14:140
irene, Catocala 22:58
irene, Speyeria 5:34
iricolor, Pteroteinon 13:146
iris, Apatura 11:163
irma, Cynea 21:166
iroides, Callophrys 14:21; 22:225, 226
Malus, Ceanothus, Cuscuta, Gaultheria, Arbutus menziesii, Chlorogalum pomeridianum. Sedum ?
iroides, Incisalia 2:42; 7:146; 10:169,170; 12:30; 13:102, 103, 231, 234; 20:160; 23:280; 24:258

Gaultheria shallon 10:170; *Arbutus* 13:103
irrorata venosata, Semiothisa 13:219
irus, Callophrys 14:21; 16:195; 19:50
irus, Incisalia 7:34; 14:127; 17:108, 109
Lupine 17:108-109
irus, Incisalias 5:70
irus, Polyommatus 12:120
isabella, Heliconius 24:299
isabellae dianasus, Heliconius (Eueides) 21:96; 22:152
isabella, Isia 2:13; 13:26
isabella, Pyrrharctia 11:148
isabellae, Poecilopsis 9:199, 200 (IG), 202
Larix 9:202
isandra, Melete 21:190
isidorus, Papilio 4:64; 15:117
ismenias, Argiolaus 13:145
ismenias, Iolaus (IA) 13:145
ismeria, Melitaea 8:141; 12:122, 130; 13:212
ismeria, Phyciodes 4:43
ismenius, Heliconius 18:14
ismenius telchinia 18:18; 21:192
isobeon, Calycopis 18:24
isola, Echinargus 10:112
isola, Hemiargus 1:9, 42; 2:37; 7:52, 146; 10: 98, 166; 12:30, 130; 13:99, 100, 225-227, 232; 16:203; 19:55, 112, 136, 137, 232; 20: 161; 21:207, 208, 278; 22:96, 97, 99, 100,103
Indigofera leptosepala 13:225-227
isola alce 2:105; 22:228, 230; 25:38, 214
isola isola 18:153, 154
Prosopis juliflora, Indigofera miniata v. *leptosepala, Melilotus indicus, Dalea pogonanthera, Albizia julibrissin, Acacia roemeriana, Indigofera lindheimeriana, Dalea pogonanthera* v. *walkerae*
isophthalma pseudofea, Brephidium 22:35
isthmia, Mechanitis 24:298
isthmia bolivarensis 22:109
italica, Eana 16:178
Ithomiidae 22:108-110
ithra, Phyciodes 21:98; 22:152
ituriensis, Diestogyna (IA) 13:137
itylius, Autochton 6:75
itylus, Autochton 21:160; 22:155
itys strigosus, Anaea 21:104; 22:153
itysalis, Udea 13:242
iulia iulia, Colaenis 21:97; 22:152

iulia iulia, Dryas 21:192
iulia hispaniola 5:56
ivallda, OEneis 9:186; 20:161; 22:60

jäckhi, Eana 16:178
JACKI, *Stallingsia (IA,IG) 17:85, 86; S1:4*
Manfreda sp.
JACKSONI, *Archichlora (IA,IG) 25:170,171,173*
jacksoni, Papilio 24:72
jacobeae, Hipocrita 18:110; 19:227
jacobeae, Tyria 11:149
jada, Dolymorpha 20:135
jada, Strymon 14:22
jaguarina, Schinia 6:53
j-album, Nymphalis 2:13; 3:17; 5:26; 6:33; 7:51; 10:55, 107; 11:164; 14:62; 15:101; 16:128, 244; 18:228; 20:124; 22:252
j-album watsoni 7:125
j-album watsoni, Aglais 15:218
j-album, Polygonia 11:47
j-album, Vanessa 12:122; 13:212; 15:218
jamaicaria, Epimecis 5:59
jamaicensis, Danaus 9:68; 15:222
jamaicensis, Erinnyis 2:86
jamaicensis, Phemiades 14:31
jamaicensis, Smerinthus 15:193; 21:27, 253
Prunus virginiana v. *demissa 21:27*
cherry, birch, poplar, willow 21:253
jamaicensis geminatus, Smerinthus 5:54
jamaicensis, Xylophanes 9:136
janais, Chlosyne 13:165, 207 (IG); 14:148; 15: 199; 21:193; 25:40
janias, Papilio 12:97
janira, Epinephile 16:104
jansonis, Pelopidas 8:95
Miscanthus (Gramineae)
januaris, Melipotis 17:226
japonica, Arhopala 7:37; 8:97
Cyclobalanopsis, Quercus (Fagaceae)
japonica, Astilbe 14:205
japonica, Dictyoploca 14:103
japonica, Hestina 7:37; 8:99; 14:196
Celtis (Ulmaceae)
japonica, Lühdorfia 8:96; 11:9; 13:199
Heterotropa, Japonasarum, Asiasarum (Aristolochiaceae)

japonica, Narathura 14:138; 18:106
Quercus acutissima
japygia suwarowius, Melanargia 3:75
jasius, Charaxes 3:76
jasminearum, Chlaenogramma 17:102; 19:189
jason, Graphium 24:304
jason, Mylon 21:164
jason, Papilio 13:22
jatrophae, Anartia 3:25; 14:188; 15:72, 217
jatrophae guantanamo 9:205
jatrophae jamaicensis 15:218
jatrophae jatrophae 21:98; 22:152
jatrophae luteipicta 15:217; 18:20, 144, 145; 21:194; 22:128
Ruellia occidentalis 18:144,145
jatrophae quantanamo 15:217
jatrophae saturata 15:218
java teutonia, Anapheis 3:40
JAVANICA, *Eressa* 8:139
jeanneli, Syntarucus 10:38
jebus, Chalybs 21:152
jesous, Azanus 10:38, 222
jezoensis, Favonius 3:80; 8:97; 13:178; 14:206
Quercus 8:97, *Quercus crispula* 13:178
jimna, Taygetis 6:98
JIVARO, *Arcas* (IA,IG) 25:89, 93, 94, 105-107
joannisi, Eana 16:178
johannae johannae, Baeotis 21:148
JOHANNI, *Phlegethontius* (IA,IG) 11:107-112
johnsonaria mollisaria, Euchlaena 13:220
johnsoni, Callophrys 13:21; 16:33, 193, 194; 18:49, 228; 20:135
Arceuthobium campylopodum ?
johnsoni, Mitoura 13:183
JOHNSTONI, *Decodes* (IG) 14:119-121
johnstoni, Platyptilia 4:29
jolas, Jolana 3:75
jole, Pap. Nymph. Gemmat. 12:197
jonasi, Shirôzua 8:97; 13:178
carnivorous on oak-aphids (*Quercus dentata, Q. acutissima, Q. serrata*) 13:178
jonasi, Thecla 3:79; 7:45, 46
aphidivorous
jonia, Thaleropis 14:196
jorulla cinctus, Rothschildia 16:61
josephina, Ascia 21:190
josephina, Pieris 12:120

jovianus fabricii, Pythonides 21:162; 22:155
juba, Hesperia 16:36, 69; 18:227; 22:94, 97, 98, 102, 130, 169
jubararia, Synaxis 13:220
jucunda, Diarsia 5:54
jucunda, Eurema 2:51; 6:75; 10:204-206
jucunda, Xanthidia 12:122
judicaël, Papilio 4:65
judith, Catocala 10:168
juditha, Catocala 24:111, 114
judithae, Agathymus 13:91, 206; 23:245; S1:36
Agave parryi S1:36
juglandis, Cressonia 10:166; 15:193; 19:189
juglandis robinsonii 21:253
Ostrya, Carya cordiformis, Carpinus, Fagus, Corylus, Alnus
julia, Colaenis 4:67
julia, Dryas 14:180; 15:71
julia cillene 25:138
julia delila 18:19; 22:75, 76
Passiflora lutea v. *glabriflora, Passiflora incarnata*
julia moderata 25:40
julia nudeola 9:209; 12:200
julia, Lerodea 13:221-227
Stenotaphrum secundatum
juliae, Agathymus 15:70; S1:40
Agave parryi-like
JULIAE, *Megathymus (IA,IG)* 11:125-127
Agave parryi ?
juno, Agraulis 15:71
juno, Dione 14:180
juno huascama 21:192
juno juno 21:97; 22:152
juno, Hemileuca 2:81
Prosopis sp.
juntura, Catocala 25:42
jurtina, Maniola 5:28; 9:145, 204; 11:165; 14:140; 17:175; 18:41; 22:127; 25:157-159
jurtina cassiteridum 5:28
jurtina iernes 5:28
jurtina splendida 5:28
jurtina, Papilio 10:110
jurtina fortunata, Epinephele 10:221
jutta, Oeneis 5:25, 26, 42; 10:120; 11:48; 14:141; 15:124; 18:236; 19:57; 20:161; 23:157-159; 25:150
jutta alaskensis 6:105; 7:124

jutta ascerta 23:159; 24:43
jutta reducta 9:141; 23:159; 24:306, 307
jutta ridingiana 16:243; 19:242; 23:103, 158
juturnaria, Sericosema 13:219
juvenalis, Erynnis 3:48; 7:146; 9:195; 10:55; 12:130, 202; 14:38; 16:36; 18:190, 191; 21: 207, 277; 22:98, 120, 130
juvenalis juvenalis 17:156; 20:35-37
  Quercus nigra, Q. fusiformis, Q. stellata, Q. marilandica
juvenalis propertius 12:31
juventus, Callimormus 21:165, 246

kali, Thecla 18:44
kalikamaha, Thecla 21:196
kallina, Doxocopa 22:153, 156
kalmiae, Sphinx 15:64, 193; 17:36; 19:189; 23:100
  Diervilla lonicera, Nemopanthus mucronatus
karschi, Acraea 13:146
karschi, Euphaedra 13:128
karwinskii, Smyrna 15:198
katahdin, Œneis 12:50
kearfottella, Acrobasis 24:254
KENDALLI, *Zamagiria* 24:249-253 (IA, IG)
  Croton fruticulosus
keneza, Taygetis 21:90
KENORA, *Lithocolletis* 24:274, 277, 280, 281 (IA, IG)
  Salix spp.
kerea, Taygetis 6:98; 21:90
kindermanni, Smerinthus 18:37
kingi, Satyrium 13:71; 14:20; 16:197; 25:143
kirbyi, Ornipholidotos 13:138
klagesi, Papilio 4:41
kleis, Clerome 13:20, 21
klugii, Dircenna 15:221; 18:17; 21:191
knysna, Zizeeria 10:38
knysna karsandra 18:107
  Amaranthus spinosus, A. ascendens
kollari, Euploea 3:39
kolyma connectens, Catacore 21:87, 101
korallion, Temenis 21:102
krausei, Mimacraea 13:146

kriemhild, Boloria 7:50; 20:114-116 (IA)
krueperi, Pieris 4:74; 10:117
KUCHI, *Melitaea (IA,IG) 15:7-8*
kuhlweini, Endrosa 3:76
kühniella, Anagasta 17:73, 22:174
kühniella, Ephestia 6:13
kuldjaënsis, Eana 16:178
kumskaka, Atrytone 9:191; 13:15
kurava septentrionalis, Nacaduba 8:98; 18:106
Bladhia (Myrsinaceae), *Ardisia sieboldii*

labdaca, Libythea 3:39; 18:74
labe, Euptychia 21:191
labradus, Zizeeria 20:134; 24:196
labruscae, Pholus 2:86; 3:78
lacaena variegata, Diaeus 21:161; 22:155
lacertinaria, Drepana 13:8
Betula (birch)
laceyi, Lithophane 25:31, 42
laceyi, Logoa 25:31, 43
laceyi, Strymon 14:22; 18:152, 153; 20:65
lachesis, Melanargia 14:139
lacinia, Chlosyne 7:146; 11:164; 12:130; 13: 165; 14:148 (IG); 15:199; 16:61; 19:113, 232; 20:203; 24:135-142
Helianthus annuus, Viguiera deltoidea, Ambrosia trifida v. *texana, Verbesina virginica, Helianthus cucumerifolius, Silphium* sp., *Viguiera dentata, Simsia calva*
lacinia adjutrix 13:224-227; 19:113; 22:229; 25:40, 214
Helianthus annuus, H. cucumerifolius, Ambrosia aptera, Verbesina virginica
lacinia crocale 19:113
lacinia lacinia 18:19; 21:193
lacinia nigrescens 19:113
lacinia quehtala 14:148
lacinia rufescens 15:219; 19:113
lacinia saundersi 21:98; 22:152
lacondones, Papilio 4:67
lacrimosa, Lepidochrysops 10:38
lacrymosa, Catocala 2:37; 10:167; 23:263
lacteata, Enchoria 13:219
lacturnus kawaii, Everes 8:98; 18:107
Vicia, Desmodium heterocarpum (Leguminosae)
lacustra, Erynnis 13:233
lacustrata, Xanthorhoe 13:218
laeta, Erora 1:42; 2:22; 3:44; 6:34; 9:146; 10:

18-24 (IA,IH), 161; 14:239; 15:92; 16:1-4; 17:43; 18:162 (IA); 25:222,295
Fagus, Corylus 16:1-4
laeta sanfordi 16:1-4
laeta, Eurema 14:52
laeta bethesba 8:96; 18:106
Cassia, Cassia mimosoides ?
laeta lineata 22:121
laeta, Holomelina 19:74
laetabilis, Paramiana 19:165
laetitia, Euryphene (IA) 13:128
laertes, Morpho 13:86
laertes laertes, Prepona 21:104
lafresnayei lafresnayei, Vettius 22:156
lafresnayei pica 21:88, 166
lagonellus, Megacraspedus 3:76
lais, Speyeria 5:34
laius, Chilades 11:167; 16:144
LAJITAENSIS, *Agathymus (IA,IG) 18:174-185; S1:45*
Agave lecheguilla
l-album, Nymphalis 11:164
l-album samurai, Polygonia 8:99
Betula (Betulaceae)
lallata, Galenara 25:270
lamae, Holomelina (IA) 19:69-74
lamarchei, Papilio 4:64
lamis lamis, Juditha 21:87, 150
lamis molpe, Peplia 18:25; 21:197
lamprospilus, Isoteinon 8:95
Miscanthus (Gramineae)
lancea, Pythonides 21:162; 22:155
lanceolata, Anthocaris 22:96, 97, 99, 100, 102
lanceolata australis 12:28
langtoni, Alypia 5:54
lankana, Rapala 20:130, 135
lanoraieensis, Callophrys 14:21
lanoraieensis, Incisalia 3:44
lansdorfi, Phyciodes 21:97; 22:152
laodamia, Hamadryas 21:103; 22:153
laodamia laodamia 21:195
laodamus, Papilio 4:41
laodice, Argyronome 14:193
laodice, Argynnis 3:75; 7:37
laodice japonica 8:98
Viola (Violaceae)
laogonus nicola, Telemiades 21:88, 161
laothoë liberia, Temenis 12:172; 18:20

laothoë bahiana 21:88, 102; 22:152
laphira, Hypothyris 21:88, 95; 22:151
lapponaria, Poecilopsis 9:200 (IG), 202
Erica, Calluna, Betula, Crataegus
lapponica, Sympistis 2:65
lara, Leptomyrina 10:39
larissa, Melanargia 10:118
larua, Taygetis 6:98; 21:90
larunda, Melitaea 18:44
larvaria, Anacamptodes 5:54
larvata, Libythea 18:44
lasiomera, Cirphis 9:57
lassauxi, Erinnyis 3:78; 25:152
lassauxi merianae 5:20; 9:136
lassauxi omphaleae 5:20; 9:136
lassia, Mylon 21:245
LASSULA, Ithome 15:89
Leucaena glauca
lasti, Charaxes 18:104
Afzelia quanzensis (toxic)
latex, Polia 5:54
lathonia, Argynnis 3:39; 11:167
lathonia, Issoria 10:221; 11:163; 14:193
latifascia, Hemileuca 2:81
Salix sp.
latifascia, Tigridia 21:104
latifasciata latifasciata, Sophista 21:161; 22: 155, 156
LATIMARGINALIS, Syllepsis (IA,IG) 24:288-290
latipalpis, Hormoschista 10:112
latipennis, Epistor 2:86
latipes, Mocis 2:S8; 10:112; 18:7
Syntherisma sp.
latreillei, Ocnogyna 4:50
laure, Chloripe 15:72; 21:195
laure lauretta, Doxocopa 21:103
laurentia, Doxocopa 21:103; 22:153
laurentina, Hesperia 6:73; 10:23, 107; 16:245
laureolus loxa, Cymaenes 21:165
lavatherae, Carcharodus 14:51
laviana, Heliopetes 19:1, 7, 32; 25:35
Abutilon abutiloides, A. incanum, Malvastrum americanum, Sida filipes, Wissadula holosericea
laviana leca, Heliopetes 14:32
laviana libra, Heliopetes 21:164
lavinia, Pieris 10:55, 97; 12:130; 16:33; 23:182
Plantago, Mimulus ? 16:33

lavinia coenia, Precis 3:17; 12:29; 13:227, 234; 18:163
Antirrhinum antirrhiniflorum 13:227
lavinia zonalis, Precis 9:209
lavinia zonalis, Junonia 17:225
lawrencei, Euphydryas 13:172
LAZA, *Cymaenes (IA,IG) 22:4, 10, 11, 13*
leander, Coenonympha 3:76
leanira, Chlosyne 17:113
leanira, Melitaea 21:271; 22:98-100, 103
leanira wrighti 12:29
lecheana, Cacoecia 13:6, 8
leda, Melanitis 6:43; 8:100; 11:167; 16:132-134 (IA), 144; 18:159 (IA)
Gramineae
leda africana 18:74
leda leda 18:108
grasses
leda, Strymon 14:22; 19:112, 232
leda ines 19:112
lederer, Thecophora 17:169
lefeburei bossardi, Phlegethontius 5:22
lefebvrei, Erebia 14:140
leilia, Asterocampa 14:190; 19:113; 22:128,230; 23:182
leilia cocles 6:33
lemberti, Callophrys 17:112, 114
lemmeri, Lithophane (IG) 23:10, 11, 13
lemnos, Anthene 10:38
lemolea, Spalgis 7:45
lemonias, Precis 16:144
lenitas, Symbiopsis (IA, IG) S4:9
leo leo, Polygonus 21:159
leo savingny 14:34
leonardus, Hesperia 9:195; 10:107; 12:38; 13:16; 14:30, 37; 19:133, 137; 23:186
leonidas, Papilio 9:129
leoninus, Thymelicus 8:95
Agropyron (Gramineae)
leopardina hilaria, Symmachia 21:87, 149
lepcha, Mycalesis 11:167; 16:144
lepigone, Athetis 3:75
lepigone, Radinogoes 2:94
lepita, Libythea 11:167
leporina, Acronycta 2:94
birches and grasses
leptotypa, Euryphene 13:128
lerna, Limenitis 21:195

leroma, Crudaria 10:39
lesbia, Colias 3:40; 7:139; 8:163
lethe, Hypanartia 21:98, 194; 22:152
leucaspis, Papilio 4:67
leuce, Eurema (Pyrisitia) 21:154; 22:154
leucitys, Symmerista 5:18
maple
leucocera, Celaenorrhinus 11:168; 16:144
leucocycla, Lasiestra 2:65
leucodyce, Aporia 11:168; 16:144
leucomelas, Actinote 21:192
leucone leucone, Nastra 21:246
leuconoe, Idea 9:76
leuconoe movia, Hestia 12:115
leuconoe riukiuensis, Idea 18:107
Cynanchum formosanum
leucophaea, Olene 4:26
Croton ?
leucosia, Nymphidium 21:88, 150
leucosphenia, Cirphis 9:57
leucostigma, Gortyna 2:93, 94
Iris pseudacorus
leucostigma, Hemerocampa 4:26; 18:150
Halesia sp.
leucotaenia, Tanaecia 13:21
leuderwaldti, Cercyeuptychia 25:15 (IA,IC,IG)
leuderwaldti, Cercyonis 25:13-16, 18
levana, Araschnia 7:37; 11:164; 14:195; 24:77; 25:163
levana obscura 8:99
Laportea, Urtica (Urticaceae)
levina, Levina 21:164
levis, Mesosemia 21:87, 146
lewisii, Papilio 12:101
lherminier, Hesperia 12:120
lherminier, Nastra 14:40; 17:157; 19:196; 21: 207, 277
lherminieri, Lerodea 9:195; 10:55
lherminieri, Pieris 12:120
libeon, Coeliades 13:146
liberia, Hesperia 14:30
libertana, Grapholithia 8:26
libocedrus achotla, Sphinx 5:22
liborius liborius, Antigonus 21:162; 22:155
libya, Pholisora 12:31; 14:32
libya lena 22:169
libya, Xylophanes 25:152
libye, Euptychia 21:191

libye, Magneuptychia 21:92
licinia, Pap. Dan. Cand. 12:197
ligea, Erebia 11:164; 14:140
ligea takanonis, Erebia 8:99
Carex (Cyperaceae)
ligeae, Schinia 25:274-277 (IA, IH, II, IP)
Machaeranthera tortifolius
lignata, Xanthorrhoe 3:76
lignosellus, Elasmopalpus 19:134
ligulella, Dichomeris 19:134
lilina, Calociasma 21:197
lilis, Melinaea 24:298
lilis dodona 15:30
lilis erica, Melinaea 15:30
lilis ezra 15:30
lilis flavicans 15:30
lilis imitata 15:30
lilis lilis 15:30
lilis messatis 15:30
lilis parallelis 15:30
lilis sola 15:30
lilius, Erynnis 17:112
limaea limaea, Pythonides 21:162
limbata, Dynamine 21:99
limbata, Nematocampa 13:220
limbata nigella, Hesperia 15:119
limitaria nigroangulata, Nyctobia 13:217
limniace, Danaus 3:39; 11:229
limnorina, Dismorphia 21:155
limpida, Papaipema 8:60
Arctium sp.
linda, Eurema 2:51
lindseyi, Hesperia 9:198; 13:239; 22:94-105
linea liberica, Adopaea 10:222
lineata, Celerio 2:86; 3:48; 7:11, 12; 10:166; 11:43; 15:193; 19:189; 21:144; 25:152
knotweed, tomato, grape, Virginia creeper, dewberries, peas, rose, elderberry, lilac, evening primrose, Russian thistle, alfalfa, boysenberry, sand lilies, garden flowers
Parasites 3:48; 7:11
lineata livornica 4:50
lineata, Hyles 25:41
lineata, Phragmatobia 21:38
lineata brunnea, Leucophlebia 21:8
lineata formosana 21:8
linella, Arivaca 24:40
lineola, Adopaea 6:92; 7:56; 9:142, 204; 10:

222; 11:166; 12:15; 16:62, 243; 17:108; 18: 116; 24:46

lineola, Thymelicus 2:105; 4:14; 10:151; 12: 173; 14:51; 19:195, 196; 20:28, 243, 244; 22:20-25, 120; 23:76, 86, 249, 250; 24:2; 25:222; form pallida 23:249,250; 22:25
Agropyron repens, Agrostis alba, Phleum pratense

lineola KUSHANA, Thymelicus (IA) 15:18

lineosa, Lerema 21:166

lingeus, Cacyreus 10:38

lingulacella, Chrysopora 4:4

linita, Leucania 9:56

lintingensis cebrene, Precis 18:74

lintneri, Gluphisia 21:36

linus, Arawacus 21:152

linus togarna, Thecla 18:23; 21:196

liodes, Emesis 18:25

liparops, Satyrium 14:20; 16:195

liparops aliparops 22:163

liparops liparops 17:152; 19:49

liparops strigosa 23:187; 24:2; 25:81, 82

liparops, Thecla 12:122

liparops, Strymon 9:11, 77; 10:107, 166; 14: 216; 15:215; 16:128; 15:108; 25:296
Crataegus, Malus, Prunus, Vaccinium, Azalea calandulacea (flowers & buds); *Ilex & Quercus*

liparops fletcheri, Strymon 16:244

liquoraria, Synchlora 13:217

liqustri, Sphinx 18:37

liria lirissa, Ectima 21:102

lisa, Eurema 2:51; 3:17; 5:55; 9:78, 208, 211; 10:23, 55; 12:130; 13:74; 14:46; 17:22, 155, 226; 18:31, 64; 21:190, 207, 277; 22:127, 132, 227, 229; 23:182, 264; 24:305; 25:37,214
albino 18:64

lisa euterpe 9:80; 18:16

lisa, Xanthidia 12:122

litaria, Apodrepanulatrix 13:219

lithosioides, Crambidia 11:148

littera, Fagitana 7:172
Aspidium thelypteris 7:172

livida, Anthene 10:38

lividus, Polygonus 14:34

lividus savigny 9:206

livornica, Celerio 4:61

livornica, Deilephila 4:72

locutia, Quinta 21:166
loewii, Polyommatus 10:117
logan, Atrytone 9:78, 191, 195; 10:55; 12:37, 41-44, 174; 13:15; 14:215; 24:243, 244
logan lagus 14:29
logan, Hesperia 17:103
lohita, Spindasis 9:14
Terminalia (Combretaceae), *Psidium* (Myrtaceae), *Xylia, Acacia* (Leguminosae), *Viscum* (Loranthaceae), *Grevillea* (Proteaceae), *Convolvulus* (Convolvulaceae), *Henslowi* (Santalaceae), *Loranthus* (Loranthaceae), *Dioscorea* (Dioscoreaceae)
loki, Callophrys 14:21; 20:135
loki, Mitoura 12:30; 13:234; 15:201
longimaculella, Ethmia 11:44
longior, Lithophane 23:13
Juniperus pachyhloea
longipennis, Autochton 21:244
longipennis, Xanthospilopteryx 13:128
longirostris, Soliana 21:167; 22:156
longstaffi, Hesperocharis (IA) 22:189, 190
longula, Callophrys 20:69, 70
lonicerae, Zygaena 18:110; 25:160
lorata, Sabulodes 13:75
lorea, Lacinipolia 5:54
lorquinaria, Elpiste 13:220
lorquini, Limenitis 3:106; 10:169; 16:33, 66, 69; 18:224; 21:225 (IA); 22:96, 98, 103; 25:278
willow, cottonwood, wild cherry, apple 18:224
lorquini burrisonii 20:172-176 (IA)
lorquini lorquini 12:29; 13:232; 20:172-176 (IA)
lorquinii, Zizera 10:222
losca, Elbella 21:157
lotus, Perichares 21:167
LOUISEAE, *Astraptes* (IA, IG) S2:6
lowii, Papilio 24:304
lubricalis, Epizeuxis 23:207-209
lubricans, Polytrema 16:144
lucens, Dasyspoudea 13:234
lucetius nubilus, Phlegethontius 5:22
lucia lucia, Lucida 21:165
lucianus pseudocrispus, Polystichtis 21:87, 149
lucida, Actizera 10:38
lucifer lucifer, Melinaea 15:33
lucifer lutzi 15:33
lucifer eryx 15:33

lucina, Hamearis 11:165
lucina, Symbrenthis 13:21, 22
lucilius, Erynnis 8:3; 9:78, 194; 10:23; 14:40, 103
lucilius afranius 13:232-234; 22:168
  Lupinus sp. ?
lucilla, Œneis 14:132
lucina, Nemeobius 24:196, 200
lucinda lucinda, Emesis 21:87, 149
lucinda fastidiosa 22:153, 156
lucinda saturata 18:25
lucius, Metacharis 21:147
lucretius, Vettius 22:156
luctuata obductata, Xanthorhoe 5:54
luederwaldti, "Cercyonis" 21:87, 90
lugubris, Epistor 2:86; 19:189
lugubris lugubris, Quadrus 21:245
luna, Actias 10:166; 14:103; 21:249-252 (IA), 255-257; 24:209-212; 25:42
luna rubromarginata 13:26
luna luna, Pierella 21:191; 22:128
luna heracles 18:17
luniferella, Oneida 5:29
lunigera, Selenephera 3:76
lunulalis, Oneida 5:29
lupina, Emesis 21:196
lupinus centralis, Epinephele 10:118
lurida, Nastra 21:165
lusca bahamensis, Perigonia 5:22
lusca major 5:22
lusca restitua 3:78
lusca tenebrosa 3:78
luscitialis, Mecyna 5:29
luscitiosa, Sphinx 15:121, 193
lucina, Hemileuca 2:81
  Spirea sp.
lutea, Japonica 3:79, 8:97; 13:175, 178; 14:204
  Quercus (Fagaceae)
lutea, Spilosoma 6:58
luteata, Eupithecia 5:54
luteata bifasciata 13:218
luteola, Euryphene 13:128
luteostriata, Diestogyna 13:137
luteizona, Hegesippe 21:157
lutosa, Rhizedra 2:94
  Phragmites communis
lyauteyi, Argynnis 10:215, 217, 221
lyca, Catagramma 21:194
lycaon, Hyponephele 14:140

lycon, Maniola 9:145
lycaon mauretanica, Epinephele 10:221
lycaste diomaea, Hypothyris 21:190; 18:16
lycaste negreta, Dynothea 15:221
lycidas, Pap. Eq. Achiv. 12:197
lyciades, Achalarus 3:17; 9:78, 194; 10:55; 12: 44, 130; 15:98; 17:22; 19:1, 27, 31, 51, 196; 21:207, 277; 23:187; 24:305; 25:35, 81
Desmodium ciliare, Lespedeza hirta, L. texana
lycides, Papilio 4:41
lycimenes, Papilio 4:41
lycimenes lycimenes, Parides 21:189
lycimnia paulista, Melete 21:155; 22:154
lycisca, Riodina 21:148; 22:153
lycopersicella, Keiferia 24:7, 8, 9 (IG)
lycophron, Papilio 4:64
lycopodiana, Sparganothis 3:36
lycormas, Glaucopsyche 8:98
Vicia (Leguminosae)
lycus, Strymon 15:215
lydia, Eurema 2:52
lydus, Bethys 15:215
lygdamus, Glaucopsyche 1:58; 5:19; 10:23, 107, 171; 11:47; 13:100, 102; 16:70; 18:41, 226; 19:167, 232; 20:131, 152; 21:123; 22:59, 120, 165; 23:149, 150; 25:241
Lathyrus couperi, L. ochroleucus, L. caroliniana, Lotus glaber, L. scoparius, Lupinus argenteus ?, L. micranthus, Vicia cracca, V. gigantea, Hedysarum boreal, Adenostoma fasciculatum
lygdamus australis 12:31; 13:232; 23:150, 151
lygdamus behrii 3:42; 10:114; 16:35; 17:218; 22:96, 97, 99, 100, 103, 104; 23:151, 152
Lupinus micranthus, Lotus glaber, Astragalus sp., Lupinus arboreus ? 23:151, 152
pupa in ant nest *3:42*
lygdamus columbia 18:186; 23:152, 153
lygdamus couperi 16:244
lygdamus mildredae 4:11
Lathyrus japonicus
lygdamus nittanyensis 10:159
lygdamus oro 2:92; 6:103; 8:65; 10:98
Lygris 25:263, 264
lysander, Papilio 4:41
lyside, Colias 12:120
lyside, Kricogonia 3:17; 9:207; 13:227, 228;

14:44; 15:72; 18:26, 163; 22:227, 229; 25: 37, 214
Porliera angustifolia 13:227, 228
lyside castalia 19:223, 224
lysimachiae, Papaipema 8:60
Lysimachia quadrifolia, L. terrestris
lysimnia, Mechanitis 13:84; 21:95; 22:151
lysimon, Zizera 10:222; 18:160
lysimon epiplatea, Nymphidium 21:88, 150
lysithous, Eurytides 14:173
lysithous, Graphium 21:88, 157; 22:154, 156
lysithous, Papilio 4:66
lyra cleadas, Lyropteryx 18:24
lytrea lytrea, Anartia 5:56

maacki, Papilio 3:62; 14:203, 206; 14:206
maackii, Papilio 5:52; 7:37; 11:9; 13:199;15:182
Phellodendron amurense, Fagara ailanthoides, orange (Rutaceae). (IA,II) 15:182
maackii tutanus 8:96
Phellodendron, Orixa (Rutaceae)
macalpinei, Agathymus 13:91; S1:36, 37
Agave related to *scabra*
MACALPINEI, *Anthoptus* (IA, IG) S2:27
macaria, Speyeria 3:106
macaira, Heliopetes 19:1, 9, 32
Malvaviscus drummondii
macaira orbigera 21:163; 22:155
macaira macaira 21:246
macarius, Nisoniades 21:161
macdunnoughi, Pero 10:168
Artemisia californica, Eriogonum fasciculatum, privit
macelhosiella, Ethmia 11:44
macfarlandi, Sandia 14:127
macfarlandi, Callophrys 14:21; 16:55-57 (IA,IH, II); 19:82, 177, 178; 20:135
macfarlanei admiralia, Graphium 25:74-76 (IA)
maculipennis, Plutella 17:78
machaon, Papilio 2:25; 3:39, 76; 4:63; 5:26; 7: 167; 8:4, 141; 9:7, 49; 10:47, 202, 215, 218; 11:9, 162; 13:151, 199; 14:5, 203; 16:143; 18:40, 41; 19:57; 22:126; 24:229, 232
fennel, rue
machaon aliaska 6:104; 7:126; 9:133; 11:36 (IA); 11:48, 49
machaon centralis 10:117

machaon dodi 9:141
machaon hippocrates 7:37; 8:96
Angelia, Cryptotaenia, Daucus, Foeniculum, Glehnia, etc. (Umbelliferae); *Phellodendron,* etc. (Rutaceae)
machaon hudsonianus 12:124 (IA); 21:136
machaon maxima 10:220
machaon saharae 10:220
machaonides, Papilio 4:64
macilentus, Papilio 5:52; 8:96; 11:9; 14:5, 203, 207
Orixa, Fagara, Zanthoxylum Poncris, etc. (Rutaceae) *8:96*
mackinnoni, Papilio 13:76
macmurtrei, Prionoxystus 21:34
macounii, Oeneis 10:107; 11:47; 16:243; 19:125; 21:136, 258-260; 23:155-157; 24:43
macris phasis, Evonyme 21:87, 102
MACROCARPAE, *Cameraria (IG) 24:86, 87*
Quercus macrocarpa
macularia crocearia, Sicya 13:220
maculata, Halisidota 2:13
maculata, Oligoria 17:157
maculata, Tegeticula 20:54
Yucca whipplei
maculatus, Pyrgus 8:95; 14:206
Potentilla (Rosaceae) 8:95
Maculinea spp. 24:196, 199, 200
maculosus, Megathymus 11:114; 12:81, 87
Manfreda maculosa 11:114
maculosus, Stallingsia 12:89 (IC); 13:90; 14: 42; S1:4
Manfreda maculosa S1:4
madagascariensis, Chrysiridia 9:65
Omphalea biglandulosa (Euphorbiaceae)
madusaria ochrearia, Euchlaena 13:220
madyes, Papilio 4:41
maelus brunnea, Melinaea 15:31
maelus cydon 15:31
maelus madeira 15:31
maelus maelus 15:31
maelus mayi 15:31
maelus purusana 15:31
maenalis obscurior, Cyrestis 24:304
maenius cocana, Melinaea 15:32
maenius hicetas 15:32
maenius juruaensis 15:32
maenius maenius 15:32

maenius pothete 15:32
maenius tarapotensis 15:32
maeonis borealis, Melinaea 15:31
maeonis maeonis 15:31
maeonis zamora 15:31
maeotis, Mesosemia 21:87, 146
maera, Dira 9:204
maera, Pararge 11:164; 14:139
maerula, Anteos 9:206; 13:62, 90; 15:198
maerula lacordairei 22:34
maerula, Gonepteryx 4:67
maesites, Strymon 9:207; 14:22
maestosa, Catocala 10:168
maestosa, Eupithecia 13:218
magdalena, Erebia 2:37; 14:132; 19:59, 129
magdalena magdalena 22:172
magellanus, Ornithoptera 13:21
magnarius ochreatus, Deuteronomos 13:220
magnifica, Cossula 12:180
magnoliata pernotata, Spargania 13:219
magnoliatoidata, Lobaphora 13:217
maha argia, Zizeeria 7:37; 8:98; 11:167; 14:136; 18:107
Oxalis (Oxalidaceae), *O. corniculata 18:107*
mahaguru niphonica, Gonepteryx 8:96; 14:52, 205; 24:214 (IA)
Rhamnus (Rhamnaceae)
mahendra, Neptis 11:167; 16:144
maia, Hemileuca 2:81; 5:71; 8:29; 10:46, 166; 14:96, 103; 15:121; 25:42
Quercus, Prunus, Corylus 2:81
maillardi, Crymodes 3:75
maja, Evonyme 21:102
mckinleyensis, Oeneis 7:124
malachitana, Eumarozia 8:102
malaya iwasakii, Megisba 18:107
Mallotus japonicus
m-album, Eupsyche 6:75; 16:199; 20:135
m-album, Panthiades 15:226-232 (II); 19:50; 25:38
Quercus sp., Astragalus ?
m-album, Strymon 2:22; 6:75; 9:207; 10:55, 163; 14:21; 15:97; 24:44
m-album, Thecla 12:122
malitiosa pecta, Synapte 14:32
malvae, Pyrgus 11:166; 14:51
manataaqua, Polites 9:78, 195; 10:23, 55; 12:38; 14:30, 216

manataaqua, Hesperia 20:241
mancinus, Erebia 11:47
mandan, Carterocephalus 11:47
mandana, Emesis 21:149
mandana mandana 21:196
mandinga, Euryphene (IA) 13:137
manea schistacea, Rapala 20:135
manica, Charaxes (IA) 13:124
manitoba, Cleora 19:189
manni, Perigonia 9:136
manni, Pieris 3:76; 14:53
manni haroldi 10:218, 220
manto, Erebia 14:140
manueli manueli, Polygonus 21:159
marathon stenotaenia, Lymnas 21:148
marcia, Phyciodes 5:126
marcida, Epimecis 5:60
marcellina, Pap. Dan. Cand. 12:197
marcellus, Graphium 6:75; 7:52; 12:130; 21:207, 277; 22:34; 24:176-188
marcellus marcellus 17:153, 154; 18:132, 192
 Asimina parviflora, Asimina triloba
marcellus, Iphiclides 4:13
marcellus, Pap. Eq. Achiv. 12:197
marcellus, Papilio 1:75; 4:14, 66, 110; 6:38, 101; 8:141; 9:208; 10:163, 164; 13:74; 15: 211
marchandi, Papilio 4:67
marcus marcus, Vettius 21:166; 22:156
mardania orientis, Euryphene 25:80
mardon, Polites 18:228; 20:254
 Festuca ovina ? 20:254
margarita, Appias 15:211
margarita, Evonyme 22:153, 156
marginalis, Lethe 8:99
 Miscanthus (Gramineae)
marginata, Barbicornis 21:88, 147
marginatus, Prodoxus 20:54
 Yucca whipplei
marginidens, Papaipema 8:60
 Cicuta maculata, Sium suave
MARGUERITARIA, *Grotella (IA,IG) 22:137-143*
mariae, Agathymus 13:91, 93, 99, 204-206 (IA); 14:42, 58; 15:67; 18:171; S1:44
 Agave lecheguilla
mariae, Megathymus 10:109; 11:115
mariae mariae, Agathymus (IA,IG) 15:19-22
mariae MICHENERI *(IA,IG) 15:19-22*
 Agave lecheguilla

mariae X carlsbadensis (IA) 13:204, 205
mariae, Papilio 12:115
marianna, Chlosyne (IG) 13:165
marianna irrubescens, Chlosyne 13:169
marina, Leptotes 5:63; 7:52, 146; 10:112; 13: 100; 17:244; 19:34, 55, 112, 232; 20:134; 21:208, 22:96, 97, 99, 100, 103, 228; 25:38, 214
marinus, Leptotes 12:30, 130; 13:232, 234; 16: 203
mariposa, Lycaena 7:43; 10:171; 18:185-188, 225; 19:61; 21:273
mariposa penroseae 22:164
maritima, Papaipema 8:60
Helianthus giganteus, H. laetiflorus
marius, Pap. Eq. Achiv. 12:197
marloyi, Erynnis 10:118
marmorata, Marmopteryx 13:218
marmorea, Iolaus (IA) 13:145
marmorea, Stugeta 13:145
marmorinata, Erilophodes 18:114
maroccana, Eana 16:178
maroccana, Epinephele 10:215, 218, 221
maroccana nivellei 10:221
marsaeus clara, Melinaea 15:32
marsaeus macaria 15:31
marsaeus marsaeus 15:31
marsaeus rileyi 15:31
marshalli, Cacyreus 10:38
marsyas damo, Thecla 18:22; 21:196
marsyas, Polygonia 15:219
marsyas, Pseudolycaena 21:151; 22:153
martha, Echinargus 7:52
marthesia, Anaea (Siderone) 21:104
marthesia manitobensis, Macrurocampa 21:37
martialis, Erynnis 6:33; 9:195; 12:130; 13:89, 100; 22:120; 25:35
martialis, Strymon 9:207; 14:20; 17:226; 25:137
Trema floridana 17:226
martini, Polyommatus 10:215, 217, 222
masariformis, Chaemaesphecia 3:75
masoni, Minois 10:114
massasoit, Hesperia 20:241
massasoit, Poanes 4:14; 9:78; 10:55; 12:41, 43, 44; 14:215; 23:186, 267, 268
massasoit hughi 14:28
massasoit suffusa 14:28
masuriensis, Pedesta 16:144

MATHERI, *Enosis (IA, IG) S2:22*
mathewi, *Thecla 21:184*
mathias, *Pelopidas 11:168; 14:51; 16:144*
mathias oberthüri *8:95; 18:105*
    Oryza, Miscanthus, Miscanthus sinensis, Imperata cylindrica, Arundo donax, Oplismenus compositus, Saccharum officinarum, Bambusa sp.
matilda, *Protitame 13:219*
matronalis, *Hesperis 18:94*
matronaria, *Epimecis 5:60*
maturna, *Euphydryas 14:194*
mauretanicus, *Thestor 10:215, 221*
mauretanicus, *Tomares 24:196*
mavors, *Thecla 18:23*
maxima, *Apharitis 10:117*
maximellus, *Schoenobius (IA,II) 17:234-236*
    Zizaniopsis miliacea
maximiana, *Euryphene (IA) 13:128*
MAYELISARIA, *Glaucina (IA,IG) 20:247-250*
mayi, *Melitaea 15:220*
mazans hayhurstii, *Staphylus 17:155; 19:1, 23, 31; 21:207, 277; 23:186; 24:305*
    Amaranthus retroflexus, Chenopodium album, C. ambrosioides
MAZAI, *Atrytone (IA, IG) S2:39*
meadi, *Cercyonis 23:167, 168, 170-173*
meadi ALAMOSA, *Cercyonis (IA,IG) 23:161-164, 166, 168*
meadi meadi *23:161, 162 (IA), 164, 165, 168*
meadi mexicana *23:161, 166, 168*
meadii, *Cercyonis 10:94*
meadi, *Colias 2:13, 7:50*
    Astragalus sp. 7:50
meadii, *Colias 5:25, 10:99, 14:44*
meadii elis *19:129*
meadi, *Coryphista 13:217; 24:220-223*
    Berberis thunbergii
meadi, *Minois 5:63; 13:91*
meadii, *Minois 7:147*
meander, *Prepona 21:104*
meander, *Rhescyntis 14:92, 103*
mechanitis mechanitis, *Athyrtis 15:27*
mechanitis oberthuri *15:27*
mechanitis salvini *15:27*
mecilentus, *Papilio 15:175*
mediodorsella, *Cameraria 25:210*
medon, *Papilio 13:22, 23*
medon innotata, *Euphaedra (IA) 13:128*

medusa, Erebia 5:28; 14:140; 18:41; 22:128
megaera, Pararge 8:143; 9:145; 14:139
megaera vividissima 10:221
megera iranica 10:118
meges, Lamproptera 9:74
meges, Leptocirus 13:22
mejicanus, Megathymus 11:129; 12:81
mejicanus, Pholisora 7:46; 13:92; 22:168
mejicanus, Turnerina 12:89 (IC); S1:28
melaena, Castalius 10:38
melaleuca, Sympisis 12:50
melanarge, Chlosyne 13:165; 21:193
Melanargia 8:174
melanchlora, Victoria 25:169
melane, Paratrytone 22:94, 97, 99, 102
melane melane, Paratrytone 12:32
melanippe, Pap. Dan. Cand. 12:197
Melanitis 24:13
melanopa, Anartia 10:54
melanops, Barbicornis 21:147
melanops, Glaucopsyche 14:138
melanops alluaudi 10:218, 222
melanthe, Limenitis 18:21
melas, Erebia 14:140

meleager, Polyommatus 9:145
melete, Pieris 7:37, 198; 13:200; 14:53
melete aglaope 13:74
melete melete 8:96
    Arabis, Raphanus, Brassica, Wasabia
melete pseudonapi 8:96
    Arabis, Raphanus, Brassica, Wasabia
meliboeus, "Thecla" 22:153, 156
melicerta, Ophiusa 2:100
melinus, Strymon 1:42, 58; 3:17; 4:12; 7:52, 146; 8:101; 9:77, 207; 10:55; 12:130; 13:99, 100; 14:20, 130; 15:227; 16:33, 69, 195, 203; 17:212; 18:149, 225; 19:112, 231; 20:131,135; 21:207, 277; 22:96-98, 120; 23:105, 106, 182, 187; 24:196
    Myrica asplenifolia 4:12, Trifolium oliganthum, T. tridentatum 8:101, Polygonum alpinum 16:33
melinus atrofasciata 13:70
    Trifolium oliganthum, T. tridentatus, Fragaria bracteata, Anaphalis margaritacea
melinus buchholzi 13:28
melinus franki 15:56; 18:151, 152; 20:160; 22:

230; 25:38, 214
Lespedeza hirta, Lupinus texensis, Phaseolus vulgaris, Callirhoe leiocarpa, Lantana macropoda, Croton monanthogynus, Diospyros texana, Tecoma stans, Cassia puberula, Indigofera texana, Passiflora incarnata, Quercus virginiana, Sesbania drummondii, Porlieria angustifolia, Hibiscus syriacus 18:151, 152
melinus humuli 10:97; 24:44
melinus melinus 17:23, 152; 22:35
Trifolium incarnatum 17:152
melissa, Danais 11:167; 24:298
melissa, Lycaeides 6:36, 41; 7:50, 51; 10:98; 14:130, 238; 19:112; 20:161; 22:129
Lupinus parviflorus 6:36
melissa annetta 6:36
melissa melissa 22:172
melissa samuelis 23:265, 266
Lupinus perennis
melissa scudderii 7:125, 198
melissa, Œneis 5:8, 26; 23:155
melissa assimilis 2:64; 5:8; 19:243; 23:103,104
melissa beanii 18:227
melissa gibsoni 19:60
melissa semidea 7:123; 10:54; 14:240; 20:161
melissa semplei 21:136; 23:104
melissa, Plebejus 7:146; 11:165; 16:70; 19:167; 18:225; 20:134
melissa melissa 12:31; 16:235
melite, Dismorphia 21:156
melite, Enantia 5:211
melite, Leptalis 13:146
mellonella, Galleria 15:102
melona meridionalis, Limenitis (Adelpha) 21:99
melpia, Mesosemia 21:87, 146
melpomene, Heliconius 17:176
melpomene melpomene 24:25
melpomene nannus, Heliconius 21:96
melsheimeri, Cicinnus 18:89; 21:34
melzeri, "Thecla" 21:87, 153
memnon, Caligo 18:18; 21:192; 20:203
memnon, Papilio 5:52; 11:9; 13:199; 25:164
orange 5:52
memnon pryeri 18:106
Citrus depressa
memnon thunbergii 8:96; 14:5; 15:175
Citrus, Poncris (Rutaceae)

menaka, Tagiades 16:144
menander, Callitaera 21:191
menapia, Neophasia 10:169; 14:238; 16:30, 70;
17:243; 18:221; 19:109, 232; 22:166
Pinus contorta 10:169, Pinus edulis ? 22:166
menapia, Pieris 21:271
menapsis saturata, Mechanitis 21:190
menas, Argiolaus 13:145
menas, Iolaus (IA) 13:145
menechmus, Jemadia 21:157
menecrates, Elbella 22:154
menelaus mineiro, Morpho 21:89
meneria, Amarynthis 21:87, 148
menestheus lormieri, Papilio 13:76
mengetana, Olethreutes 2:65
menippe, Anteos 21:155; 22:154
menippus, Mylon 21:162, 245; 22:155
menophilus menophilus, Melinaea 15:32
menophilus orestes 15:32
menophilus zaneka 15:32
meolans, Erebia 14:140
MEPHISTO, *Neperigea (IA,IG) 22:137-140*
mera, Strymonidia 13:178
Rhamnus
mercurius, Proteides 13:90
mercurius mercurius 21:158; 25:36
mercurius sanantonio 14:35
meridianus, Erynnis 13:91
meridionalis, Calosaturnia 12:17, 160
merione, Ergolis 25:146
Ricinus communis
merione merione (IA) 18:160
merione, Eurytela 25:146
meris, Pap. Pleb. Rural. 12:197
meris arsis, Lasaia 21:148
mermeria, Taygetis 6:97; 18:7, 116
mermeria tenebrosus, Taygetis 21:90
merops, Marpesia 21:195
merops, Sphinx 25:152
merrickata, Papaipema 8:60
Podophyllum peltatum
merus, Strymon 14:206
merus, Strymonidia 8:97
Rhamnus (Rhamnaceae)
mesentina, Pap. Nymph. Gemm. 12:197
meskei, Hesperia 4:44, 62; 21:207, 277
messalina, Eurema 2:51; 9:80, 206
messapus, Eicochrysops 10:38

metaleuca, Euptychia 18:17; 21:191
metanemaria, Elpiste 13:233
metea, Hesperia 4:44; 5:70; 9:192; 10:55; 12: 39; 13:16; 17:108, 109; 19:215-219; 20:241
Andropogon scoparius ? 19:215-219
metea licinus 9:193
metea licinius 13:89
metatoides, Zale 24:3
metelkana, Rhyparioides 2:94; 3:76
palustrine plants
metella, Synale 21:166
methymna, Lepidochrysops 10:38
meton, Rekoa 21:152
meton, Thecla 18:23; 21:196
metrodorus metrodorus, Phocides 21:158
MEXICANA, *Carystoides* (IA, IG) S2:37
mexicana, Eurema 2:37, 51; 7:52, 147; 9:191; 10: 100, 163, 166; 12:130; 14:44; 15:72, 198; 17: 154; 19:112, 231; 21:208; 22:34, 229; 25:37, 214
mexicanaria, Phaeoura (gynand.) (IA,IG) 23:274,275
MEXICANUS, *Polythrix* (IA,IG) S2:10
micacea, Hydroecia 2:94
palustrine plants
micans, Amphipyra 3:75
michaelis, Careana 3:79
micheneri, Agathymus 23:245; S1:46, 47
Agave lecheguilla S1:47
microdamus, Papilio 4:66
microgamma, Autographa 18:236
microgamma montana, Syngrapha 5:54
micronympha, Catocala 14:177; 23:5; 24:108, 110, 114; 25:43
micropis, Borolia 9:57
middletoni, Hamearis 21:87, 146
midea, Anthocaris 6:42, 99, 100; 7:41, 42; 8: 141; 21:207, 277; 22:34
Arabis (rock cress) 6:99, 100
midea, Anthocharis 2:22; 3:48
parasites 3:48
midea midea 25:37
mikado, Parnassius 8:96
Corydalis (Papaveraceae)
milberti, Aglais 11:164; 12:50
milberti, Nymphalis 2:22; 4:13; 5:26, 69; 6:73, 105; 9:15; 10:107, 169; 12:41; 14:179, 238; 16:32, 244; 17:107, 109; 18:223; 19:113; 21: 136; 22:120; 25:140

Urtica 16:32
milberti subpallida 10:97
milberti, Vanessa 12:120
milhauseri, Hoplitis 3:75
milionia, Euaspa 16:144
milleri, Recurvaria 8:91
Pinus contorta murrayana
miltiades, Antirrhea 21:191
militaris, Haploa 21:38
mimas, Pap. Pleb. Urbicolae 12:197
mimas, Nisoniades 21:164
mimosae, Epamera 10:39
mimosae, Iolaus 20:135
mincia, Limenitis 21:98; 22:152, 156
MINEIRA, Adlerodea (IA,IG) 22:14, 16, 18, 19
mineus, Mycalesis (IA) 18:159
minima, Copaeodes 13:75, 225-227; 17:156; 20:102
Cynodon dactylon 13:225-227
minima, Pseudomya 11:148
minima, Teriomima 13:138
minimus, Cupido 9:204; 11:165; 14:136; 24:196
ministra, Datana 25:43
minnehaha, Plebius 7:51
minor, Viola 21:164
minor minor, Staphylus 21:161
minori, Papilio 16:140
minualis, Anepischetos 10:112
minuta, Agoroea 11:159
minuta, Ethmia (II,IF) S6:30
Cryptantha intermedia
minuta, Melitaea 19:231
minuta minuta, Poladryas 25:40
minyas, Eumaeus 18:22; 20:135
mira, Catocala 12:201; 24:110, 114
miranda, Troides 9:75
mirifica, Euliphyra 6:43
miscitata, Laspeyresia 20:252, 253
Pinus ponderosa, P. jeffreyi
miselioides, Phosphila 2:S8
Smilax rotundifolia
miserabilis, Callophrys (IG) 20:68-70
misippus, Hypolimnas 7:169; 8:99; 9:206; 18:74, 108; 24:72; 25:80
Ficus 8:99, *Portulaca oleracea* 18:108
misippus misippus 5:56
misturata, Eupithecia 13:217
mitchellii, Euptychia 12:41-46 (IA); 14:209-226 (IA,II); 19:148, 153-159; 20:43, 44; 24:1,43

Carex alopecoidea, Scrirpus atrovirens, Carex cephalophora 14:209-226
mithradates thraso, Achlyodes 22:155
mithrax, Chiomara 21:164
mittrei, Argema 9:66
Eugenis (Myrtaceae), *Weinmannia* (Cunoniaceae), (Mimoseae)
mixturata, Coenonympha 7:124; 15:222
mixtus, Charaxes 13:124
mizon, Pero 13:220
mneme, Melinaea 15:30
mnemopsis, Melinaea 15:31
mnemosyne, Parnassius 4:15, 16; 13:199
mnemosyne athene 4:74
mnestra, Erebia 14:140
mniszechii, Satyrus 10:218
modesta, Liptena 13:138
modesta, Pachysphinx 1:34; 5:17; 15:193; 21:254
Populus tremuloides, aspen, willow 21:254
modius, Thoon 21:246
MOECKI, *Macroglossum (IA) 23:256, 257*
moera alluaudi, Pararge 10:221
moera meade-waldoi 10:221
moeschleriana, Tortrix 4:29; 12:50
mohammed, Sloperia 10:215, 217, 222
mohave, Philotes 8:8; 20:131; 23:52, 57
Eriogonum pusillum 23:52, 57
molliculata, Lygris 25:262-264
Physocarpus opulifolius
molpodi, Appias 15:211
molops, Papilio 4:66
mona, Barbicornus 21:147
moneta, Dione 22:152
moneta poeyii 21:192; 23:177, 179 (IA), 180, 184 (1st U.S.)
monetifera, Euherrichia 19:189
monima, Eunica 8:25; 9:206; 13:62-63; 15:72, 199; 18:229
monima, Pap. Nymph. Gemm. 12:197
monoco, Hesperia 20:241
montana, Chlosyne 14:148
montana, Cupressus 13:233
MONTANUS, *Decodes (IG) 14:118, 119*
montanus, Erynnis 8:95; 14:51, 204
Quercus (Fagaceae)
montensis, Plebeius 10:222
MONTEZUMA, *Mellana, (IA,IG) S2:41*
montezuma, Papilio 4:40

montezuma, Parides 21:189
montinus, Argynnis 20:241
montinus, Boloria 2:86
montium, Melitaea 14:194
monuste, Ascia 3:17, 18; 7:13, 14; 9:206, 209, 211; 10:74; 14:48, 156; 17:226; 22:166; 24:77; 25:186
  Cakile edentula 17:226
monuste monuste 5:55; 21:154, 155, 189
monuste phileta 7:14; 17:155; 22:34, 25:36, 137
morbidalis, Chytolita 12:132; 23:206
mori, Bombyx 20:207, 217-225; 22:204
moriqua, Azanus 10:38
mormo, Apodemia 7:52, 147; 13:100, 233, 238; 15:145-171; 16:57, 70; 18:49, 226; 20:160; 21:272; 24:196
  Eriogonum 18:226
mormo cythera 14:238; 15:154 (IA)
mormo deserti 9:148; 15:165 (IA); 17:205; 20:134
  Eriogonum inflatum 15:165
mormo DIALEUCA (IA) 15:167
  Eriogonum hastatum ?
mormo langei 15:158; 22:93, 96, 97, 99, 100, 102, 105; 23:278
  Eriogonum latifolium 15:158
mormo maxima 15:168
mormo mejicanus 15:156; 19:112; 22:163
mormo mormo 15:148, 171 (IA), 214; 22:96, 97, 99, 100, 102, 105
mormo TUOLUMNENSIS (IA) 15:159
mormo virgulti 12:29; 15:160 (IA); 17:205; 20:134
  Eriogonum fasciculatum v. *foliolosum, E. elongatum* 15:160
mormonia, Speyeria 2:5; 5:31; 19:61; 25:51
mormonia arge 16:31
mormonia eurynome 2:92; 10:95; 14:184; 22:160
mormonia luski 19:107, 114
mormonia opis 19:61
mormonia washingtonia 18:222
morphea, Argynnis 12:120
morpheus, Caradrina 13:77
morpheus, Heteropterus 4:72; 11:166
morpheus, Papilio 12:120
Morpho 25:223
MORPHO, *Symbiopsis* (IA, IG) S4:17

morrisii, Boloria 15:221
morrisonarius, Pero 5:54; 13:220
morrisonella, Heterographis 13:233
morrisoni, Eupsilia 24:3
morrisoni, Ochlodes 9:148
morrisoni, Stinga 22:169
morsei, Leptidea 8:96
Vicia (Leguminosae)
morsei major 18:110
morvus boisduvali, Anaea 18:22
morvus stheno, Anaea (Memphis) 21:105
mossii, Callophrys 14:21
mossii, Incisalia 13:101- 103
Sedum spathulifolium, S. oreganum
mossii schryveri 10:97
mothone messenina, Melinaea 15:32
mothone mothone 15:32
motschulskyi, Ypthima 8:99
Digitaria (Gramineae)
motya, Hecaerge 15:216
motya, Libytheana 15:216
mozambica, Spindasis 9:14
Vigna (Leguminosae)
muiri, Gyros 13:232
muiri, Mitoura (I) 6:95, 96
muiri, Thecla 6:95
mulciber, Euploea 16:144
MULLERI, *Papaipema 10:139, 140; 12:208*
Helianthus strumosus 10:139, 140
multicaudata, Papilio 7:146; 12:101; 18:132-133, 220; 19:109, 231; 22:94, 97, 99-102; 25:36
Fraxinus sp., F. pennsylvanica v. *subintegerrima, Ligustrum vulgare, Prunus demissa, Ptelea trifoliata*
multicaudatus, Papilio 9:137; 10:99; 11:17, 224; 12:101, 103-113; 13:91, 100; 14:238; 16:69; 17:243; 18:40; 22:126. 12:103-113 (II)
Prunus demissa, P. virginiana, P. americana, P. capuli, P. cerasus, Ptelea angustifolia, P. trifoliata, Ligustrum lucidum, L. vulgare, Fraxinus "viridis", F. oregonia, Amelanchier sp., Fragaria, Rosa, Umbellularia, Foeniculum vulgare
multilinea, Leucania 9:56
multiscripta, Eupithecia 13:218
multiscripta canadensis, Cerura 21:37
multistrigata, Eupithecia 13:218
multivagata, Entephria 13:219

mulucha, Callicista 21:153; 22:153
mulucha, Thecla 18:23
muticum, Calephelis 23:187
mundana, Nudaria 2:94
mundula, Ommatochila 19:133
munitata, Xanthorhoe 12:50; 13:218
murrayi, Crymodes 4:28
    Lathyrus maritimus
musa, Eressa 8:135
musa, Eurema 21:154; 22:154
muscerda, Pelosia 2:94
muscosa, Cargolia 18:115
muscosa, "Euptychia" 21:92
muscosa, Godartiana 22:151
muscosa, Neodesmodes 18:115
muscosa, Phlegethontius 25:152
mushaellus, Neozephyrus 3:79
mustelinalis, Mecyna 5:29; 13:233
mustelinus, Hepialus 5:54
muticum, Calephelis 1:42; 7:119; 10:166
muticum, Lephelisca 12:41, 42-44; 22:35
myia, Phyciodes 18:20
mylitta, Pap. Nymph. Gemm. 12:197
mylitta, Dynamine 18:21; 21:194
mylitta mylitta 21:99; 22:152
mylitta, Phyciodes 3:106; 7:146; 10:96; 13:207 (IG); 16:32, 69; 17:243; 18:223; 19:113, 231; 20:161; 22:98, 99, 103
    Cirsium 16:32
mylitta barnesi 7:51
mylitta mata 18:228
mylitta mylitta 22:35
mynesalis, Tetanolita 19:133
myops, Paonias 15:193; 19:189; 21:254
    Prunus pensylvanicus, P. virginiana ?, P. serotina ?, Amelanchier 21:254
myops macrops, Calasymbolus 5:22
myrina, Pap. Nymp. Phaler. 12:197
myrina atrocostalis, Boloria 5:54
myrina terrae-novae 4:11
myrina tollandensis 8:64
myrinna, Vanessa 21:98; 22:152
myrissa, Argynnis 12:120
myrmecophila, Pseudodipsas 6:43
myrodora, Cosmosoma 11:148
myron, Ampeloeca 10:166; 19:189
myron cnotus 5:23
myron mexicana 5:22

myron, Darapsa 15:193
myrrha, Libythea 3:40; 16:144
myrtea, Charis 21:196
myrtilli, Anarta 2:94
Calluna vulgaris
myrto, Nymphalis 12:120
mys, Zariaspes 21:164, 246
mys cytis, Euselasia 21:87, 146
mysis aestiva, Delias 22:121; 23:280
mysis mysis 22:121
mystic, Hesperia 20:241
mystic, Polites 7:54; 10:55, 159; 12:46; 16: 245; 19:196; 20:239, 240
mystic sonora 13:233
mythra, Yphthimoides 21:91

nadia, Ethmia S6:38
Phacelia ramosissima v. *suffrutescens, P. tanacetifolia*
nais, Apodemia 7:146; 18:76, 77 (IA,IG); 19: 112, 232; 22:35
napaea, Boloria 14:193; 19:60, 61
napaea, Litholomia 15:121
napaealis, Pyrausta 5:29
napi, Pieris 2:6; 3:17, 39; 4:28; 5:18, 19, 26; 6:105; 7:5, 37; 8:3, 4; 9:145; 10:18, 28 (IA), 107, 171; 11:162; 12:7, 42, 44; 13:100; 14: 215 (IA, 239; 16:5-19 (IA), 105-127 (IA); 18: 51, 91-100, 110; 19:110, 232; 22:126, 223; 23:113; 25:6, 9, 10, 11, 160, 161
Alliaria petiolata, Armoracia rusticana, Barbarea, Biscutella laevigata, Brassica, Cardamine hirsuta, C. pratensis, C. amara, Diplotaxis tenuifolia, Cardamine trifolia, Erysimum officinale, Nasturtium officinale, Rapa, Reseda, Roripa silvestris, Raphanus raphanistrum, Sinapis, Sisymbrium
napi blidana 10:220
napi-bryoniae 12:7
napi castoria 15:212; 17:111, 116
napi hulda 7:125
napi japonica 8:96
Arabis, Cardamine (Cruciferae)
napi MACDUNNOUGHII 8:75; 10:100; 14:46, 238
napi marginalis 18:221; 25:8, 10, 11
napi mogollon 9:141
napi napi 14:53

napi nesis 14:53
napi oleracea 4:11; 5:43; 14:46; 16:245; 24: 77-81 (IA); 25:8, 10, 11
napi pseudomelete 8:96
Arabis, Cardamine
napi pseudonapi 2:92
napi segonzaci 10:218, 220
napi venosa 15:212; 22:94, 97, 99, 100, 102
narcaea, Heliconius 13:84
narcosius, Pap. Pleb. Urbic. 12:197
nareda, Ypthima 11:167; 16:144
narses, Lasaia 21:196
nasica, Epimecis 5:59
naso bada, Parnara 18:105
Oryza sativa, Bambusa sp.
nastes, Colias 2:13; 5:25, 42; 11:12; 14:52; 16:46; 18:47, 227; 19:60
nastes alaskae 7:125
nastes cocandicides 15:213
nastes streckeri ab. *palliflava 9:141*
nastes subarctica 2:64; 9:141; 15:214
natalensis, Spindasis 9:14; 10:39
Vigna, Mundulea (Leguminosae)
natalii, Satyrus 12:6
nausicaa, Speyeria 5:33
nausithous, Maculinea 14:138
neaerea arge, Pyrrhogyra 21:103
nearchus, Antigonus 21:162, 245
nebris, Papaipema 8:60
Ambrosia artemisiifolia, Arctium sp. 8:60
nebulosa, Catocala 10:168
nebulosa, Yphthimoides 21:91
necopina, Papaipema 8:60; 10:139, 140
Cacalia tuberosa, Helianthus divaricatus
neda, Eurema 6:77; 9:208
neda, Pieris 12:120
negra, Junonia 12:194
negundana, Archips 10:154
nehemia, Pseudopieris 22:154
nehemia nehemia 21:156
neis, Autochton 21:160
nelita, Papaipema 8:60
Rudbeckia laciniata
nelsoni, Callophrys 14:21; 16:33, 193; 17:111, 245; 18:228; 20:135; 24:196
nelsoni muiri 22:96, 97, 99, 100, 102
nelsoni, Mitoura 6:95 (IA), 96; 8:101; 12:30; 15:201

Libocedrus decurrens ?, *Thuja plicata*
nelsoni muiri 6:95 (IA), 96
nise nelphe, Eurema 2:51
nemesis, Calephelis 12:29
nemesis, Lephelisca 13:225-227; 19:112
Clematis henryi, C. drummondii
nemesis, Siderone 4:67
nemesis nemesis 5:56
neocypris, Phoebis 21:154; 22:154
neogama, Catocala 10:167; 24:108, 110, 115
neophanes, Myelois 17:39
neophilus, Papilio 4:41
neophilus eurybates, Battus (Parides) 21:156
neoptolemus, Xylophanes 3:78; 5:23; 25:152
nephalion, Battus (Parides) 21:89, 156; 22:154
nephalion, Papilio 4:41
nephele, Minois 6:33
nephele, Satyrus 9:188
nepheleptena, Papaipema 8:60; 23:48
Chelone glabra 8:60
NEPHEOS, *Exoteleia* (IG,II) 21:9-11, 15-21
Pinus resinosa, P. mugho, P. sylvestris
nephrasyntheta, Papaipema 8:61
Eryngium yuccifolium
neptaria, Semiothisa 13:219
neptunus, Papilio 9:75
nereis, Pierella 22:151, 156
nereus, Amblyscirtes 13:91
nerippe, Fabriciana 14:193
nerissa, Huphina 3:39; 18:160 (IA)
nero, Appias 9:74
nero palawanica 24:304
nero, Cobalopsis 21:166
nero, Greta 21:191
nervana, Eana 16:178
nesimachus, Dichorragia 14:203
nesimachus ishigakianus 18:108
Meliosma sp.
nesimachus nesiotes 8:99
Meliosma (Sabiaceae)
nessus, Amphion 10:166; 15:64; 23:100, 192, 194; 25:41, 85
Epilobium angustifolium 15:64
nessus floridensis, Amphion 5:23
nessus, Celotes 13:91, 99, 224-227; 19:1, 6, 32, 133, 137; 25:35
Abutilon incanum, Althaea rosea, Sida filipes, Sphaeralcea lobata, Wissadula amplissima

neumoegeni, Agathymus 12:89(IC); 13:206; S1:35
Agave parryi
neumoegeni, Chlosyne 24:226
neumoegeni, Hemileuca 2:81; 14:103
Polygonaceae
neumoegeni, Megathymus 7:147; 8:77-87 (I); 12:82
neumoegeni stephensi 12:32
neumoegeni, Xanthothrix 3:74
Chaenactis glabriuscula
neurica, Archanara 2:93, 94
neurona, Plebejus 17:244
Eriogonum wrightii
nevada, Hesperia 10:100; 16:36; 20:254; 22:169
grass 16:36
nevada, Thorybes 16:35
nevadae, Hydriomena 13:234
nevadata, Eupithecia 13:218
nevadensis, Chionobas 12:194; 21:271
nevadensis, Eana 16:178
nevadensis, Hemileuca 2:81; 14:103
Populus sp. 2:81
nevadensis, Œneis 10:171; 16:64, 70; 18:222; 21:259; 23:157
nexa, Phragmitiphila 2:94
nexilis, Eutricopis (IA,IC,IF,II) 24:151-156
Antennaria sp.
ni, Trichoplusia 17:67, 70; 19:132, 133
nicaea paulla, Eurybia 21:87, 146
nicea, Stibochiona 11:167; 16:144
nicias, Aricia 14:137
nicippe, Eurema 1:42; 2:22, 51; 3:17; 6:33; 7:146; 9:148; 10:200; 12:95, 96, 121, 130; 13:99; 14:46; 15:198; 17:226; 19:112, 232; 20:102; 21:190, 207, 277; 22:227, 229; 23:182, 187; 25:37, 214
Cassia corymbosia 2:22, 51
nicippe nicippe 17:22, 155
Cassia occidentalis, C. bicapsularis
nicippe, Pap. Dan. Cand. 12:197
nicomedes brazus, Telemiades 21:164
nigra, Aporophyla 2:94
Rumex sp.
NIGRA, *Laspeyresia (IA,IG) 20:251-253*
Pinus agacahuite (cones)
nigrescens, Chlosyne 13:165
nigricans, Hylesia 14:103
nigricaria, Glena 13:220
nigrilineella, Eriphia 15:82

nigristrigella, Gracilaria 25:202
nigritus, Phlegethontius 5:21
NIGROCOLA, *Cynea* (IA,IG) S2:31
nigrofimbria, Xanthoptera 2:S8
Syntherisma sp.
nigrum, Empetrum 4:11
nihana ryukyuensis, Pithecops 18:107
Desmodium caudatum, D. tashiroi
niitakanus, Neozephyrus 3:79
nikias, Speyeria 5:32
nilus, Calephelis 21:148; 22:153
nimbice, Catasticta 15:199
nimbice nimbice 22:126
niobe, Argynnis 10:217
niobe, Fabriciana 11:163; 14:193
niobe taura 10:118
nipalicus, Spindasis 16:144
niphon, Callophrys 14:21
niphon niphon 17:153
niphon, Incisalia 7:34; 10:55
niphon clarki 10:107; 16:244; 20:41; 22:26
niphonica, Erebia 8:99
NIPIGON, *Lithocolletis* (IA,IG) 24:274, 276-278
Populus balsamifera
nippia, Symbiopsis (IA, IG) S4:20
nireus, Papilio 25:271
nireus lyaeus 13:76
nirmala, Erebia 11:167; 16:144
nise, Pap. dan. Candid. 12:197
nise, Eurema 15:72; 22:229
nise nelphe 15:214; 21:190
nise nise 15:214
nise perimede 2:51; 9:209; 18:16
niscana, Argyrotaenia 13:232
nishikaze, Neozephyrus 3:79
nissa, Rapala 16:144
nitra, Papilio 4:63; 22:167
nivalis, Erebia 14:140
nivalis, Lycaena 14:128; 15:107; 16:34, 69; 18: 225; 20:161
Polygonum douglasii 18:225
nivalis browni 2:92; 10:98; 17:26; 22:165
Polygonum douglasii 17:26
nivaria, Euryphene (IA) 13:128
niveicosta, Schinia (IA,II) 2:42; 20:29-33
Palafoxia linearis 20:29-33
nivella, Heliopetes 14:32
niveicornis, Chondrolepis 18:74

nivescens, Agrodiaetus 14:138
niveus, Acentropus 8:23; 9:69-73
Elodea, Ceratophyllum, Potamogeton 8:23;
Elatine americana, Ceratophyllum 9:69-73
niveus, Pap. pleb. Urbicol. 12:197
nivigcrata, Lobophora 5:54; 23:4
nobilis, Papilio 13:76
noctua bipuncta, Noctuana 21:245
noctuella, Nomophila 13:86,233; 19:133; 24:172
noctuiformis BREDINI, Cautethia 24:267 (IA,IG)
noctuiformis noctuiformis 24:268, 270 (IA)
nokomis, Samia 8:2
nokomis, Speyeria 2:5; 22:160
nokomis apacheana 23:62-68; 25:45, 59-52 (IA)
nokomis coerulescens 15:198; 19:114; 25:51
nokomis nigrocaerulea 25:49, 51
nokomis nitocris 23:271-272; 25:44, 47-52 (IA)
nokomis nokomis, Speyeria 19:114; 22:128, 132;
25:44-52 (IA)
nokomis wenona 25:51
nomion, Parnassius 15:210
nomius, Hemeroplanes 3:78
nomius, Papilio 3:39
nora kanoi, Nacaduba 18:107
Entada phaseoloides
norna, Œneis 5:26; 15:22
nostrodamus, Gegenes 10:222
notataria, Eufidonia 8:3; 23:7
notatis, Afrida 6:53
notatus, Colias 15:213
nouna, Teracolus 4:72; 10:215
novascotiae, Argynnis 6:38
nubicola, Metaxmeste 13:241
nubiculata, Semiothisa 13:219
nubigena, Euphydryas 12:165
nubilis, Euparthenos 2:S8
Robinia pseudoacacia
nubilofasciata, Hydriomena 13:218
numata, Heliconius 24:299
numenes, Charaxes (IA) 13:124
numeria, Yphthimoides 21:91
numilia esite, Catonephele 21:194
numilia penthia 21:100
numitor, Ancyloxypha 5:69; 7:54; 9:40, 78, 195;
10:55; 12:130; 13:74; 14:37, 215; 16:245; 17:
156; 19:196; 20:230; 22:169; 24:243, 305
unidentified grass 5:69
Zizaniopsis miliacea 20:230

nuptialis, Catocala 15:124
Amorpha
nyassae, Spindasis 9:14
Entada, Acacia (Leguminosae)
nycteis, Chlosyne 14:179; 23:192, 193
nycteis nycteis 19:48; 25:40, 81, 82
nycteis, Melitaea 11:38-41; 12:46, 130; 15:112, 219; 21:207; 277; 22:120; 24:305; 25:141
Actinomeris alternifolia 11:39
nycteis drusius 10:96; 16:244; 21:136
nycteis nycteis 17:23; 20:241
nycteis reversa 10:107
nycteis, Phyciodes 2:22; 9:78
nycteis drusius 2:92
nycteis ab. *milburni 12:208*
nyctelius, Hesperia 12:120
nyctelius, Nyctelius 17:157
cultivated rice (pest, Cuba)
nyctelius nyctelius 21:167
nyctimus, Catonephele 18:20; 21:194
nympha, Taygetis 6:98; 21:191
nysa, Amblyscirtes 13:90, 91, 221-227; 19:180; 22:231; 25:34
Stenotaphrum secundatum 13:221-227
nyseus, Talicada 16:131-134; 18:163
nyssaecolella, Actrix 24:255

oakesii, Thecla 25:97
oaxacana, Leucania 9:57
obelus, Thecla 21:152
oberthueri, Papilio 4:66
obliqualis, Eublemma 19:133
obliquus, Pholus 3:78
obscura, Catocala 24:109, 110, 115
obscura, Erinnyis 2:86; 3:78; 13:66; 17:226
obscura jamaicensis 5:22, 23; 9:135
obscura obscura 5:23
obscura socorrensis 5:22
obscura stheno 5:22
obscura, Thoressa 8:95
Arundinaria, Pleioblastus, Sasa (Gramineae)
obscuripennis, Nothris 4:50
obscurus narva, Anasfrus 21:164
obsoleta, Heliothis 6:53
obstipata, Percnoptilota 12:171
obtusa, Pelosia 2:93, 94; 3:75
OCARIA *24:56*

occidentalis, Colias 16:69; 17:115; 18:220
Melilotus alba, Vicia angustifolia, Lupine
occidentalis, Pero 13:220
occidentalis canaster 13:234
occidentalis, Pieris 10:100; 14:48; 15:212; 18:221; 20:49; 21:135
occidentalis, Synclita 13:242
occiduaria, Anagoga 13:220
occiduaria, Itame 13:219; 25:267
occulta, Eurois 24:3
occulta, Phlegethontius 25:152
occultus pacificus, Phlegethontius 5:22
ocellata, Dyops (IA,II,IG) 16:47-54
ocellatus, Smerinthus 18:37
ocelleus, Euchromius 19:134
ocellinata, Semiothisca 23:4, 7
ocelloides, Megisto 21:92
ocepete, Epistor 25:152
ochosalis, Pyrausta 5:29; 13:242
ochracea, Coenonympha 13:74; 14:127; 17:253
ochracea, Ochlodes 14:51
ochracea rikuchina 8:95
Oplismenus (Gramineae)
ochracea, Yphthimoides 21:91
ochrearia, Aspilates 3:76
ochroptena, Papaipema 8:61
Humulus lupulus
ochsenheimer, Pieris 15:212
ochus, Phlegethontius 3:78
ocirrhoe ocirrhoe, Pareuptychia 21:90; 22:151
ocola, Panoquina 9:79; 17:148, 157; 21:167, 207, 277; 25:34
cultivated rice ? 17:157
ocrisia, Ocaria 24:56, 57 (IG), 60
octavia, Pap. Nymph. Phaler. 12:197
octavia, Precis 24:77
octo, Amyna 10:112; 15:111
octomaculata, Polythrix 15:72; 21:278
octomaculata octomaculata 21:159
oculigera, Dyops 16:54
ocypete, Epistor 2:86; 25:152
ocypore, Emesis 21:196
ocypore zelotes, Emesis 21:149; 22:153
oda, Heliophorus 11:167
odata, Chaetoprocta 16:144
odia, Historis 12:172
odilia trebius, Cymaenes 14:31; 21:246
odius, Historis 21:195

odius odius, Historis 21:104
odius, Papilio 13:66
odora, Erebus 2:22, 34, 86; 3:63; 4:13, 46; 11:205; 12:118; 13:17, 65, 85; 15:121
Cassia fistula 2:22, *Acacia decurrens*, *Cassia fistula*, *Ficus trigonata*, *Gymnocladus diocus*, *Pithecollobium unguiscate*, *Saman* sp. 13:17, *Acacia delbotta*, *Prosopis pubescens*, *Acacia greggi*, *Prosopis juliflora* 13:65
adult in Wisconsin 2:34
odorata, Erebus 17:227
oedippus, Coenonympha 7:37; 14:139
oedippus annulifer, 8:100
Cyperaceae
oedipus, Coenonympha 3:75
oeme, Erebia 9:204; 14:140
oenomais, Anaea 21:195
oenone, Brenthis 15:219
oenone, Melitaea 20:241
oenone, Phyciodes 15:219
oenotrus, Erinnyis 3:78; 25:152
oetus, Cercyonis 7:124; 14:132; 17:114; 18:221; 19:114; 23:167, 168, 170, 171, 172, 173
oetus charon 10:94; 14:238; 23:165
oetus oetus 23:165
oetus phocus 23:165
oileus, Lasaia 21:148
oileus oileus, Pyrgus 21:246
oileus orcus 21:163; 22:155
oileus philetas 22:230
oileus scamander, Caligo 21:192
olane, Ogyris 20:134
olena, Aeria 21:94; 22:151
oligyrtis, Patricia 15:27
olivacea, Lacinipolia 20:51
olivacea, Schinia 19:105, 106
Sida physocalyx, *Sphaeralcea lindheimeri*
oliviae, Hemileuca 2:81
olympia, Euchloe 2:22; 6:99-102; 10:23, 159, 163, 165; 11:162; 12:130; 15:113; 16:25; 21:208; 22:252; 25:64
Arabis (rock cress) 6:99, *Arabis viridis* 10:165, *Arabis glabra* 25:64
olympia rosa 13:89; 15:113
olynthus, Eudamus 12:122
omrina, Heliopetes 21:164
onaca, Vettius 21:246
oneko, Hesperia 20:241

ononis, Heliothis 6:53
onopordi, Pyrgus 14:51
onopordi fulvotincta, Hesperia 10:222
ONTARIO, *Lithocolletis* (IA,IG) 24:273-275, 277
    Populus tremuloides, P.grandidentata
ontario, Strymon 10:166; 13:71; 14:21
ontario autolycus, Euristrymon 18:150; 25:38
    Quercus virginiana, Quercus laurifolia
ontario ontario 16:195, 197; 19:50; 24:2; 25:80
opaca, Liptena 13:138
opalina, Pantoporia 11:167; 16:144
opalinus, Anthene 13:145
opella, Holomelina 19:75
ophidicephalus, Papilio 9:2; 10:35; 11:155
ophionalis, Microtheoris 13:242
opinatus, Charaxes 13:124
opis, Cynadra (IA) 13:137
OPLERELLA, *Adela* 23:212, 215, 216, 235 (IA), 238
oporana, Cacoecia 10:156
Opsiphanes 24:13
optica goya, Typhedanus 21:160
optilete, Vacciniina 11:165; 14:137
optilete daisetsuzana, Vaciniina 8:98
    Vaccinium (Ericaceae)
optilete yukona, Plebejus 11:51
orabilis, Papilio 4:67
orates, Ogyris 20:134
orbiculosa, Oxytripia 3:76
orbifera, Magusa 24:3
orbis browni, Phoebis 5:56
orchamus, Xeniades 20:228
orchamus orchamus 21:167
orcidia, Thecla 18:23
orciferaria, Aspilates 2:65
orcinus, Udranomia 21:158
orea, Epiphile 21:101; 22:152
oreas, Polygonia 10:169
oreas silenus 22:98-100, 103, 104
oregona, Leucania 9:57
oregonensis, Philotes 8:8
oregonia, Papilio 16:149
oregonica, Heliothis (IA,IF,II) 25:1-6
    Castilleja spp.
oregonius, Papilio 18:65; 22:53-56 (IA,IH,II)
    Artemisis dracunculus, Foeniculum vulgare
orellana, Papilio 4:41
orgia, Panthiades (Parrhasius) 21:152

orientalis, Favonius 3:80; 8:97; 13:178
Quercus spp. (Fagaceae)
orion, Scolitantides 14:138; 24:196, 199
orion jezoensis 8:98
Sedum (Crassulaceae)
orion, Typhedanus 21:160
origines, Polites 19:196; 20:241; 24:305
orillata, Semiothisa 5:54
orithya, Precis 8:99; 16:132-134 (IA), 144; 18:107, 163
Justica procumbens, Antirrhium (Scrophulariaceae)
orithya albicncta 22:121
orithya evarete 6:74
orithya madagascariensis 18:74
orithyia, Precis 11:167
orius orius, Naevolus 21:166, 246
orizaba, Rothschildia 14:103
orma, Mopala 13:146
ornatrix, Utetheisa 8:2
ornatrix bella 17:231-233; 22:197-203
ornatrix ornatrix 17:231-233
ornatrix stretchii 17:231-233
ornithogalli, Prodenia 15:111
ornythion, Papilio 4:64; 13:195, 199; 15:56
orono, Hedone 9:192
orsedice, Iratsume 3:79; 8:97; 13:178; 14:204
Hammamelis (Hammamelidaceae)
orseis, Phyciodes 8:3
orthosilaus, Graphium 21:157
orthosilaus, Papilio 4:66
ortygnus, Oenomaus (1st U.S.) 18:189-190
ollari, Amblyscirtes 13:91; 25:34
oslari, Anisota 14:103
oslari, Satyrium 14:21
osseana, Phalaena 16:176
osseana darvaza, Ablabia 16:176
osseana niveosana 16:176
osseana osseana 16:176
osseana niveosana, Eana (IG) 16:182, 190
osseana osseana (IG) 16:183, 187
ostenta, Holomelina 19:74
ostrogowichii, Cleogene 3:76
otacilia, Anthene 10:38
octacilia benadirensis, Anthene 13:145
otho, Wallengrenia 10:55; 13:221-227; 14:29; 16:174; 21:207, 277; 23:192, 194; 24:244; 305

Stenotaphrum secundatum 13:221-227
otho egeremet 9:195; 17:109; 20:160
otho otho 17:21, 108, 109, 156
otis, Zizeeria 11:167; 16:144
otis alope, Zizina 8:98
Lotus, Indigofera, Lespedeza (Leguminosae)
otis riukiuensis 18:107
Melilotus suaveolens, Kummerowia striata, Lespedeza cuneata, Vigna marina
oto, Greta 21:191
otolais, Pyrrhogyra 16:20
otolais neis 15:91; 18:20; 21:194
otrere, Anaea (Memphis) 22:153, 156
otreus, Pap. Pleb. Urb. 12:197
ottoe, Hesperia 12:130; 14:57; 22:169
ottomana, Erebia 14:140
ottomana, Zeuxidia 13:21
ou, Autographa 13:26
outis, Cogia 13:89, 90
ovata, Acronycta 10:88
ovinia edwardsi, Atrytonopsis 14:28
oxione squalida, Euryphene (IA) 13:137
oxygala luteopallens, Aletia 9:56
oxygala oxygala, Aletia 9:56
oxygramma, Autographa 15:121
oxynius, Papilio 4:64

pachalica, Hypolycaena 18:74
pacifica, Aeria 21:190
pacifica, Catagramma 21:194
packardella, Gracilaria 4:5
sugar maple
packardella, Rhagea 24:255
packardi, Cosymbia 4:26
Bidens ?
packardiana, Tortrix 7:28
Zenodochium citricolella, Choristoneura fumiferana
pacuvius, Erynnis 9:15
Ceanothus fendleri
pacuvius lilius 18:228
padella, Hyponomeuta 5:30
pahaska, Hesperia 10:100; 13:100, 233
pahaska williamsi 14:30
palaemon, Carterocephalus 5:26, 54; 6:105; 10:23; 11:43, 47, 166; 18:227

palaemon mandan *10:35*
palaemon mesapano *14:32, 239*
palaemon satakei *8:95*
Gramineae
palaeno, Colias *5:25; 8:163; 11:163; 14:52*
palaeno aias *8:96*
Vaccinium (Ericaceae)
palaeo chippewa *6:105; 7:125; 11:48; 19:58, 60*
palaeogama, Catocala *22:234-236; 23:263; 24:108, 110, 114*
palamedes, Papilio *3:25; 4:64; 9:205, 206, 211; 13:26, 74, 197, 199; 16:198; 17:22; 22:34*
Persea borbonia, Magnolia glauca, Sassafras, Glabraria aestivalis, Nectandra, Misanteca
palamedes palamedes *17:153; 18:135, 136*
Persea borbonia *18:135,136*
palatka, Atrytone *4:62*
palawanicus, Papilio *21:50; 24:304*
palegon, Rekoa *21:152; 22:153*
palegon, "Thecla" *24:196*
palegon, Thereus *24:60*
palemon, Pap. Pleb. Rural. *12:197*
palemon, Cacyreus *10:38*
palemon mesapano, Carterocephalus *16:245*
pales, Boloria *3:75; 7:123; 14:193*
palla, Chlosyne *13:207 (IG); 14:184; 16:31, 69; 17:113; 18:223*
Aster, Castilleia *16:31*
palla whitneyi *17:112*
palla, Melitaea *2:92; 19:173, 175 (IA; 22:98, 99, 100, 103*
palla calydon *10:96; 15:220*
pallene rogersi, Colotis *(IA)* *13:137, 138*
pallescentella, Acedes *23:192, 195*
pallicolor, Protitame *13:234*
pallida, Perigonia *3:78*
pallida, Phyciodes *22:162*
pallida benuncas, Erynnis *10:222*
pallidissima, Pieris *8:75*
pallidivittalis, Orenaia *13:241*
pallifrons, Eana *16:179*
palmerii, Apodemia *19:112*
palmerii marginalis, Apodemia *9:148*
palmira, Eurema *2:52, 67*
palmira palmira *9:80*
paludata, Carsia *2:94*
Vaccinium oxycoccos
paludata boreata *6:34*

palustris, Athetis 2:94
palustrine plants
palustris, Sesia 2:94
Euphorbia aquatica
pamina, Automeris 14:103
pamira KOTANDARI, *Karanasa 15:10, 11 (IA); 18:78 (IA)*
pamira TWOMEYI, *Karanasa 15:11, 12 (IA)*
pamphiloides, Coenonympha 15:222
pamphilus, Coenonympha 9:145, 204; 11:164; 14: 139; 15:222
pamphilus arenosa 10:221
pamphilus latevittata 10:221
pampinaria, Anavitrinella 13:220; 23:192, 194
pan denticulata, Hemeroplanes 5:22
panagaea, Turanana 10:117
panamensis, Symbiopsis (IA,IG) S4:15
panariste, Consul 25:19
pandion, Papilio 12:194
pandora, Coloradia 14:103
pandora, Damora 10:118, 194
pandora seitzi, Dryas 10:221
pandorus, Pholus 11:43; 15:193
pandrose, Erebia 14:140
panicifoliella, Cyloplasis 4:4
panic grass
pannonica, Nola 3:76
pannonica, Porphyrinia 3:76
panopus, Euthalia 13:21
panopus, Symphaedra 13:21
panoquin, Panoquina 17:108, 109; 20:241
panoquinoides, Panoquina 17:226
panoquinoides errans 12:32
pansa, Chamaelymnas 21:87, 147
pantherata pantherata, Anetia 5:56
panthonus, Papilio 4:41
panthonus jaguarae, Battus (Parides) 21:156
paon, Papilio 4:64
paphia, Argynnis 7:37; 9:145; 11:10, 163; 14: 194, 204; 15:43, 59; 18:41, 110; 22:128
paphia reisha 8:98
Viola (Violaceae)
pavanus, Heliconius (Eueides) 22:152, 156
PARA, *Dardarina (IA,IG) 22:2-4,6-8*
paradisea, Abantis 9:129
paradoxa, Papilio 9:74
paraeca, Limenitis 21:195
paranympha, Phulia 15:119

parasita, Ocnogyna 3:76
parce, Hemeroplanes 3:78; 25:152
paria, Eutychide 21:246
parmeno diversa, Gnophodes 18:74
Parnassius 22:204
parta, Catocala 13:182; 24:109, 111, 115
parthenoides, Melitaea 9:204
parthenos, Parasemia 5:54
pasiphae philippina, Epinephele 10:221
passova practa, Passova 21:158
pastor, Strymon 14:22; 15:56
patala, Euthalia 16:144
patilla, Ithomia 21:190
patricia, Lepidochrysops 10:38
patrobas, Elbella 20:227
patrobas, Gyrocheilus 7:147; 19:232
patrobas tritonia 19:114
patronaria, Epimecis 5:60
paucipuncta, Apodemia 21:87, 149; 22:153
paula, Oleria 18:15, 16
paulella, Aroga 13:232
paulina, Appias 3:39
pausianus, Papilio 4:66
pavana, Lycaena 11:167; 16:144
pavon, Asterocampa 15:216
pavon, Chlorippe 13:90; 15:72; 17:37; 21:195
pavonia, Saturnia 12:17; 14:103; 18:36
pavonii, Elzunia 15:30
pawnee, Hesperia 12:37-40; 14:57
  Leptoloma cognatum 12:37-40
pawnee montana 14:30
pearsalli, Stamnoctenis 13:218
pearsalli, Venusia 13:219
peckius, Polites 9:195; 10:55; 12:46, 130; 14: 216; 16:245; 22:130; 25:82
pedaliodina extincta, Anisochoria 21:162
pedrona pedrona, Phyciodes 21:87, 97
pegala, Minois 6:74
pegala, Cercyonis 6:74; 11:165; 12:41, 130; 14: 132; 16:69; 19:179, 180; 21:208; 22:127, 132; 23:100, 169-172, 192, 193; 24:305; 25:14
pegala alope 23:166; 24:2; 25:14 (IC, IG)
pegala ariane 18:221; 19:161; 20:161; 22:98-100, 104; 23:166, 169, 173
pegala boopis 14:238; 22:99, 103; 23:166, 169, 173
pegala damei 23:166, 169, 173
pegala ino 23:166

pegala maritima 22:127
pegala nephele 12:38; 15:93; 16:243; 20:12, 161; 21:136; 22:120
pegala olympus 10:94; 19:114
pegala pegala 13:74; 17:23, 150; 20:161, 186-188; 22:180-182; 23:166
pegala texana 18:137; 23:166, 169; 25:14,41 (IC)
pelaus, Papilio 4;¢4
peleides, Morpho 12:118; 14:130; 15:71,199; 20:203; 25:299
Machaerium riparium (Leguminoaeae)
peleides montezuma 18:18; 20:97; 21:192
peleides, Papilio 4:64
pelias saturnus, Charaxes 13:124
pelidne, Colias 2:13; 5:8,42; 12:122; 13:212; 16:245
Vaccinium sp.
pelidne skinneri 7:50
Vaccinium sp. 7:50
pelidne labradorensis 5:9; 19:242; 23:39
pellenea, Actinote 13:83; 21:96; 22:151
pellenis, Nymphalis 12:120
pellucida, Polytremis 8:95; 14:51
Arundinaria, Shibataea. Pleioblastus, Oryza, Miscanthus, etc. (Gramineae) 8:95
pelope, Nymula 21:87, 150
pelopidas, Mylon 21:162
pelops, ANTILLEA (IC,IG) 12:162-164
pelota, Pyrrhopyge 21:157; 22:154
pendulinaria, Cosymbia 5:54; 23:4, 7, 8
pendulinaria griseor 13:217
peneia, Euphyes S2:43
penelea, Posttaygetis 21:90
penelea, Taygetis 6:97
penidas, Telemiades 21:164
penitalis, Pyrausta 12:180; 17:227
pennatus, Symbiopsis (IA,IG) S4:18
pensilis, Lacinipola 8:101
PENTAS, Symbiopsis (IA,IG) S4:14
penziana, Eana 16:178
penziana, Tortrix 16:177
perditalis, Legna 8:27
Cephalanthus ?, Scirpus cyperius
perezi, ATLANTEA (IC, IG) 12:162, 163
perezi, Synchloe 12:162
perfida, Mellana 21:167
perfracta, Hydriomena 13:218
perfumosa, Aseptis 13:234

perfusca kootenaiata, Eupithecia 13:218

perigenes, Vidius (1st U.S.) 18:214; 20:35,36

Stenotaphrum secundatum

perimale scyllara, Cepora 22:121

peristera, Paulogramma 21:100

perkinsae, Leptotes 9:80

perlata, Campaea 5:54; 13:220

permagnaria, Ceratonyx 21:13

pernyi, Antheraea 21:257; 24:209

perquiritata, Anomogyna 5:54

perrhebus, Papilio 4:41

perrubralis, Pyrausta 5:29; 13:242

perse, Melitaea 19:114

persea, Meliteae 10:117, 118; 14:194

persephone, Chazara 10:118

persephone, Prodryas 1:42

perseus, Mycalesis (IA) 18:159

persicana, Archips 5:29

persimilis, Diagora 16:144

persius, Nisoniades 20:241

persius, Erynnis 8:3; 9:195; 14:38,103; 16:69; 17:111; 18:227; 21:207,277; 22:94,97,99,102

persius avinoffi 7:126; 11:48

persius frederickii 10:100; 14:33; 22:168

perspicua, Datana 21:36

peruviana, Zellerina 11:178

petasitis, Hydroecia 2:94

Petasites hybridus

petiveranus, Heliconius 18:18; 21:193; 22:34

petreus, Athena 3:25

petreus, Pap. Eq. Achiv. 12:197

petreus, Marpesia 20:122; 22:162; 25:39,185,186

petreus petreus 21:99

petreus thetys 9:205

petreus, Megalura 12:200

petropolitana, Pararge 14:139

petrus, Heliopetes 21:164

phaedrus, Lycaena 10:120

phaeocapna, Zale 24:3

phaesyla, Cybdelis 21:102; 22:152, 156

phaeton, Euphydryas 2:37; 4:43; 6:38; 9:78; 10:23,46,55,166; 12:41,42,43,165; 14:179,215,240; 21:208; 22:120; 23:48,188; 25:141

Chelone glabra, Aureolaria flava, Pentstemon hirsutus 2:37; 10:46; 23:48

phaetontea, Argynnis 12:120

phaetusa, Dryadula 21:97,192; 22:152

phaetusa phaetusa, Justinia 21:88, 166

phainis, Papias 21:165
phalaecus, Papilio 4:40
phalaenodectes, Myrmonyssus 9:55
phalaenoides cama, Helias 21:245
phalaenoides palpalis 21:163
phalantha, Atella 3:40; 11:167; 16:144
phaleros, Cycnus 21:152
Phalonia sp. 13:232
phantasia, Euryphene 13:128
phaon, Graphium 13:198, 199
phaon, Papilio 4:66
phaon, Phyciodes 9:191; 10:166; 12:130; 14:186; 17:150,225; 18:143,144; 20:161; 21:207,208, 278; 22:128,230; 25:40,214
    Phyla nodiflora 18:143,144
phaozalis, Herculia (II) 13:67
    dead leaves of Lombardy poplar
pharnaces, Papilio 4:64
pharella, Pharneuptychia 21:91; 22:151
pharte, Erebia 14:140
pharsalus, Acraea 13:146
phaselis, Sarmientoia 21:160
phasiana manga, Melinaea 15:32
phasiana phasiana 15:32
phedima, Melanitis 13:200; 14:139; 16:144
phedima oitensis 8:100
    Miscanthus, Zea, Coix, Setaria, Saccharum
phegea, Syntomis 11:148; 18:110
pherecydes, Pap. Eq. Achiu. 12:197
pheres, Lycaena 22:62
phestus reductus, Papilio (IA) 25:76, 78
phiale flavomaculata, Eurema 21:87, 154
phiale majorina, Eurema 21:154; 22:154, 156
phidippus, Amathusia 13:19; 18:101-104 (II)
    Cocos nucifera, Areca catechu, Corypha elata, Nipa fruticans, Musa sapientum 18:101-104
phleas, Lycaena 14:136
philea, Phoebis 1:70; 2:22; 9:206,207; 14:44; 15:72; 18:15; 21:137,190,207,208,278; 22:34
    Cassia corymbosia, Cassia fistula 2:22
philea philea 21:155; 22:154
philemon mathata, Leuchochimona 21:146; 22:153
philenor, Battus 6:75; 12:130; 13:195; 15:23, 71; 18:149; 19:109,229,232; 21:207,209,277; 22:34,200; 23:191,192; 24:305; 25:222,278
philenor hirsuta 12:204; 22:94,97,98,100-102
    Aristolochia 12:204
philenor philenor 17:22,154; 18:130,131; 25:36,214

Aristolochia longiflora 18:130,131
philenor, Papilio 3:17,25; 4:14,41; 6:33,101; 7:146; 9:143; 10:36,164
Aristolochia 4:14
philetas, Papilio 4:41
philetes adela, Perichares 14:31; 21:166; 22:156
philetes dolores (IA) 25:20-22
philippinensis, Hypolimnas 13:22
philippus, Hypolycaena 10:39; 20:135
phillone, Nymula 21:150; 22:153
phillyra phillyra, Phyciodes 21:193
philodice, Colias 2:13,25,60; 3:17,48; 5:25; 6:73,108; 7:139-145; 8:3,76,163; 9:78; 10: 55,107,111,143,166; 11:12,163,207; 12:120; 14:42,215; 15:43,93,112,213; 16:129,130,245; 17:76; 18:149,227; 19:110 (IA, 182,232; 21: 207,277; 22:120; 23:257-260; 24:305; 25:108
Trifolium sp. 2:13
philodice eriphyle 16:30
philodice vitabunda 7:125
philolaus, Graphium 21:188
philolaus, Papilio 4:66; 12:118,128,171; 24:233
philyra, Neptis 14:196
philyra excellens 8:99
Acer (Aceraceae)
phragmatidicola, Leucania 9:56, 57
phranza moreelsi, Euryphene (IA) 13:128
phranza phranza (IA) 13:128
phronius, Paryphthimoides 21:92
phrutus, "Thecla" 22:154
phocata, Psychophora 4:27, 29
phocion phocion, Perichares 5:55
phoebe, Melitaea 9:145,204; 10:215; 14:194
phoebe dorae 10:118
phoebe punica 10:221
phoebe scotosia 8:99
Cirsium, *Serratula* (Compositae)
phoebus, Chrysophanus 10:215, 222
phoebus, Parnassius 10:140; 19:203
phoebus catullius 10:142
phoebus smintheus 18:220
Sedum
phoebus sternitzkyi 24:189
pholus, Darapsa 10:166; 15:64,193; 19:189; 23: 100
hackberry 15:64
pholus, Lycomorpha 2:88; 11:150
phorbas, Pholus 3:78

phorcas ansorgei, Papilio 13:76
phosphorus, Papilio 4:41
photinus, Papilio 4:40
phlaeas, Lycaena 2:25; 5:26,27; 7:37; 8:125-128; 9:145; 10:55; 11:165,167; 12:43; 16:59-69,144; 18:110; 20:134; 24:196,305
phlaeas americana 5:27; 6:108; 12:38,202; 15:93; 16:136; 19:103; 21:207,208,278; 22:129; 24:2
phaeas daimio 8:97; 17:106
Rumex (Polygonaceae) 8:97
phaeas feildeni 11:52
phaeas hypophlaeas 19:61
phaeas phaeas (IA) 17:105, 106
Rumex acetosella
phlaeas pseudophlaeas, Chrysophanus 10:222
phlegia, Stalachtis 21:150; 22:153, 156
phloeas f. *eleus, Lycaena 10:117*
phyllophora, Heptagrotis 5:54
phylace, Amblyscirtes 13:91
phylaeus, Hylephyla 3:17; 9:191, 195
phyleus, Hylephila 2:22; 5:55; 12:130; 13:221-227; 14:31; 15:98; 17:156,226; 19:55; 21:207; 22:35,94,97,98,102,130,228,230; 23:186; 25:34,213
14:31; 21:277
Stenotaphrum secundatum 13:221-227
phyleus phyleus 21:166,247; 22:156
phyllira, Apantesis 4:26
Oenothera sp.
phyllis, Agrodiaetus 14:137
phyllus phyllus, Vettius 21:166
phylo oiclus, Bolla 21:164
phylo pullata, Bolla 21:245
piasus, Glaucopsyche 25:240-246
piasus catalina (IA) 25:242, 244
piasus daunia (IA) 25:241, 243, 244
piasus piasus (IA) 25:242, 244
piasus sagittigera (IA) 25:242, 244
piasus TOXEUMA *(IA) 25:244, 245*
piasus, Lycaena 21:272
piasus, Phaedrotes 16:37; 18:226
Lupinus 16:37
piasus daunia 10:98
piasus, Scolitantides 22:98
piasus daunia 22:165
Lupinus argenteus ?
piatrix, Catocala 24:109, 111
picta, Arachnis 13:237

picta hampsoni 13:237
picta picta 13:237
picta, Ceramica 23:42
picta, Phyciodes 13:91; 18:142,143; 19:114; 22:162
Siphonoglossa pilosella 18:142, 143
picta canace 25:214
pictipennis, Heliolonche (IA,IF,II) 25:287-292
Malacothrix glabrata
piera, Haetera 22:109
piera piera 24:16, 17
pieridoides, Theope 21:88, 150
Pierella 24:15
pierus, Aloeides 10:38
pigmalion, Pap. Pleb. Urbicol. 12:197
pigmalion batabano, Phocides 14:35
pigmalion hewitsonius 21:158; 22:155
pigmalion mancinus 14:35
pigmalion okeechobee 22:35
pigmalion pyres 5:55
pilumnus, Papilio 4:64; 5:16; 13:197,199; 15:198
pima, Anthocaris 19:112
pinastri, Sphinx 3:76
pinensis, Ellema 19:206
piniarius, Bupalus 25:159
piniata, Paraphia 5:54
pinifoliella, Exoteleia 21:10 (IG), 18, 20, 21
jack pine
pinus, Christoneura 7:57
pion, Allosmaitia 24:60, 61
pirithous, Tarucus 10:222
pirus, Butleria 10:100
pisonis kicaha, Itabalia 21:189
pitheas, Catagramma 21:194
pithecium, Phobetron 4:26
Ilex cassine ?
pithys, Adelpha (IA,IG) 24:292,293,296
pithyusa, Anaea 21:195
pizzaro, Papilio 4:41
plagiata, Olene 5:54
plagiata, Orgyria 21:37
plagiobothrae, Ethmia (II) S6:25
Plagiobothrys nothofulvus, P. tenellus
plantaginis, Nemeophila 11:149
plantaginis, Parasemia 17:177
platella, Homosassa 24:40
pleasure heredia, Limenitis 21:98; 22:152
plebeculata, Epirrhoe 13:219

plebia, Atreides 19:189
plebeja, Atreides 10:166
plenoscripta, Eupithecia 13:218
plexaure, Danaus 21:93
plexippus, Danaus 3:17,18,39; 4:14,45; 5:37; 6:15,69; 7:28,41,146,168; 8:123,124,162; 9:16,77,119,143,191; 10:44,55,95,111,169, 171; 12:130; 15:24,112; 16:30,229,243,249; 17:178,225,243; 18:41,47,165,221,230; 19: 114,137,180,182,232; 20:45,129,235; 21:31, 32,207,277; 22:63,64,98-103,120,127,204; 23:182,254; 24:298,305; 25:164,278,294
Asclepias curassavica, A. tuberosa rolfsii, A. syriaca, Sarcostemma clausa 25:294
plexippus megalippe 9:121
plexippus nigrippus 15:221
plexippus plexippus 8:27; 15:222; 17:150; 20: 102; 21:191; 22:34; 25:41, 294
plexippus portoricensis 5:56
plisthenes, Hypoleria 22:151, 157
PLUMASATA, *Anepia (IA,IG) 21:235-240*
plumigeraria, Coniodes 13:220
plumbilinea, Epimecis 5:59
plumosata, Itame 13:219
pluto, Xylophanes 2:86; 3:78; 13:78; 25:152
pocahontas, Hesperia 20:241
podalirius, Graphium 13:199; 15:58
podalirius, Iphiclides 6:14,26,40; 9:145
podalirius, Papilio 8:143; 10:215,218,220
Prunus 10:218
podana, Archips 10:154
podana, Cacoecia 17:48
podarce, Lycaena 12:194; 21:271
poeyi, Nannoparce 9:136
poeyi haterius 5:22, 23
poggei, Muschampia 10:118
pola, Melitaea 19:206,207 (IA); 21:273
pola arachne 14:238; 10:96; 19:114
polaris, Boloria 2:65; 5:8; 11:48,49; 19:243; 23:113-115 (IA)
polaris groenlandica 5:8; 19:242
polata, Dasyuris 2:65
polax, Cyclogbypha 21:87, 163
polibete, Doleschalia 13:22
polibetes, Panthiades (Parrhasius) 21:152
polibetes, Thecla 21:196
policenes, Graphium 25:273
polingi, Euristrymon 16:201-203

polingi, Agathymus 22:177; S1:53
Agave schottii S1:53
polingi, Strymon 13:91; 14:22
polingi, Megathymus 10:109
Agave schottii
polingi, Protoqyqia 2:42
polios, Callophrys 14:21
polios, Incisalia 7:33-35; 10:98; 11:47; 14:127; 22:26; 24:258, 260
Arctostaphylos uva-ursi 7:33-35
polixenes, Œneis 11:52; 19:60; 21:136; 23:113, 155
polixenes brucei 15:223
polixenes peartiae 19:60
polixenes katahdin 22:34
poltius, Limenitis (Adelpha) 22:152, 156
polybius, Phocides 15:72
polybius lilea 15:114
Psidium guajava
polybius phanias 21:158; 22:155
polycles polycles, Papilio 18:106
Citrus depressa, Toddalia asiatica
polychloros, Nymphalis 9:15; 11:164; 14:195; 18:40, 110; 22:128
polychloros, Eugonia 10:215
polychloros algirica 10:221
polyctor, Papilio 11:167; 16:144
polyctor polyctor, Polyctor 21:161; 22:155
polydorus dampieriensis, Papilio 25:79
polydorus manus, Pachlioptera (IA) 25:75, 79
polydorus queenslandicus, Pachlioptera 22:121
polydamas, Battus 15:198; 21:207,209,189,277; 22:34; 25:186
polydamas lucayus 18:130; 25:36
Aristolochia elegans 18:130
polydamas polycrates 5:55
polydamas polydamas 21:156; 22:154
polydamas, Papilio 4:41; 12:128
polydamas jamaicensis 7:170
polydamas lucayus 9:206
polymnia casabranca, Mechanitis 21:95; 22:151
polymnia lycidice, Mechanitis 21:190
polymnia polymnia, Mechanitis 22:109
polymniae, Papaipema 8:61
Polymnia uvedalia
Polyommatus 8:174
polyphemus, Antheraea 20:47; 25:42
polyphemus, Telea 2:22; 3:48; 10:45 (IA), 166;

12:50 (diapause 2:22; parasites 3:48)
polyphemus luna, Morpho 18:18; 20:95
polystictus, Papilio 4:41
polytes, Papilio 3:39; 11:229; 16:144; 17:163-169 (IA,II); 24:72
polytes cyrus 13:76
polytes ledebouria 21:50
polytes X Papilio helenus 18:116
polytes X Papilio protenor 17:163-169 (IA,II)
Poncirus trifoliata, Citrus natsudaidai
polyxena, Eryphanis 21:93
polyxena, Zerynthia 4:74
polyxenes, Papilio 1:42; 4:13,63; 5:26; 7:47; 9:7,8,78; 10:55,99; 12:130; 13:99,196,199; 14:5; 17:180; 23:182; 24:229,232,305
Petroselinum crispum, carrot 9:7,8
polyxenes americus 15:211
polyxenes asterius 4:110; 6:101; 9:49 (I); 13:74; 15:111; 16:203; 17:22,153; 18:133,134; 19:109,182,232; 21:207,277; 22:34,120,173-176,229; 23:258-260; 25:36,150,214
Daucus pusillus 17:22, Anethum graveolens, Ruta graveolens, Thamnosma texana, Daucus Carota, D. pusillus 18:133, 134
polyxenes forsythae 6:76
polyxenes polyxenes 21:189
polyxenes X Papilio xuthus (IA,II) 13:151-164
polyzelus, Papilio 4:41
polyzelus polyzelus, Parides 18:14; 21:189
pometaria, Alsophila 7:127; 8:140
pomona, Catopsilia 3:39; 6:69; 18:106
Cassia siamea 18:106
pomona pomona 22:121
pomonaria, Poecilopsis (IG) 9:200-202
Quercus, Crataegus, Malus
pomonella, Carpocapsa 5:30
pompeius, Hesperia 12:120
pompeius, Pompeius 17:37; 21:166,247
poohella, Arivaca 24:40
populi, Laothoe 18:36-41
populi, Limenitis 11:163; 14:196, 206
populi jezoensis 8:99
Populus (Salicaceae)
populi, Smerinthus 1:34
populiella, Lithocolletis 24:274, 277, 279, 280 (IA, IG)
Populus alba
porcelaria, Protoboarmia 4:26

Verbena canadensis ?
porcelaria indicataria 5:54
porcus, Xylophanes 5:23
porcus continentalis 3:78; 5:23; 25:152
porrectella, Plutella 18:94
porsenna, Polyommatus 20:241
portia, Anaea 9:80
portlandia, Lethe 6:32; 10:55,107; 12:41,130; 15:97; 17:109; 19:180; 22:120; 23:100; 24: 144,145,147 (IA),305
portlandia anthedon 17:107; 25:81,82
portlandia borealis 16:243
portlandia portlandia 17:149; 25:145, 146
postpuncta, Pompeius 21:167
potaro, Cobalopsis 21:165; 22:156
potatoria, Cosmotriche 2:94
Cares sp., Dactylis sp., Luzula sp.
potatoria, Philudoria 18:36-41
POTOSIENSIS, Atrytone (IA, IG) S2:40
potrillo potrillo, Cabares 21:244
POWELLELLA, Coptodisca 25:194,195,197,199,(IA,IG)
Quercus agrifolia, Q. wislizenii
poweskeik, Oarisma 24:46
praeclara, Catocala 12:201; 23:5; 24:108,111,114
praefica, Prodenia 4:25
Sabatia gracilis
prasinana britannica, Bena 5:28
pratensis, Phyciodes 20:246
pratti, Parabasis (IC) 10:15-17
praxinoe, Dismorphia 18:14, 15; 21:189
precationis, Autographa 15:111
prema, Alesa 21:146
premnas, Wallengrenia 21:166; 22:156
pressus, Anaplectoides 5:54
preussi, Euphaedra 13:128
preyeri, Artopoëtes 3:79
priamus, Ornithoptera 24:88-105
priamus admiralitatis (IA) 25:76, 79
priamus bornemanni 25:79
priamus euphorion 22:121
priamus poseidon 23:71 (II); 25:62, 63
priamus urvilleanus (II) 23:69-71,73,75,76
Aristolochia tagala
priamus richmondius, Troides 16:99
Aristolochia praevenosa, A. elegans (toxic)
prieuri, Satyrus 10:215, 218
prieuri kebira 10:221
primordialis, Anageshna 23:192, 194

primordialis, Geshna 19:133
procerus, Polythrix 21:278
proclea, ANTILLEA *12:164*
procne, Goniurus 19:53
procne, Urbanus 19:53,54; 20:35,38,39; 21:159, 244; 22:155
Cynodon dactylon 20:35-39
progne, Pap. Nymph. Phalerat. 12:197
progne, Polygonia 2:59; 6:73; 10:107,166; 12: 130; 15:99; 16:244; 21:207, 208, 278; 25: 139, 141
prolata, Melipotis 17:226
promethea, Callosamia 10:166; 18:31; 20:47; 22: 185,186; 25:234 (gynandromorph 18:31)
Prunus
promethea, Hyalophora 7:58; 13:215; 15:103
proneus, Battus (Parides) 21:89, 156
proneus, Papilio 4:41
pronoe, Erebia 14:140
propertius, Erynnis 13:100-102; 14:33; 16:70; 17:111,112; 18:227; 20:160; 22:94,97-102
propinqualis, Rivula 24:3
propinquilinea, Colocasia 5:54
propulsata, Lygris 13:218; 25:264
Ribes
proserpina, Apatura 19:200
prosope prosope, Cupha 22:121
protenor, Papilio 5:52; 11:9,167; 14:207; 15: 180 (IA,II); 16:144; 17:76,163-169 (IA); 13:199
Citrus junos, C. unshiu, Poncirus trifoliata, Xanthoxylum piperitum, Skimmia japonica, Fagara ailanthoides, Evodia rutaecarpa, Fortunella japonica 15:180
protenor demetrius 7:37; 8:96; 14:5; 15:175
Citrus, Poncris, Zanthoxylum (Rutaceae) 8:96
protenor liukiuensis 18:106
Citrus depressa
protenor X *Papilio helenus 18:116*
proterpia, Eurema 2:51; 14:44; 15:72; 18:16; 19:112; 21:190
protesilaus, Graphium 21:189
protesilaus protesilaus 21:88, 157
protesilaus, Papilio 4:66; 12:171
proteus, Eudamus 3:25
proteus, Urbanus 3:17; 5:55; 9:193; 13:225; 14:34; 15:100; 19:1,28,31; 24:47-51,246; 25:36,294
Phaseolus limensis, P. vulgaris, Clitoria

ternata *19:1-31*, *Desmodium tortuosum 25:294*
proteus proteus, Goniurus 17:155
proteus proteus, Urbanus 21:159,244; 22:155
proto gigas, Sloperia 10:222
protoclea, Charaxes 13:124
protodamus, Papilio 4:66
protodice, Pieris 2:105; 5:26; 7:146; 9:148; 10:100; 12:122,130; 13:99,100; 14:215,238; 15:198,212; 16:30,69; 17:243; 19:110,180,232; 20:49; 21:207,277; 22:94,97,98,102,120,126,227,229; 23:182,187,189-190 (IA),271; 24:77,245,305
Cruciferae 16:30
protodice protodice 17:155; 25:36,214
protomedia, Melitaea 8:99
Valeriana (Valerianaceae)
proxima consimilis, Hypoleria 21:87, 94
pruna, Bryoptera 18:115
pruna, Cargolia 18:115
pruna, Pteroteinon 13:146
pruni, Fixsenia 24:196
pruni, Strymonidia 11:166; 13:178; 14:138; 15:228; 20:135
Prunus 13:178
pruni jezoensis 8:97
Prunus (Rosaceae)
pryeri, Artopoëtes 8:97; 13:178
Ligustrum ibota (Oleaceae)
pryeri, Neptis 8:99; 14:196
Spiraea (Rosaceae)
pryeri, Paraneptis 14:204
psamathe, Dismorphia 21:155; 22:154
pseudargiolus, Argus 12:122; 13:212
pseudargiolus, Celastrina 13:102; 22:129
pseudargiolus, Lycaenopsis 2:25; 3:48; 7:146; 9:78; 10:162,169; 11:47; 12:130
Spiraea douglasii ? 10:169
pseudargiolus echo 2:42; 8:101
Spiraea discolor 8:101
pseudargyria, Leucania 8:3; 9:56
pseudoastrologa, Annaphila 19:237
Phacelia minor
pseudobryoniae, Polygonia 11:50
PSEUDOCERCYONIS *25:13, 16-19*
PSEUDOCOPIVALERIA *20:84*
pseudofea, Brephidium 9:209,210; 10:74
pseudolongula, Callophrys 20:70
Pseudolycaena 25:91

pseudomorpha, Eurema 21:154
pseudonapi, Pieris 8:75
pseudospretella, Borkhausenia 4:29
pseudothyreus, Madoryx 13:78
pseudozeritis, Chloroselas 13:145
psidii, Eutachyptera 13:24; 18:164 (error)
psidii pytho, Xanthocleis 21:95
psyche, Thecla 12:122
pteridis rubra, Diacrisia 8:101
    Plantago lanceolata
pterisii, Papaipema 8:61
    Pteridium aquilinum
pterophoralis, Zellerina 11:178
pudicaria, Epimecis 5:60
puellaria, Epimecis 5:59
pulcherrima, Eutelia 19:190
pullata, Gnophos 3:76
PULVERIPLAGA, *Archichlora (IA,IG) 25:170,173*
pulverosalis, Gabara 24:3
pulverulenta, Antigonus 13:91,223-227; 15:73
    Abutilon wrightii, Wissadula holosericea
pulverulenta, Systasea 19:1,9,21,32; 25:35
    Abutilon wrightii, A. abutiloides, A. incanum, Wissadula holosericea, W. ampliss-ima, Sphaeralcea angustifolia
pumila, Atychia 3:76
pumila, Nymphidia 12:122
punctiferella, Adela 23:212,216,231,236 (IA),238
punctilia, Cogia 21:164
punctomacularia, Philedia 13:220
punctularia, Nemoria 13:234
punctum, Chiomara 21:163; 22:155
punctum, Hyala 3:75
punctum, Zygaena 18:110
pupilla, Symbiopsis (IA,IG) S4:11
pupillata, Dyops 16:54
pupula, Eustixia 22:157
    Lepidium virginicum, Brassica oleracea, Brassica nigra
pura, Crambidia (1st Canadian) 24:227
puritana, Taygetis 6:98
purpuralis, Zygaena 18:110
purpurana, Archips 10:154
purpurea, Adela 23:212,215,216,219,233 (IA),237
purpurifascia, Papaipema 8:61
    Aquilegia canadensis
puspa ishigakiana, Celastrina 18:107
    Rosa wichuraiana ?

puspa, Lycaenopsis 11:167; 16:144
puspa umenois, Celastrina 8:98
Pieris (Ericaceae), *Glochidion* (Euphorbiaceae), *Rosa* (Rosaceae)
puta, Hasodima 18:115
pustularia, Itame 20:44
puziloi, Luedorfia 11:9
puziloi, Lühdorfia 13:199
puziloi yessoensis, Lühdorfia 8:96
Asiasarum (Aristolochiaceae)
pygas thamyras, Callicore 21:100; 22:152
pygas splendens, Callicore 21:100
pygmina, Tapinostola 2:94
pylades, Thorybes 7:146; 9:78,194; 10:55,107; 12:130; 13:99,232,233; 14:34; 16:70,245; 18:169,226; 19:1,26,31,51,180,196; 21:207,277; 22:98,120; 24:305; 25:81,82
Astragalus nuttallianus, Desmodium paniculatum, Rhynchosia texana 19:26,31
pylades albosuffusa 13:91; 19:1,27,32; 25:35
Rhynchosia texana 19:27, 32
pylaon iranica, Plebejus 10:117
pyracmon, Euptychia 15:222; 22:159
pyrame, Callidula 21:101
pyramidoides, Amphipyra 23:207-209
pyranthe, Catopsilia 3:39; 18:106
Cassia obtusifolia, C. occidentalis 18:106
pyranthe minna 6:69
pyrene, Ixias 3:39
pyri, Saturnia 18:36
pyrina, Zeuzera 9:142
pyrippe, Mesene 21:149
pyrrha, Actinote 21:96; 22:151
pyrrha, Perrhybris 13:84
pyrrhus, Xylophanes 3:78

quaderna, Erora 7:147; 10:23; 19:232
quaderna quaderna 16:1-4
quadra, Oenistis 3:76
quadraria, Apodrepanulatrix 13:219
quadratus, Papilio 4:41
quadrifasciata, Pseudoscada 21:87, 94
quadrilinearia, Itame 13:219
quadrilineata, Adelocephala (II) 15:233
Gleditsia triacanthos
quadrilineata, Scopula 5:54

quadristrigella, Opostega 17:237-242; 20:232-234 (IA)
quantius, Haywardina 21:92
quassa, Thecla 21:152
quehtala, Chlosyne 13:165
quenseli, Apantesis 12:50
quenselii, Orodemnias 3:76
quenselii gelida, Callarctia 21:38
querciella, Coptodisca 25:195-199 (IA)
Quercus emoryi
quercifolia alnifolia, Gastropacha 18:35-41
quercivora, Wagimo 8:97
Quercus (Fagaceae)
quercivoraria, Besma 4:26
Quercus phellos ?
quercus, Lasiocampa 13:24; 18:37; 25:162
quercus, Marumba 3:76
quercus, Quercusia 24:196
quercus, Thecla 11:166; 15:228; 20:130, 135
quercus, Zephyrus 10:215
quercus callunae, Lasiocampa 17:175
quercus iberica, Zephyrus 10:221
quieta constricta, Archanarta 2:65
quino, Melitaea 22:61
quinquelineraria, Scopula 13:217
quinquemaculatus, Phlegethontius 3:74; 10:166; 14:102; 15:111, 193; 19:189
quinquepunctata, Ithome 15:88
Coccoloba uvifera
quinquepunctella, Aristotelia 19:134
quiteria meridionalis, Opsiphanes 21:92; 22:151
quiteria quirinus 21:192

rachelae, Poecilopsis 9:199,200 (IG),202
dwarf willows 9:202
radiola pusillus, Callimormus 21:164
rafflesiae, Euschemon 2:60
Wilkiea (Monimiaceae)
ragalis, Thecla 18:22
rainieri, Boloria 14:158
ralla, Anathix 15:121
rama, Amblypodia 11:167
ramburi, Plebeius 10:217
RAMIREZI, *Dalla (IA,IG) S2:23*
RAMSTADTI, *Bomolocha (IA) 21:125, 126*
randa, Heliopetes 21:164
ranta rancida, Pellicia 21:161

rapae, Pieris 2:105; 3:39; 4:43; 5:53; 7:5,6, 37,39,48; 8:88; 9:58, 77, 145, 204; 10:18, 25, 55, 100, 107, 169; 11:162; 12:29, 44, 130, 174; 13:200; 14:79; 15:93, 101, 110; 16:30, 69, 128, 245; 17:22; 18:41, 94, 110, 221; 19:55, 110, 182; 20:13-27 (IA), 102, 217-225; 21:207, 277; 22:34, 94, 97, 98, 102, 120, 126, 132, 191, 211-224; 23:40, 77-93 (IA); 24:125-134 (IC), 224, 305; 25:9, 144, 145, 165

Arabis sp., Barbarea vulgaris, Cruciferae, Hesperis matronalis 7:48, 8:88, 16:30

rapae aestivus 15:212

rapae crucivora 8:96; 14:53; 18:106

Arabis, Brassica oleracea, B. juncea, Cleome, Rorippa indica, Raphanus, Wasabia

rapae iranica 10:117

rapae mauretanica 10:220

rapae rapae 14:46, 53

raphaelis, Coreana 3:79; 8:97; 13:178

Fraxinus (Oleaceae) 8:97

rastrata, Eana 16:178

ravidula, Emesis 21:149

ravocostaliata, Eupithecia 13:218

ravus, Thecla 21:152

rawsoni, Calephelis 25:37

reakirtii, Anthocaris 3:48; 6:106; 8:10; 9:118

rectifascia, Eudrepanulatrix 13:219

rectifascia, Taygetis 6:98

rectilinea, Limacodes 19:190

RECTILINEATA, *Archichlora 25:170,173-176 (IA,IG)*

recurvalis, Hymenia (Zinckenia) 13:86; 17:227

reevesii, Eryphanis 22:151, 156

reflexus, Autochton 22:155

refusana, Phaneta 25:284-286 (IA,IG)

regalis, Evenus 21:151

regalis, Citheronia 8:29; 14:96, 103

regalis joiceyi, Elzunia 15:30

regalis regalis, Elzunia 15:29

regis, Ornithoptera 24:89-105

relicta, Catocala 12:201; 13:75; 23:2,4,5 (IA); 24:108, 111, 113, 114

remingtoni, Agathymus 20:181; S1:49

Agave utahensis ?

REMINGTONI, *Megathymus (IA,IG) 11:117-118*

Agave utahensis ?

remus, Moeris 21:165, 246; 22:156

renata disaffecta, Euptychia 12:172
reniculella, Dioryctria 13:242
renunciata, Hydriomena 5:54
repressa, Epimecis 5:60
repugnalis, Anticarsia 17:226
residua, Catocala 24:108, 110, 115
resimachus, Dichorragia 14:206
resta, Xylophanes 3:78
resumens, Pachylia 25:152
retecta, Catocala 22:234-236; 24:108,110,115
rethon, Agathymus S1:53, 54
Agave sisalana
rethon, Megathymus 11:116
Agave sisalana
reticulata, Caloptilia 25:202
reticulata, Zopyrion 21:162
rex, Papilio 9:129; 13:76
rezia, Phanes 22:155
rhadamantus, Papilio (Ornithoptera) 13:76
rhadinaria, Ceratonyx 21:13
rhamni, Papilio 12:121
rhamni, Gonepteryx 9:145; 11:163,168; 14:52,204, 207; 15:58; 16:144
rhamni maxima 8:96
Rhamnus (Rhamnaceae)
rhamni meridionalis 10:220
rhamni TIANSHANICA *24:218-220 (IA)*
rhesus, Yvretta 5:63; 10:100; 22:165,169
rhodia, Mesosemia 21:146
rhoeo, Dircenna 21:94; 22:151
rhodope, Actinote 22:152
rhodosticta, Papilio 4:64
RIBA, *Cymaenes (IA,IG) 22:12-15*
ribearia, Itame 13:219; 25:266-270 (IC)
ribensis, Diestogyna (IA) 13:137
RICEI, *Agathymus (IA,IG) 20:163-167; S1:42*
Agave parryi-like; "Maguey" type of *Agave*
richardsoni, Anartia 2:65; 4:28
ridingsella, Adela 23:212,215-219,233 (IA),237
Parthenocissus quinquefolia ? 23:233
ridingsi, Neominois 7:51; 10:94; 18:47; 23:155
ridingsi stretchi 22:172
rielana, Eana (SUBEANA) *16:177*
rigida, Papaipema 8:61
Zizia aurea, Heliopsis helianthoides, Helianthus strumosus
rigidana, Rhyacionia (IA, IG) 14:231

rileyana, Heteropacha 18:89; 21:35
rimicola, Eriogaster 18:37
rimosa congratulans, Isognathus 5:21,22
rimosa inclitus 5:21,22
rimosa jamaicensis 5:21,22; 9:136
rimosa molitor 5:21,22; 9:136
rimosa papayae 3:78; 5:21,23
rimosa rimosa 5:21,22
rimosa wolcotti 5:21,22
rimosa woodi 5:21
RINDGEI, *Agathymus* (IA,IG) 18:180-185; 20:184, 185; S1:46
Agave lecheguilla
RIOBALSENSIS, *Chlosyne* (IA,IG) 14:152-154
ripartii, Agrodiaetus 9:145; 14:138
rita, Philotes 9:40; 20:131
rita coloradensis 22:165; 23:52,58
Eriogonum effusum
rita elvirae 23:52, 58
Eriogonum plumatella
rita pallescens 23:52, 58
Eriogonum sp.
rita rita 21:141; 23:52, 58
riukiuana, Ypthima 18:108
Gramineae
rivularis, Limenitis 14:196
rivularis, Paraneptis 14:204
robinsonii, Ctenucha 21:273
rogatalis, Hellula 10:112
rogeri, Papilio 4:64; 20:203
rogersi, Desmolycaena 13:145
rogersi, Kedestes 13:146
rohria dyrta, Lethe 16:144
rohria nilgiriensis, Lethe (IA) 18:160
roma, Polythrix 21:160
romanovi, Tomares 10:117
rosalia rosalinde, Sais 21:87,95; 22:151,156
rosana, Archips 10:154
roseata, Aemilia 21:273
roseola, Amolita 24:3
roseosuffusella, Aristotelia 19:134
rosita, Chlosyne 13:165; 14:148
rosita BROWNI (IA) 14:150-152
rossi, Anatoli 20:134
rossi, Byrdia 12:50
Potentilla tridentata
rossii, Erebia 2:64; 5:41; 17:8; 19:57
rossii gabrieli 7:124

rougeouti, Eresina (IA) 13:138
roxelana, Pararge 3:76
ruatensis, Xenophanes 13:33
ruberata nevadae, Hydriomena 13:218
rubi, Macrothylacia 18:36
rubi, Callophrys 10:215; 11:166; 14:138; 15:227;
20:130,134; 24:196
rubi fervida 10:222
rubicunda, Dryocampa 14:91, 103
rubicunda, Melitaea 23:176
rubidus, Lycaena 14:128; 18:225
Rumex salicifolius 18:225
rubidus sirius 10:98; 19:112; 20:161
rubifera, Diarsia 6:37
rubra, Hyalophora 25:68-73
rubricata, Euptychia 13:221-227; 16:203; 18:137;
19:114, 232; 25:41
Stenotaphrum secundatum 13:221-227
rubricata, Megisto 7:146
rubricoma, Acronicta 24:3
rubrocostata, Euryphene (IA) 13:128
rubromaculata, Liptena 13:138
rubromarginaria, Xystrota 13:217
rudkini, Papilio 8:3,7; 9:7,51,148; 17:7; 18:65
Daucus carota, Thamnosma montana 9:7
Foeniculum vulgare 18:65
ruficillata, Mesoleuca 5:54
rufimago, Panopoda 23:206
rufofascialis, Noctuelia 3:106; 13:242
rufofusca, Strymon 14:22
rufofusca, Thecla 21:153
rumina, Zerynthia 10:215
rumina africana 10:218, 220
rumina ornatior 10:220
rundiapicana, Eana 16:178
ruptifasciata, Timochares 9:193; 13:90
ruptistrigella, Gracilaria 25:202
ruralis, Notarcha 13:8
Urtica dioica
ruralis, Pyrgus 10:100; 16:35; 18:227
Potentilla drummondii ?, P. tenuiloba 16:35
ruralis macdunnoughi 14:32
ruricola, Atrytone 7:146; 9:78, 195; 10:55,101;
12:41; 14:215
ruricola metacomet 16:245
RUSEA, *Perisama (IA,IG) 24:10-13*
ruslana, Argynnis 7:37; 8:98
Viola (Violaceae)

ruslana, Argyronome 14:193
RUSPOSORIA, *Keiferia (IC,IG)* 24:6,7,8
russeliata, Eupithecia 5:54
rustica, Protoparce 2:86
rustica, Phlegethontius 3:78; 19:189; 25:152
rusticus cubanus, Phlegethontius 5:21,22,23
rusticus dominicanus 5:21,22,23
rusticus harterti 5:21,22
rusticus postscriptus 5:21
rusticus rusticus 5:21,22,23
rutulus, Papilio 2:42,92; 4:64; 9:46,139; 10:99; 12:28,103-113; 13:74,100,194,199,232; 14:33; 16:29,45,66; 17:243; 18:220; 19:232; 22:94, 97,98,102
Salix lasiolepis, S. lasiandra, Populus tremuloides, P. angustifolia, Alnus viridis, Platanus, Prunus, Liriodendron tulipifera
rutulus X Papilio glaucus 9:46-48
Liriodendron, Syringa
ryphea phidile, (Memphis) Anaea 21:104; 22:153

saadi, Coenonympha 10:118
sabini, Psychophora 4:27, 29
sabuleti, Polites 9:148; 18:50,227; 22:94.97,98, 102,170
sabuleti chusca 14:29
sabuleti mardon 14:30
sabuleti sabuleti 12:31
sabuleti tecumseh 13:232; 14:29; 16:36; 18:50
saepestriata, Japonica 3:79; 8:97; 13:178
Quercus acutissima, Q. serrata, Castanea
saepestriata, Thecla ? 13:175
Quercus acutissima
saepiolus, Plebeius 2:92; 5:54; 6:103,105; 7:50, 125; 10:98; 11:48; 14:127; 16:34,244; 18:49, 226; 19:171; 20:134
Trifolium hybridum 16:34
saepiolus gertschi 11:2
saepiolus insulanus 1:58; 10:170; 18:185-188
Lupinus ? 10:170
saepiolus saepiolus 5:4
saepium, Satyrium 14:20; 16:33,69; 17:111,243; 18:224; 20:135,160; 22:96,97,99,100,103,105
Ceanothus velutinus 16:33
saepium, Strymon 10:97; 13:235
saepium, Thecla 9:188

safitza, Mycalesis 18:74
sagana, Damora 14:194
sagana liane, Argynnis 8:98
Viola (Violaceae)
sagaris satnius, Phaenochitonia 21:149
sagata, Boloria 5:63
sagittigera, Lycaena 12:194; 21:272
salapia, Cargolia 18:115
salapia, Hasodima 18:115
salentia, Parthenos 12:116; 14:155 (siamese IA)
salenus, Synapte 24:212
SALICACEANA, *Aphania (IA,IG) 11:27, 28*
salicivorella, Lithocolletis (IA,IG) 24:274,277, 279
Salix alba
SALMOCOLOR, *Dasypyga (IA,IG) 24:250,251,253*
salome, Eurema 15:198; 23:188
salome limoneus 2:51
salona, Strymon 15:215
salonina, Hypoleria 21:94; 22:151
salvini, Papilio 4:67
salvini, Taygetis 6:98
samarcandae, Eana 16:179
samoset, Amblyscirtes 4:26; 14:27; 20:241; 24:304
Sorghastrum sp. 4:26
samoset, Hesperia 20:241
samuelis, Lycaeides 6:41
sandace, Zopyrion 21:245
SANDRAELLA, *Lithocolletis 25:196,203,205,207, 208,209 (IA,IG)*
Quercus agrifolia, Q. wislizenii
SANGOANA, *Archichlora 25:170,172,173,174 (IA,IG)*
sanguinalis, Pyrausta 17:48
sanguinea, Schinia 6:53
sanguinella, Gracilaria 25:202
sankara, Neptis 11:167; 16:144
sannio, Diacrisia 11:149
santa monka, Myrinia 21:164
saphirina, Diestogyna (IA) 13:137
saphirinus, Favonius 3:80; 8:97; 13:178
Quercus, Q. dentata
sapho, Heliconius 21:193
sappho, Catocala 4:31; 13:75
sappho, Rebelia 3:76
sara, Anthocaris 6:100-103,106; 7:146; 8:10; 16: 30,70; 18:221; 19:232; 22:94,97,99,100,102
Descurainia pinnata 6:106, mustard 16:30
sara inghami 19:112; 22:166

sara julia 10:99
sara reakirtii 2:42
sara sara 12:28; 13:232
sara julia, Euchloe 1:42
sara veraepacis, Heliconius 21:193
sarae apseudes 22:152, 156
sarae thamar 21:87, 88, 96
sardonyx, Phasis 10:38
sareptana, Epibactra 3:76
sareptensis, Cryptolechia 3:76
sarita inanoides, Euphaedra 13:128
sarpedon, Zetides 16:144
sarpedon, Papilio 13:22
sarpedon, Graphium 9:74; 11:9; 13:199; 21:50
sarpedon morium 18:105
Cinnamomum japonicum, Litsea japonica, Machilus thunbergii
sarpedon nipponum 8:96
Cinnamomum, Machilus, Neolitsea (Lauraceae)
sassacus, Hesperia 10:55, 159; 19:196; 20:241; 25:81, 82
sassacus manitoboides 14:30; 21:136
satakei, Papilio 8:96
Phellodendron, Orixa (Rutaceae)
satanaria, Ceratonyx (IA) 21:11, 12, 13
satellitia, Pholus 2:86; 3:78
satellitia pandorus 19:189
satellitia posticatus 5:22
satellitia satellitia 5:22
satevis aurantia, Melinaea 15:30
satevis crameri 15:30
satevis satevis 15:30
saturata, Anartia 15:217
saturnus, Callimormus 21:165; 22:155
saturnus servus, Saturnus 21:166
satyr, Artines 21:165
satyrata, Eupithecia 5:54
satyroides, Macusia 21:151
satyrus, Vanessa 19:205
satyrus, Polygonia 7:146; 10:96, 107, 169; 13:73; 16:69; 17:243; 18:223; 20:161; 22:98, 99, 103, 252; 25:140
nettle 18:223
satyrus marsyas 16:244; 19:113
satyrus tampa, Timochreon 21:162
saucia, Peridroma 13:86; 22:200
saucia, Phaloesia (1st U.S.) (IA) 24:301-302
saundersiana, Oncocnemis 24:3
savanna alaudinus, Passerculus 20:73

savigny, Hesperia 12:120
scabra, Plathypena 15:110; 19:133
scamander, Papilio 4:65; 21:88
scamander grayi 21:157; 22:154,156
scanda, Erebia 11:167; 16:144
scapha, Prolimacodes 21:33
schadenbergi, Ptychandra 13:20
schakra, Pararge 16:144,145
schellbachi, Speyeria 5:33
schmidti, Monima 3:76
schönmanni, Eana 16:178
schrenckii menalcas, Aranda 8:99
    Carex (Cyperaceae)
schulziana, Olethreutes 4:29
sciata, Papaipema 8:61
    Veronicastrum virginicum
scintillans, Epamera 13:145
scintillans, Iolaus (IA) 13:145
scintillanus, Neozephyrus 3:79
scipio, Erebia 14:140
scirpicola, Leucania 9:56
scolopacea, Epimecis 5:59
scolymus, Anthocaris 7:37; 8:96; 14:52,204
    Arabis, Brassica, Cardamine, Rorippa
scopas, Thecla 18:23; 21:196
scopulaepes, Hemeroplanis 17:227
scribonia, Ecpantheria 24:227
    Polypodium aureum
scripta, Habrosyne 5:54
scriptura, Pyrgus 13:231; 22:94,97,99,102,165,
    168
scrupulosa, Synedoida 13:234
scudderella, Lithocolletis 24:277,278 (IG)
    Salix
scudderi, Colias 7:50; 8:163; 10:99; 14:42;
    19:202
    Vaccinium 7:50
scudderi ruckesi 22:166
scudderi, Pamphila 11:151
scudderi, Plebeius 5:8
scudderi aster 5:9
scutata, Phlegethontius 3:78
scutellariaeella, Gnorimoschema 4:4
    Scutellaria sp.
scutosa, Canthylidia 6:53
scylax, Melinaea 15:30
scylla, Ethmia (IF,II) S6:12
    Collinsia heterophylla

sebaldus, Pap. Pleb. Urbicol. 12:197
sebrus, Cupido 9:204; 14:136
secedens, Polia 5:54
secundaria, Drepanulatrix 13:219
seitzi, "Thecla" 21:87, 153
sejona, Phyciodes 21:97; 22:152
sejuncta, Coeliades 18:74
selene, Argynnis 2:6
selene, Clossiana 11:163
selene, Boloria 5:8,26; 6:105; 9:78; 10:55; 12: 41,43; 14:215,230; 17:171; 18:222; 19:62; 20:103; 25:141
Viola nephrophylla
selene alboquina 5:26
selene atrocostalis 5:8,26; 10:107; 16:243; 18:233; 19:242
selene marilandica 5:26
selene myrina 5:26
selene nebraskensis 5:26
selene selene 14:193
selene terraenovae 5:26; 23:36, 37
selene tollandensis 1:42; 5:26; 7:49; 10:95; 14:158,184; 15:115
selenophora ishiana, Tacoraea 18:108
Mussaenda pubescens
selima, Doxocopa 21:103
selima selima, Callicore 21:100; 22:152
selira, Rapala 16:144
semele, Eumenis 9:145; 18:110
semele, Hipparchia 11:165; 14:140
semele algirica, Satyrus 10:221
semialbata, Cargolia 18:114
semialbata, Neodesmodes 18:114
semiargus, Cyaniris 3:39; 9:204; 11:165; 14:137; 24:196
semiargus, Polyommatus 4:74
semiargus maroccana, Glaucopsyche 10:222
semiatrata, Eustroma 13:218
semicana, Patriciola 24:255
semiclarata, Babta 13:219
semidea arctica, Œneis 9:141
semiferana, Archips 10:154
SEMIFUNERALIS, *Syllepsis (IA,IG) 24:287,288*
semilugens, Ethmia (II) S6:40
Phacelia calthifolia, P. crenulata
semitenebrella, Ethmia S6:51
Cercocarpus parviflorus, C. ledifolius
semperi, Papilio 13:23

semperinus, Papilio 13:76
sempiternus dilloni, Echelatus 4:72
sempiternus sempiternus, Anastrus 21:245
sempiternus simplicior 21:163
sena, Heliophorus 11:167; 16:144
senex, Comacla 2:94; 3:75
    Jungermannia sp.
sennae, Phoebis 3:25,40; 13:240; 15:199; 19:232; 22:166,228,229; 23:182; 24:305
sennae eubule 4:25; 8:162; 10:100; 17:22,154; 20:102,215; 21:137,207,277; 22:34,182; 24:118,119; 25:37,214
    Cassia nictitans ? 4:25; Chamaerista cinerea
sennae marcellina 12:28; 18:15; 21:190
sennae sennae 5:56; 21:155; 22:154
sennanensis, Agrodiaetus 14:137
separata, Schinia (IA,IF) 25:177-181
    Artemisia tridentata, A. cana ?
separatus melaena, Sphinx 5:22
sephyrus, Plebejus 3:76
septentrionalis, Danaus 3:39
septentrionella, Adela 23:212,215,216,229,235 (IA),238
    Holodiscus discolor 23:229
sepulchralis, Citheronia 19:189; 23:25
sequax, Pseudaletia 9:57
sequoiae engelhardti, Sphinx (II) 21:27-31
    Juniperus californica
serapio, Thecla 18:23
serena, Catocala 10:168; 24:110,115
serotina, Erebia 13:104
serpa paraena, Limenitis 21:87,99
serpentifera, Zellerina 11:178
serratissimalis, Crocidophora 8:102
serratulae, Pyrgus 11:166; 14:51
serrulata, Cosymbia 13:233
serville, Papilio 4:67
sessilis, Lasaia 13:90
sesostris, Papilio 4:41
severini, Euryphene 13:128
sexmaculata, Semiothisa 8:3
sexmaculata incolorata 13:219
sexta, Manduca 25:41
sexta, Protoparce 2:86
sexta, Phlegethontius 19:189; 25:152
sextus, Phlegethontius 5:21; 11:43; 10:166; 14:102
sextus jamaicensis 5:22

sextus lucia 5:22
SHAKTI, *Paralasa (IA) 15:13*
shasta, Plebeius 7:51; 10:98; 16:35; 17:114; 18:49
shandura APSARA, *Melitaea (IA) 15:6*
shastella, Gracilaria 25:202
sheridani, Callophrys 10:97; 14:21; 20:134; 24:258
sheridani newcomeri 18:224
SHIVACOLA, *Hyponephele (IA) 15:16*
shurtleffi, Heterogenea 21:33
sicelis, Kirrodesa 14:139
sicelis, Lethe 7:37; 8:99
Pleioblastus (Gramineae)
sideraria, Scopula 13:217
sidonis, Lethe 16:144
sidus, Epamera 10:39
sidus, Eupsilia 19:183
sidus, Iolaus 20:135
sierrae, Gabriola 13:220
sierrae, Hydriomena 13:218
sifanica, Boloria 11:163
sifanica numida, Hesperia 10:222
signata, Wagimo 13:178
Quercus spp.
signata signata 8:97
Quercus (Fagaceae)
signatus, Wagimo 3:80
silaceata albolineata, Diactinia 5:54
silas, Argiolaus 10:39
silas, Iolaus 20:135
silphii, Papaipema 8:61
Silphium spp.
silvester, Adopaea 9:204
silvestris, Cercyonis 18:221; 22:98-100,103
silvestris, Minois 3:106; 4:70; 12:29
silvicola, Carterocephalus 11:166
simaethis, Strymon 14:22
simaethis sarita 15:56
similalis, Loxostege 6:108; 19:133,135
similis, Euptychia 21:191
similis, Catocala 12:202; 24:108,111,114; 25:42
similis, Pamphila 18:44
similis, Porthesia 6:42
similis similis, Celaenorrhinus 21:160
similis similis, Radena 18:107
Tylophoratanakae
simius, Amblyscirtes 13:91; 22:170
simius, Yvretta 10:100
simplex, Adopaeoides 13:91

simplex, Autographa 13:26
simplex, Comacla 11:148
simpliciella, Chalceopla 23:231,232
simplicius, Pap. Pleb. Urbic. 12:197
simplicius, Urbanus 19:53,54; 21:159,244; 22:155
simulans, Œneis 9:141
sinapis, Leptidea 9:145,203; 11:162; 14:53;18:110
sincera, Anomogyna 5:54
sinensis, Pelopidas 11:168; 16:144
singulella, Adela 23:212,215,216,229-231,236,238 (IA)
    Gilia capitata, G. achillaefolia 23:236
sinuosa, Cirphis 9:57
sirenia nitida, Mesosemia 21:87,146
siris, Polites 14:30
sisymbrii, Papilio 11:47
sisymbrii, Pieris 7:146; 10:100; 11:47; 16:30; 17:112; 18:221; 19:110,232; 21:121 (IA); 22:105
    Cruciferae 16:30
sisymbrii sisymbrii 12:29
sita, Caduga 14:204,207
sita niphonica 18:107
    Hoya carnosa, Marsdenia tinctoria, Asclepias curassavica
sito, Thecla 21:196
siva, Callophrys 14:21; 16:20,57; 24:228
siva, Mitoura 7:147; 8:7; 9:4; 10:97; 13:100; 15:91; 19:112,232
siva juniperaria, Mitoura 15:201
smaragdinus, Chrysozephyrus 14:206
smaragdinus, Neozephyrus 3:79; 8:97; 13:178
    Prunus (Rosaceae)
SMALLI, *Symbiopsis* (IA,IG) S4:12
smilacis, Thecla 12:122
smintheus, Parnassius 10:141,171; 13:194-201; 16:70
smintheus olympiana 9:141
smintheus sayii 10:99
smintheus sternitzkyi 9:141
smithi, Epicausis 9:66
smithi, Megathymus 11:114; 12:81; 14:59
smithi, Stallingsia S1:3,4
    Manfreda maculata ?
smithiae, Lymnas 21:148
snowi, Lycaena 7:50; 14:128
snowi, Ochlodes 16:234
snowi, Thyreion 6:53

SOBRA, *Morys (IA,IG) 22:12,14,15,18*
socrates socrates, Pyrrhopygopsis 21:167
socus socus, Epargyreus 21:158
sodalis, Sodalia 21:165
somnulentella, Bedellia 4:5; 19:135
Ipomea 4:5
sonoma, Psaphida 20:84
sonoma, Pseudocopivaleria (IA,IG) 20:84-88
sonora, Polites 16:36; 18:228
grasses ? 16:36
sonora siris 11:43
sonora utahensis 10:100
sonorae, Melitaea 21:273
sonorella, Anderida 24:255
sonorensis, Apodemia 12:194; 21:272
sonorensis, Lycaena 12:194; 21:272
sonorensis, Philotes 9:40; 10:172; 12:31; 17:202,216-223 (IA,IG); 19:95-101; 20:50; 22:105; 23:52,54,59,60; 25:241
Sedum, Dudleya cymosa setchellii?, Stonecrop
sophia, Precis 18:74
sophocles, "Thecla" 21:152; 22:154
sophorae laurentii, Brassolis 21:92; 22:151
sophus sophus, Euryphene (IA) 13:128
sorana, Callicore 21:101; 22:152,156
sordida, Catocala 10:168; 22:234-236;23:262; 24:108, 110, 114
sororius, Hemileuca 2:60,81
sosana, Epiblema 19:136
sospeta, Xanthotype 8:3
sosunga, Godyris 21:191
sosybia, Euptychia 3:25
sosybius, Euptychia 6:76
spadaria, Doryodes 19:190
spaldingi, Philotes 8:8; 22:165; 23:52,59
Eriogonum racemosum
sparganii, Archanara 2:94
Typha latifolia, Phragmites communis
spatiosa, Euphaedra (IA) 13:128
speciosa, Palaeacrita 21:57,58
speciosa, Philotes 8:8; 23:52,59
Oxytheca perfoliata, O. trilobata, Eriogonum reniforme 23:52,59
speciosissima, Papaipema 8:61
Osmunda cinnamomea, O. regalis, O. claytoniana
sperryorum, Diastictis 13:242
Speyeria 25:247-256
Viola, V. papilionacea, V. priceana 25:253

spinetorum, Callophrys *14:21; 16:70,193; 17:245; 18:48,224; 20:135; 22:96,97,99,100,102,164*
Arceuthobium ? 16:193
spinetorum, Mitoura *1:42; 7:147; 10:97; 12:14; 13:100,183,233*
Arceuthobium campylopodum, A. campylopodum f. *divaricatum, A. vaginatum* f. *crytopodum, A. americanum 12:14*
parasite - *Apanteles n. sp. 12:14*
spini, Strymonidia *20:130,131,135; 24:196*
spinosae, Eupanychis *6:53*
spitzi, Udranomia *21:87,158*
splendens, Citheronia *16:62*
Gossypium thurberi, walnut
splendens, Lycomorpha *11:150*
splendor, Arcus *(IA,IC,IG) 25:87,90,93,94,103, 104,105*
spurina, Strymon *14:22*
squanda, Telemiades *21:164*
stalachtioides, Apodemia *21:149*
stallingsi, Celaenorrhinus *24:212*
STALLINGSIA *12:87*
statilinus, Eumenis *9:145*
statilinus, Satyrus *10:215*
statilinus rungsi *10:221*
statira, Pap. Dan. Cand. *12:197*
statira, Phoebis *18:15; 21:154; 22:154*
statira floridensis *9:207,211*
Ecastophyllum ecastophyllum
statira hispaniolae *5:55*
statira jada *21:190*
stator, Noctuana *21:245*
stauderi, Carcharodus *10:222*
steinbachi, Papilio *4:41*
stelenes, Metamorpha *9:206, 210; 14:188; 15:199*
stelenes biplagiata *21:194; 25:137*
stelenes stelenes *5:56; 21:98; 22:152*
steneles, Metamorpha *15:72*
steneles biplagiata *18:20; 25:39*
stellata, Actizera *10:38*
stelligera, Haywardina *21:92; 22:151, 156*
stenodamus, Papilio *4:66*
stenoscelis, Papaipema *8:61*
Woodwardia virginica, Dryopteris noveboracensis
stephensi, Agathymus *12:179, 186; 23:244-248; S1:51, 52*
Agave deserti S1:52
stephensi, Satyrus *19:161-164*

stephensi, Megathymus 10:109; 20:53
stercorea, Ancylostomia 24:254
stevenaria, Gnophos 3:75
sthenele, Minois 10:113
sthenele, Cercyonis 15:222; 16:69; 17:111; 19: 232; 23:166-168,170-173
sthenele masoni 23:165
sthenele oetus 16:31
sthenele paulus 23:165
sthenele sthenele 23:165
sthenele silvestris 17:243; 23:165
stictomenes stictomenes, Vehilius 21:165
stigmata, Æromachus 16:144
stigmaticus, Grais 14:40; 15:72; 25:35
stigmaticus stigmaticus 21:163; 22:155
stigmella, Acrobasis 24:254
stirius, Erebia 13:104; 14:140
stolida, Grammodes 3:75
stomoxyformis, Synanthedon 3:75
storax storax, Parphorus 21:246
stramentacealis, Zanclognatha 3:75
strataria, Biston 9:199
stratocles, Papilio 13:22
streckeri, Megathymus 18:45; 12:180,186,189; 13:93, 100; 22:171,177,178; S1:23
    Yucca baileyi ?, Y. glauca ? 12:189
streckerianus, Papilio 4:41
strenua, Symbiopsis (IA,IG) S4:8
strenuana, Epiblema 19:134
stretchii, Catocala 22:58
striana, Archips 10:154
STRIATA, PEBA *(IA,IG)* 22:4,8,9,11
strictomenes illudens, Vehilius 21:246
strictomenes strictomenes, Vehilius 22:155
strigifinitella, Neurobatha 19:135; 25:200,201
strigilis, Protambulyx 3:78; 25:152
strigilis strigilis 13:78
strigillata, Praefaunala 21:92
strigosa, Clostera 21:35
strigula, Rhyacia 2:94
    Erica cinerea
stubbendorfii, Parnassius 11:9
stubbendorfii hoenei 8:96
    Corydalis (Papaveraceae)
stulta, Parigonia 3:78
styx, Erebia 13:104; 14:140
suaveola, Callophrys 10:117

subalbida, Epimecis 5:60
SUBARGENTANA, *Eana (Ablabia) 16:176, 183 (IG)*
subatomaria, Hypagyrtis 13:220
subcolorata, Eupithecia 13:218
subcostulata subcostulata, Papias 21:165
subdifferens, Dyops 16:54
subdita, Lasionycta 12:50
subflexa, Heliothis 6:53
subfuscata, Scopula 13:217
subgrisea subgrisea, Morys 21:166
subhastata stygiata, Eulype 5:54
sublinea, Heliopetes 14:32; 15:72; 21:205
sublivens, Lycaeides 6:36 (♀desc.)
submedialis, Mecyna 5:29
subnata, Catocala 24:111,113,114
subnebulosa, Parasemia 4:27,28
subhastata albodecorata, Eulype 13:219
subota, Phyciodes 21:193
subpallida, Asterocampa (IA) 19:113
subplicana, Epinotia 13:234
subpunctata, Pherne 13:233
subrotata, Heterocampa 21:37
subsolana yagina, Lycaeides 8:98
    Vicia
substriataria danbyi, Epirrhanthis 13:220
subtentyris, Euryphene (IA) 13:137
subterranea, Agrotis 18:5
SUBTILIS, *Lithophane (IA,IG,II) 23:10-13,15-18*
    Juniperus californica
SUBTROPICA, *Rhyacionia (IA,IG) 14:231-236*
    Pinus elliottii v. *elliottii, P. taeda, P. palustris, P. tropicalis*
sugitanii, Celastrina 8:98
    Æsculus (Hippocastanaceae)
suk, Virachola 13:145
sulphuraria, Hesperumia 13:220
summandosa, Pareuptychia 21:91
superba, Annaphila 3:74
    Gilia lutea
superba, Metamorpha 21:194
superior, Anisochoria 21:162
surima, Actinote 21:95; 22:151
susanna, Stalachtis 22:153,156
swammerdamella, Nemophora 23:214
sybillina, Chrasmarcha 10:77
syces cubensis, Pachylia 5:22
syces insularis, Pachylia 2:86; 5:22

syces syces, Pachylia 5:21
sycorax, Papilio 9:75
syla, Thecla 16:144
syllius pelagia, Melanargia 10:221
sylvanoides, Ochlodes 9:148; 11:151; 14:28,40; 16:37,69; 18:227; 20:160; 22:94,97-102; 23:278
sylvanoides napa 10:100
sylvanoides sylvanoides 12:32
sylvaticus, Thymelicus 8:95
Gramineae
sylvestris, Thymelicus 18:110
sylvestris sylvestris, Adopaea 22:130
sylvicola, Panoquina 21:167
sylvicola sylvicola 21:247
sylvicola isshikii, Carterocephalus 8:95
Gramineae
sylvina, Yphthimoides 21:91
sylvinus, Satyrium 14:20,130; 16:33; 18:48, 224; 20:135; 22:163
willows 16:33
sylvinus itys, Strymon 19:112
sylvinus sylvinus, Strymon 12:30
sylvo, Episcada 21:94; 22:151
syma, Limenitis 22:152,156
SYMBIOPSIS S4:4
syncellus syncellus, Thecla 18:23; 21:196
synemonistis, Pemphigostola 9:66
syntomoides, Syntomeida 11:148
syrictus, Pyrgus 9:206

tachasara, Stolidoptera 3:78
tages, Erynnis 11:166; 14:51
tagis mauretanica, Anthocaris 10:218,220
taikosama, Aloeides 10:38
taiwanus, Neozephyrus 3:79
takanonis, Spindasis 7:45; 8:97; 9:13; 15:246
fed orally by host ants
talayra, Thecla 21:196
talboti, Anthene 13:145
tamarinde, Opsiphanes 21:192
tamarindi, Opsiphanes 17:37
tamasea, Elzunia 15:29
tameamea, Vanessa (IA,II) 22:111-114
Pipturus albidus
tamyris, Marela 21:164
tamyroides, Marela 21:164

tanais, Symbiopsis (IA,IG) S4:19
tanna, Urbanus 21:244
tantalus clavipes, Sesia 5:22
tantalus eumelus 5:22
tarania, "Thecla" 21:153; 22:154
tarda, Lithophane 23:13
  Juniperus pachyhloea
tarquinius, Feniseca 1:42,56; 4:26; 7:45; 9:78; 10:55,107; 12:130; 15:92; 16:244; 20:152; 22: 120; 24:196
  Viburnum dentatum (pupa on stem of)
tarricina, Tithorea 21:190
tarricina bonita 15:28
tarricina duenna 15:28; 24:299
tarricina pinthias 15:28
tarricina tagarma 15:28
tarricina tarricina 15:28
tasso, Papilio 4:64
tatalis, Pyrausta 13:233
tatila, Eunica 12:25,130; 13:62,63
tatila tatilista 6:33; 9:206; 22:35
tau, Aglia 4:13; 14:103; 18:36
  beech 4:13
taunayi, "Thecla" 21:87,153
taxila, Neozephyrus 7:37; 13:178; 14:138
  Alnus spp. 13:178
taxila japonicus 8:97
  Alnus (Betulaceae)
taxila monticola 8:97
  Alnus
taxila regina 8:97
  Alnus
taxiles, Poanes 10:100
taygete, Œneis 5:26; 11:49; 23:155
taygete fordi 19:58
taygete hanburyi 2:64
taygete taygete 19:243; 23:103
tehachapina, Speyeria 17:244
tehuana, Anaea 20:203
telamonides, Papilio 1:75; 12:194
telata tyrtaeus, Monca 21:246
teleboas, Archimestra 24:207
telekii, Athetis 3:76
telekii, Psodos 3:76
telemachus, Erynnis 22:168
telesilaus, Graphium 21:157
telesilaus, Papilio 4:66
teleus, Urbanus 15:72; 21:159,244; 22:155

telicanus, Syntarucus 10:38; 14:136; 18:74
tenedia, Emesis 21:149,196
tenella, Eurema (Pyrisitia) 21:154; 22:154
tenuidactylus, Pterophorus 19:133
tepida, Graptolitha 15:121
TEPHRINA, *Lithophane* (IA,IG) 23:12,13,14
teras, Lepricornis 21:147
tereas, Archonias 22:154
terlooi mooseri, Arctonotus 5:22
terlootii, Neophasia 19:100; 22:60
terminus terminus, Mycalesis 22:121
terpsichore terpsichore, Lyropteryx 21:146;22:153
terrestris, Argyreuptychia 21:87,92; 22:151
tersa, Xylophanes 2:86; 3:78; 10:166; 19:189; 25: 41, 152
tesselata, Lycaena 16:104
tessellana, Olethreutes 2:65
tessellaris, Halisidota 2:36; 4:26; 8:3
Carpinus caroliniana 4:26
tessellum tersa, Muschampia 10:118
testata, Lygris 25:264
tethys, Daimio 7:37; 14:51,203
tethys daiseni 8:95
Dioscorea (Dioscoreaceae)
tethys tethys 8:95
Dioscorea
tetonia, Speyeria 5:34
tetrica, Mesosemia 18:24; 21:196
tetrio, Pseudosphinx 19:227; 25:152
tettensis, Abantis 9:129
teuthras erubescens, Cosmosoma 13:83
texana, Horama 11:148; 25:42
texana, Megathymus 11:113
texana, Mocis 2:S8
Syntherisma sp.
texana, Phyciodes 12:130; 13:226, 227; 18:144; 19:113, 232; 20:156, 161; 21:208; 22:162
Beleperone guttata, Dicliptera brachiata, Jacobinia carnea, Ruellia occidentalis, R. drummondiana 13:226, 227; 18:144
texana seminole 17:151; 19:48
texana texana, Phyciodes (Tritanassa) 22:230; 25:40, 214
texanae, Amblyscirtes 13:91, 92
texanus, Megathymus 12:180, 189; 13:91-93, 99; 18:45; S1:24
Yucca baileyi ?, Y. glauca ? 12:189

texanus leussleri S1:24
Yucca glauca
texanus texanus S1:24
Yucca glauca
TEXOLA *12:161*
textula, Sisyrosea 21:33
thadana, Biblis 12:120
thalestris, Phoebis 5:56
thamyra, Taygetis 21:90
thoas brasiliensis, Papilio 22:154
tharos, Phyciodes 2:22; 5:126; 6:33,105; 7:125; 8:3; 9:77,194; 10:55,96,107; 12:130; 14:186; 15:93; 16:244; 17:180; 18:228; 19:104,133, 137,180; 21:207,277; 22:120,128,132,228,230; 23:192,193; 24:77,305
tharos pascoensis 2:92
tharos pulchella 15:219; 19:113
tharos tharos 17:23,150; 18:142; 20:102; 22:35; 25:40,141,143,144
Aster prealtus 18:142
tharossa, Argynnis 12:120
thasus, Cremna 21:87, 146
thasus thasus, Mysoria 21:157
theano, Erebia 5:26; 19:57
theano alaskensis 7:124; 19:61
theano ethela 2:37
thelephassa, Chazara 10:118
themis, Euptychia 18:17
themis aureola, Euphaedra (IA) 13:128
themisto, Thyridia 21:94
themistocles, Polites 9:78,195; 10:55,100; 12: 38,120,130; 14:216; 16:245; 18:228; 19:180, 196; 21:207,277; 24:305
themistocles turneri 17:113
theon, Olyras 15:27
theon, Pellicia 21:164
theodora theodora, Chalodeta 21:148
theodora, Hamearis 21:87,146
theona, Melitaea 7:146
theona bolli 13:226,227; 15:56; 19:114
Leucophyllum texanum, L. frutescens
theona theckla 19:114
theona, Thessalia 21:193
theona bollii 25:40
theona ethela, Erebia 7:49
theona theona, Chlosyne 18:19
theophrastus, Tarucus 10:38,219-222
camel-thorn 10:219

thera eratosthenes, Melinaea 15:33
thera thera, Melinaea 15:32
Thereus 24:58,59
thermaea, Zethera 13:20,21
thermesia, Dismorphia 22:154
thermesia thermesia 21:155
thermus valgus, Phocides 21:160
thero, Phasis 10:38; 24:196
thersamon kurdistana, Lycaena 10:117
thersites, Agrodiaetus 14:137
thersites, Lysandra 9:145; 20:134
thersites, Papilio 4:64; 7:39
thersites, Polyommatus 10:222
theseus, Dynamine 18:21; 21:194
theseus justitiae, Morpho 20:95-101
thespis, Tarucus 10:38
thetis, Curetis 20:135
thisbe, Panara 21:87,147
thius, Callicista 21:153,154
thoas, Papilio 4:63,64
thoas autocles 13:195,199; 18:14; 21:189
thoas melonius 7:170
thoasa gerona, Limenitis 21:99; 22:152
thoë, Lycaena 6:108; 8:102,128; 10:23,98,112; 11:165; 12:130; 14:215; 16:244; 20:130,134; 21:207,208,278; 22:120; 24:196,305
thoë X *Lycaena phlaeas americana 16:235*
thomasi, Hemiargus 9:207
thore scandinavica, Boloria 14:193
thore jezoensis 8:98; 14:193
    Viola (Violaceae)
THORPELLA, *Adela 23:212,215,216,226-228,234 (IA), 238*
thrasibulus thrasibulus, Cycloglypha 21:163
thraso, Achlyodes 13:90; 14:33
thraso tamenund 14:33; 19:1,20,32; 25:35
    Zanthoxylum fagara 19:20-32
thraso thraso 21:163,245
thyastes, Papilio 4:67
thyatiroides, Dasylophia 21:36
thyelia salvini, Xylophanes 5:22
thymbraeus, Papilio 4:66
thymetus thymetus, Phyciodes 21:97; 22:152
thyodamas, Cyrestis 11:167; 14:195; 16:144
thyodamas ishigakiana 18:108
    Ficus septica, F. erecta (Moraceae)
thyodamas mabella 8:99; 18:108

Ficus septica, F. erecta
thyra, Aloeides 10:38
thyrsis, Gangara 16:131-134; 18:74,163
thysa, Belenois 25:80
thysbe, Hemaris 4:12; 10:166; 25:41
Vibarnum cassinoides 4:12
thysbe, Poecilmitis 10:39; 24:196
tianschanicus SHIVA, Parnassius (IA) 15:4
tigrina, Amplypterus 3:78
tigrinaria, Euchlaena 13:220
TILDENI, Hesperia (IA) 9:196-198
tiliae, Mimas 18:37
timais, Xanthopastis 25:138
timberlakei, Ethmia S6:53
Phacelia ramosissima v. *suffrutescens*,
Phacelia ramosissima
timias, Papilio 4:41
timon, Pygaera 3:75
tiphon, Coenonympha 14:139
tiphon mackenziei 9:141
tircis, Chamaelymnas 21:88,148
tiridates, Charaxes (IA) 13:124
titan cubensis, Sesia 5:22
titani, Xylophanes 3:78
titania, Boloria 5:26,27; 6:105; 7:50; 11:48, 49; 14:230; 19:59,62; 20:103; 23:113
titania boisduvalii 5:26; 19:242
titania grandis 5:26,27; 6:105; 7:125; 14:230; 16:243; 18:237; 21:136
titania helena 5:26; 10:96; 14:184
titania ingens 5:26
titania montina 5:26; 10:54
titania montinus 15:221
titania rainieri 5:26; 18:222
Viola sp.
titania, Catagramma 21:194
titania, Catocala 3:51
tithia, Dynamine 21:99; 22:152
tithonus, Maniola 9:145
tithonus, Pyronia 11:164
titus, Chrysophanus 14:127; 15:227; 16:195; 18: 48,224; 20:135
titus mopsus 21:277
titus, Harkenclenus 22:129,132; 23:187; 25:82
titus mopsus 21:207; 25:296
titus watsoni 25:37
titus, Satyrium 14:20; 16:199
titus, Strymon 1:58; 4:13; 9:77; 10:55,97,107;

12:43,130; 14:215
titus immaculosus 13:92
titus mopsus 6:107
titus titus 6:107
tityrus, Epargyreus 3:59; 7:54
tityrus, Heodes 11:165
tityrus, Lycaena 14:136
tityus, Hemaris 18:41
toddi, Boloria 5:26,69; 7:50,123,125; 9:54,77; 10:55,96,107,160; 11:163; 12:43; 14:215; 19: 104; 20:103; 22:120; 25:141
toddi ammiralis 5:26; 18:228; 20:114 (IA)
toddi jenistae 5:26
toddi toddi 16:243; 19:242
TODDI, *Oncocnemis (IA,IG) 22:135-137,139*
tolimus robigus, Anastrus 21:164
tolosa, Napeogenes 18:16; 21:190
topazata, Stamnodes 13:218
toralis, Timora 6:53
torquatus, Papilio 4:64; 20:203
torquatus polybius 21:157; 22:154
torrefacta, Apatelodes 18:89; 21:35
torrentium, Borolia 9:57
torris, Thecla 21:152
toxea, Eumenia 12:119,120
toxeus, Achalarus 14:37-56; 19:1,27,28,31; 25:35
Pithecellobium flexicaule 19:27-31
transcaucasica, Melitaea 14:194
transversata, Sabulodes 23:206
trapeza, Papilio 4:66
trayja, Metamorpha 21:98; 22:152
trebellius martinus, Tagiades 18:105
Dioscorea alata, D. cirrhosa
trebula, Calycopis 21:196
tremuloidiella, Lithocolletis 24:274-277 (IA,IG)
Populus tremuloides
trianguliferata, Neoterpes 13:220
TRIANGULIFERA, *Syllepsis (IA,IG) 24:288-290*
triangulum, Pholus 3:78
tribulella, Anacostia 24:40
tricolor, Hemileuca 2:81
Cercidium sp. 2:81
trichostomus, Crambus 2:65
trifasciata trifasciata, Timochares 21:163,245; 22:155
trifolii, Discestra 10:201
trifolii, Lasiocampa 13:24; 18:36-41
trifolii trifolii, Scotogramma 9:56

trigrapha, Adela 23:212,215,216,220,221,234,237 (IA)
Linanthus androsaceus, L. bicolor 23:221
trilinearia, Plataea 13:220
trimeni, Crenis 18:74
trinotata, Eilicrinia 3:76
triopas, Papilio 4:41
tripunctata, Acanthopteroctetes (IA,IG) 23:137, 139,140,142,146
tripunctata, Taygetis 6:97
tripunctus theogenis, Cymaenes 14:31; 21:165
triquetra, Mithras 21:152
trisignata, Harrisimemna 1:83 (wood-boring)
Virburnum lentago, ash, privet
tristis, Erynnis 22:94,97,98,102; 23:278
trite, Phoebis 21:156
tritona, Acronicta 5:54
trivena, Limenitis 16:143
trivia, Melitaea 14:194
trivia robertsi 10:118
trixus, Pap. Pleb. Urb. 12:197
trochilus, Gorgopas 21:164
trochilus, Zizeeria 11:167; 16:144
troglodyta troglodyta, Anaea 5:56
troilus, Papilio 3:17,48,64; 4:64; 5:16; 6:5,6; 9:77,143; 10:55,164; 12:130; 13:74,197,199; 15:23; 17:180; 21:207,277; 22:120; 24:305; 25:36
troilus ilioneus 3:25; 9:206; 13:26,74; 17:153; 18:136,137; 20:102; 22:34
Lindera pubescens, Sassafras albidum, Persea borbonia, Magnolia 18:136,137
troilus troilus 17:22; 22:34
trojana, Trogonoptera 21:49-52; 24:303
tros, Battus (Parides) 21:156
truncata ochrofuscaria, Dysstroma 13:218
tryhana istria, Vinius 21:164
tryxus, Xenophanes 21:162,245; 22:155
tuberculum Schinia 6:53
tullia, Coenonympha 4:72; 5:26; 9:204; 10:23; 11:164; 16:69; 17:158; 19:159; 21:1
tullia ampelos 17:113; 18:221; 21:1-7 (IA)
tullia californica 16:31
tullia eryngii 21:6
tullia EUNOMIA *(IA) 21:1-7*
tullia inornata 8:102; 16:243; 17:158
tullia mono 17:112
tullia nipsisquit 1:58

tullia ochracea 10:94
tullia yukonensis 11:52
tullia semperi, Discophora 24:304
tullia, Pythonides 21:162
tulliola, Everes 21:151
tullius lasius, Perophthalma 18:24
tullius tullius 21:87, 146
tuneta, Arcas (IA,IC,IG) 25:90-94, 100-107
tuolumnalis, Pyrausta 5:29; 13:242
turfosalis, Schrankia 2:94
TURNERINA *12:84*
tutana, Aricoris 21:150
tyndarus, Erebia 5:27; 14:139
tyndarus callias 2:37
tyndarus, Xylophanes 3:78
typhae, Phragmitiphila 2:94
  Typha latifolia, T. angustifolia
typhon, Methionopsis S2:32
tyrtaeus, Lerodea 13:90
tytia niphonica, Caduga 8:98
  Marsdenia, Metaplexis (Asclepiadaceae)
TZOTZILI, *Pyrrhopyge (IA) S2:2*

ubaldus, Azanus 10:38
uhleri, Œneis 3:26; 10:94; 14:127; 23:160
ulsterata, Philobia 13:219
ultramarinus, Favonius 3:80; 8:97; 13:178, 180
  Quercus crispula, Q. dentata
ultronia, Catocala 3:64; 12:201; 23:5,100;
  24:108,110,113,114 (predator 3:64)
u-lucida PARABUS, *Quadrus 22:3,6,155*
ulysses gabrielis, Papilio (IA) 25:75,78
umbra, Cremna 18:24
umbra, Pyrrhia 2:S8; 24:234
  Rhus typhina
umbrosaria, Pseudoboarmia 8:3
umbrosaria nigrovenaria, Nepytia 13:220
unangulata intermediata, Euphyia 13:219
uncas, Hesperia 10:100; 13:91,92
uncas uncas 25:213
uncinata, Taygetis 6:98
uncula, Eustrotia 2:94
  grasses
undulata, Calocalpe 5:54; 11:19
  willow
undulata, Hydria 13:217

undulatus, Typhedanus 21:159,244
undulella, Epichnopteryx 3:76
undulosa, Ceratomia 4:26; 10:166; 15:193; 19:
189
Fraxinus sp. 4:26
ungemachi, Polyommatus 10:222
unicalcararia, Apodrepanulatrix 13:219
unicolor, Leptalina 8:95; 14:204
Miscanthus, Imperata, Setaria (Gramineae)
unicolor, Psyche 18:37
unicornuis, Schizura 4:26
Aronia arbutifolia ?
unifascia, Cisthene 21:38; 24:277
unifascialis, Pyrausta 5:29
unijuga, Catocala 12:201; 23:5; 24:108,111,114
unimaculella, Ithome 15:81
unipuncta, Leucania 3:51; 4:26
Senecio sp. 4:26
unipuncta, Pseudaletia 9:56; 13:26; 19:132,133;
22:200; 25:42
unipuncta antica 9:57
unitaria, Lytrosis (IA) 23:3,4
Urania sp. (Migration) 4:67
urentis, Abrostola 24:3
ursaria, Plataea 13:232
ursula, Leucania 8:3
ursus, Megathymus 3:36; 10:1-8 (IA,IG); 18:117;
S1:24,25
Yucca schottii S1:25
ursus violae, 14:27
urticae, Aglais 7:37; 8:143; 9:15,204; 11:164;
13:3
urticae connexa 8:99
Laportea, Urtica (Urticaceae)
urticae, Nymphalis 3:39; 4:61; 14:195; 18:40;
22:128
urticae, Vanessa 9:15
urticae maxima 3:75
urticaeae, Eugonia 16:104
urticaria, Xanthotype 4:25; 8:3
Salix sp. 4:25
utahensis, Polites 7:50
utha, Pyrrhocalles 14:31
uvydixa, Thorybes 15:225

vacciniae, Pseudotamila 6:53

vacuna, Doxocopa 21:103
vagans, Olene 2:13; 23:42
Populus tremuloides 2:13
vaillantina, Egybolis 18:74
valentina, Taygetis 6:98
valeriana, Thorybes 15:225
valerius valerius, Morys 21:88,166
vallina colophonia, Hypothyris 22:109
VALLISCOLA, Acronicta (IA,IG) 22:133,134,137,139
VALVERDIENSIS, Agathymus (IA,IG) 20:181-185; S1:49,50
Agave lecheguilla S1:50
vanensis, Agrodiaetus 14:137
vanillae, Agraulis 2:37; 3:17,25; 7:146; 10:163-166; 11:163; 15:112; 17:227,228; 18:229; 19:33,34,114,232; 20:53,215; 21:207,277; 22:228,229; 23:182; 25:139,141
Passiflora incarnata 17:227,228
vanillae incarnata 12:204; 18:19; 21:192; 22:98,99,103; 25:41,214
vanillae maculosa 21:97; 22:152
vanillae nigrior 9:206,211; 17:23,150; 22:34
Passiflora incarnata 17:23
vanillae, Dione 3:52; 4:67; 12:130
vanillae insularis 5:56
vanillae, Papilio 12:121
vansomereni, Argiolaus 13:145
vansomereni, Iolaus (IA) 13:145
vardhana, Lycaenopsis 11:167; 16:144
varia, Halpe 7:37; 14:205
varia, Mellicta 14:194
varia, Thoressa 14:51
varia varia 8:95
Arundinaria, Pleioblastus, Sasa (Gramineae)
variicolor, Microceris 21:87,158
varuna, Rapala 20:135
vau-album, Polygonia 14:179,195,206
vau-album, Nymphalis 15:218;23:187; 25:140
vau-album j-album 22:120; 23:100,193
vaucheri, Coenonympha 10:215,218,221
vaucheri annoceuri 10:221
vasiliata, Earophila 13:219
VEADEIRA, Lerema (IA,IG) 22:12,14,15,18
vecina najoides, Pellicia 21:164
veleda veleda, Eprius 21:246
velutina, Calephelis 18:24; 21:196
venata, Ochlodes 9:204; 14:51
venata herculea 8:95

Miscanthus (Gramineae)
venatus, Ochlodes 11:166
venusta, Hypanartia 21:194
venustissima, Pacidara 11:214
venustus, Tmolus 21:152
venusulella, Depressaria 3:76
verma, Lethe 11:167; 16:144
VERNA, *Phaneta (IA,IG) 25:285,286*
verna, Polites 9:40,78,195; 10:55; 12:43; 14:216
verna, Pompeius 19:196; 21:207,277; 24:305; 25: 81,82
verna sequoyah 14:29
vernata, Paleacrita 7:127
versicolor, Darapsa 15:193
versicolor versicolor, Mimoniades 21:157; 22:155
versicolora, Endromis 4:13; 18:36
white birch 4:13
vertumnus, Papilio 4:41
vesta, Phyciodes 13:226,227; 20:161; 22:230; 25:40,214
Siphonoglossa pilosella 13:226,227
vestalis vestalis, Leucochimona 21:196
vestigiata, Paryphthimoides 22:151
vestris, Atrytone 18:50
vestris, Euphyes 14:28; 18:227; 19:180; 21:207, 277; 24:305
vestris metacomet 14:28; 25:34
vestusta, Hemerocampa (II) 7:153
vetulana, Eana 16:179
vexillata, Epimecis 5:59
VEYTIUS, *Staphylus (IA,IG) S2:18*
vialis, Amblyscirtes 9:195; 12:130; 16:70,245; 18:50,227; 19:1,31; 21:207,277; 24:305
Cynodon dactylon 19:1,31
vianna, Anisochoria 21:87,163
viardi, Eana 16:179
viardi, Pieris 15:198
viator, Atrytone 12:44
viator, Poanes 10:23; 12:41; 13:75; 14:216; 15: 98; 20:229; 23:186
Zizaniopsis miliacea 20:229
vibex, Polites 13:74; 15:110
vibex catilina 21:166; 22:156
vibex brettoides 13:90; 17:156
vibex praeceps 19:1,31; 25:34
Cynodon dactylon, Stenotaphrum secundatum
vibex vibex 20:241
vicrama, Philotes 8:8; 14:138; 24:196

victoriae, Ornithoptera 23:69,72 (II),74,75,76
Aristolochia tagala
victoriae epiphanes 25:61
victorinus victorinus, Papilio 21:189
vidius, Vidius 21:165
vidleri, Erebia 18:227; 19:94
vidua, Catocala 23:5; 24:111
vierecki, Atrytonopsis 7:146;13:99,100; 19:220
viertlii, Boarmia 3:76
villiersii, Battus 15:210
vinula, Cerura 6:5; 13:3
vinula, Dicranula 4:13
viola, Charaxes 9:129
viola kirki 13:124
viola phaeus (IA) 13:124
viola, Speyeria 5:34
VIOLAE, Megathymus 10:4-8 (IA,IG); 13:93,99; 14:40; 15:68; 18:117; S1:25
Yucca torreyi, Yucca baccata ?, Yucca treculeana ?
violella, Viola 21:161; 22:155
violescens, Iaspis 21:87,152
violetta, Charaxes 13:124
virbius hersilia, Cobalus 22:156
virescens, Heliothis 4:26; 6:53; 15:121
Rhexia mariana ? 4:26
virescens, Urbanus 21:160; 22:155
virgaureae, Heodes, 9:204; 11:165
virgaureae, Lycaena 14:136; 20:134
virgilia, Taygetis 6:98; 21:90; 22:151
virgilius, Theope 21:197
virginaria, Epimecis 23:192,194
virginiana virginiana, Hyposcada 21:190
virginica, Diacrisia 2:13; 13:75
virginiensis, Anisota 25:84,85
Quercus nigra
virginiensis, Calephelis 6:75; 14:128; 19:49; 21:281
virginiensis, Erycina 15:214
virginiensis, Lephelisca 9:206; 20:160; 25:137
virginiensis, Pieris 5:26; 8:3; 10:18-28 (IA,IH), 55,159; 14:46; 22:126,132,252; 24:45; 25:6-11, 222
Dentaria diphylla 10:25-28; 25:6-11
virginiensis, Tharsalea 8:128
virginiensis, Vanessa 3:17,18,48; 5:26; 7:39, 147; 9:77; 10:55,97,107; 12:130; 13:73; 14: 186; 16:32,233,244; 17:23,151,225; 18:223;

19:119,173,182,232; 20:42; 21:194,207,277; 22:35,98,99,103,120; 23:182; 24:305; 25:39, 140,214
Gnaphalium palustre 16:32; parasites 3:48
virginiensis brasiliensis, Vanessa 21:98; 22:152
virgo, Callogonia 3:76
virgo, Platychasma (IC) 17:33,34
viridana, Tortrix 13:5,6
viridescens, Eana 16:178
viridipennata, Mesothea 13:217
viridis, Callophrys 10:115; 20:134; 21:113,114; 23:280; S5:3-41 (IA,IF,II)
Eriogonum latifolium latifolium, Lotus scoparius 21:113,114; S5:3-41
viridis, Ceramidia 22:187
viridis, Hesperia 13:91,100; 16:203; 19:1,31; 25:34
Cynodon dactylon, Lolium perenne 19:1-31
viridis vista, Niconiades 21:247
viterboana, Urbanus 21:244
viterboana alva 22:155
vitis, Pholus 3:78; 25:152
vitis hesperidum 5:22
vitis vitis 5:22
vitreus, Phanus 21:158
vitrinus, Cocytius 2:86; 9:135
vitrinus musgravi 5:22; 9:135
vitrinus vitrinus 5:22
vogelii, Philotes 10:215,217,222
Erodium cheilanthifolium 10:217
voigti, Karanasa (IA) 15:9,10
volumna intricata, Evonyme 21:87,102
vulcanus, Spidasis 9:14
Plectronia (Rubiaceae), *Zizyphus* (Rhamnaceae), *Allophylus* (Sapindaceae), *Clerodendron* (Verbenaceae)
vulneraria, Sphacelodes 25:139
vulpinus, Ablepsis 22:155

w-album, Strymonidia 11:166; 13:178; 20:131, 135; 24:196
Ulmus 13:178
w-album fentoni 8:97
Ulmus (Ulmaceae), *Prunus* (Rosaceae)
walkerata, Dysstroma 5:54; 13:218
walkeri, Amphimoea 25:152

walkeri, Apodemia 13:90
walkeri, Samia 3:106; 5:16
walkeri advena 9:142
walshii, Papilio 1:75
walsinghami, Schinia 25:181-185 (IA,IF,II)
Chrysothamnus nauseosus
walterorum, Calosaturnia 14:103
walterorum, Saturnia (correction) 12:160
WALTORUM, *Saturnia* (IA,II,IG) 12:17-25
Arctostaphylos sp., Rhus laurina (lab.)
wanda, Acronycta 12:208
warscewiczii, Papilio 4:65; 15:117
washingtonalis, Phlyctaenia 4:29
washingtonalis, Udea 13:242
WATSONAE, *Victoria* (IA,IG) 25:169,170,175
wedah, Pseudergolis 11:167; 16:144
weidemeyerii, Limenitis 2:92; 10:97; 14:188, 238; 19:232; 22:215-227; 25:278
fm. *fridayi* (IA) 21:223-227
ab. *nigerrima* (IA) 21:215,226
ab. *sinefascia* (IA) 21:216,217,226
weidemeyerii angustifascia 8:130; 21:214-228 (IA)
Salix
weidemeyerii LATIFASCIA (IA) 21:214-228
weidemeyerii nevadae 17:112; 21:214-228 (IA)
weidemeyerii nevadae X *Limenitis lorquini* 17:112
weidemeyerii oberfoelli (IA) 21:214-228
weidemeyerii sinefascia 19:113
weidemeyerii weidemeyerii (IA) 21:213-228
westwoodi, Euptychia 21:90
weymeri, Taygetis 6:98
wiedemeyerii, Limenitis 7:146; 20:159,161
wilsonensis macdunnoughi, Sericosema 13:219
WILSONI, *Mysoria* (IA,IG) S2:3
wilwerthi, Euryphene (IA) 13:128
wimico, Panoquina 14:27
WINDI, *Epargyreus* (IA,IG) S2:4
WINDI, *Windia* (IA,IG) S2:17
WINDIA S2:16
wingina, Hesperia 20:241
WISLIZENIELLA, *Cameraria* (IA,IG) 25:195,203, 208-210
Quercus wislizenii, Q. wislizenii v. *frutescens, Q. agrifolia*
woodgatei, Hesperia 7:147
wrefordi, Spialia 13:146
wrighti, Lephelisca 20:134
wrightii, Melitaea 3:106; 4:70

xami, Callophrys 14:21; 16:203; 18:85-89; 20:212
Echeveria gibbiflora, Sedum allantoides, Dudleya saxosa ?
xami, Mitoura 13:90
xanthaphes, Niconiades 21:167,247
xanthica, Perisama 24:12
xanthochlora, Cyphanta (IC) 17:33,34
xanthocles, Heliconius 23:105
xanthoides, Lycaena 17:212,244; 19:171; 20:161; 22:96-98,100,103; 24:305
xanthoides dione 10:98,112; 11:165
xanthoides, Tharsalea 12:130
xantholepis, Eudyops 16:54
xanthomelas, Nymphalis 7:38; 11:164; 14:195, 205; 18:40
xanthomelas japonica 8:99
Celtis (Ulmaceae), *Salix* (Salicaceae)
xanthometata, Mellilla 23:192,194
xanthopleura, Papilio 4:65
xanthus, Pyrgus 14:32; 22:168
xanticles, Papilio 4:66
xenana, Taygetis 6:97
xenia erythra, Lymnas 21:148;22:153
xerces, Glaucopsyche 10:113-115; 12:94; 23:153,154
Lotus glaber, Lupinus arboreus, L. micranthus, Astragalus menziesii 23:153,154
xerces, Lycaena 22:60
xicaque, Paramecera 19:114
xuthus, Papilio 4:110; 5:52; 7:37; 8:96; 11:9; 13:199; 14:5-18,207; 17:70, (IA,II) 14:6-18
Citrus, Fagara, Phellodendron, Poncris, Zanthoxylum, Poncitrus trifoliata (Rutaceae)
xuthus X Papilio hippocrates (IA,II) 13:151-164; 14:5-18
xuthus xuthus 18:106
Zanthoxylum beecheyanum, Fagara ailanthoides
xylina, Lygris 13:218; 25:264
xynias, Papilio 4:66
xypete caerulescens, Euphaedra 13:128
xypete cyanea 13:128

yama, Lethe 16:144
yarigadakeana, Lycaeides 8:98
Astragalus, Hedysarum (Leguminosae)
ydatodes, Afrida 6:53

yehl, Poanes 17:156
yojoa, Strymon 14:22; 21:196
youngi, Erebia 11:50; 19:57; 19:60
youngi herscheli 11:49
youngi rileyi 7:124
yphthima, Taygetis 6:97
yphthima pacta, Yphthimoides 21:91
yreka, Pieris 20:52
yuasai, Favonius 8:97; 13:178
Quercus acutissima
yuccae, Eudamus 12:122
yuccae, Megathymus 7:146; 11:113,137-141 (IA); 12:89 (IC),175-183,186; 13:89,92; 14:27
Yucca filamentosa, Y. smalliana, Y. aloifolia, Y. gloriosa, Y. flaccida, Y. louisianensis
yucca arizonae 20:172
yuccae buchholzi 13:28; S1:9
yuccae coloradensis 13:99; 17:94; 19:89; 20:172; 22:177,178
Yucca intermedia v. *ramosa 13:99*
yuccae ELIDAENSIS *(IA) 20:170-172*
Yucca intermedia v. *ramosa*
yuccae KENDALLI *(IA) 19:83-89*
Yucca constricta, Y. rupicola, Y. pallida, Y. necopina
yuccae LOUISEAE *(IA) 17:90-99; 19:89*
Yucca thompsoniana
yuccae martini 12:177,178; 20:53
Agave deserti
yuccae MAUDAE *(IA) 20:168-170*
Yucca schidigera
yuccae navajo 13:93; 22:171
yuccae REINTHALI *(IA) 17:89-99; 19:89*
Yucca louisianensis
yuccae REUBENI *(IA) 17:87,88; 19:89*
Yucca baccata, Y. elata, Y. torreyi ?
yuccae stallingsi 13:89; 17:94; 19:89
Yucca arkansana
yuccae WILSONORUM *(IA,IG) 11:129,130; 13:90; 17:97; 19:89*
Yucca treculeana ?
yuccae WINKENSIS *(IA) 19:85-90*
Yucca campestris
yuccae yuccae 14:37; 19:51; S1:9
Yucca smalliana
yuma, Ochlodes 9:148; 11:151-154; 12:203; 14:28; 22:94,97,99,100,102; 23:278
yuma scudderi 22:170

zabulon, Hesperia 12:122
zabulon, Poanes 9:195; 10:55; 19:217; 21:207, 277; 24:305
zacynthus, Papilio 4:41
zaddachi, Euphaedra (IA) 13:128
zagreus, Papilio 4:65; 5:16
zaida zaida, Neptis 16:144
zangis, Calisto 9:80
zarex, Autochton 21:160,244; 22:155
zarucco, Erynnis 13:71,72; 14:176; 17:148,156; 21:207,277
Daubentonia, Sesbanea grandiflora, Robinia pseudoacacia
zarucco funeralis 12:31; 13:235; 25:35
zarucco zarucco 17:21,22
Robinia pseudoacacia, Sesbania exaltata
zatima, Spilosoma 6:58
zea, Heliothis 18:6; 19:132,133; 22:200
zebina, Strymon 14:22
zela ares, Emesis 19:112
zelicaon, Papilio 2:25; 3:64; 4:63; 5:26; 6: 102; 7:43,44; 9:10; 10:169; 13:74; 16:29,45, 66,70; 18:220; 19:173; 20:53; 22:94,97,98, 102; 23:279; 24:229,231,232
Heracleum lanatum 10:169, Cymopterus terebinthinus 16:29, Lomatium grayi, L. triternatum, Pteryxia terebinthina 18:220
zelleri, Acanthopsyche 3:76
zelleri, Parnara 10:215
zelleriella, Ethmia 11:45
ZELLERINA proposed instead of *Stenoptycha 11:178*
zelmira, Lycaena 12:194; 21:271,272
zelmire, Pap. Dan. Cand. 12:197
zenobia, Thysania 1:42; 2:34; 4:13; 13:65
zephodes, Ephyriades 14:33
zephyralis, Choristostigma 13:234
zephyria, Automeris 19:82
zephyrus, Polygonia 10:97,169; 14:186; 16:32, 70; 18:223; 19:113,173,232; 20:161; 24:44
Ribes viscosissimum 16:32, Ribes cereum 18:223
zephyrus, Vanessa 19:205
zerene, Speyeria 2:5; 5:31; 7:56; 16:81-97 (IA); 17:111,112; 19:172; 22:98-100,103
zerene carolae 15:221
zerene conchyliatus 17:113
zerene cynna, Speyeria 5:33
zerene garrettii 18:222
zerene gloriosa 17:114

zerene hippolyta 17:116
zerene malcolmi 23:64
zerene platina 5:33; 10:95
zerene sinope 2:92; 10:102; 14:182
zerene zerene 16:31
zeroe, Polyommatus 21:273
zetes, Papilio 4:41; 25:142
zetides, Battus 25:142
zeuxo zeuxo, Phasis 20:131
ziczac, Notodonta 10:201
ziha, Thecla 16:144
zilpa, Chioides 25:36
zilpa, Eudamus 15:72
zippora, Taygetis 6:97
zitenius, Melanitis 16:144
zobeidah, Caradrina 10:201
zohra monticola, Cigaritis 10:222
zolicaon, Papilio 8:3
ZOLUS, *Ouadrus* 22:2,3,5,6 (IA,IG)
zonalis, Junonia 12:194
zonalis, Precis 8:1
zonara, Euryphene (IA) 13:137
zonaria, Graphium 5:55
zonaria, Nyssia 9:201
zonaria, Papilio 4:66
zosine, Ogyris 20:134
zuni, Arachnis 13:236
Melilotus alba, pigweed, Chinese elm
ZURITUS, *Staphylus* (IA,IG) S2:19
zurkvitzi, Thecla 21:153
zygia epone, Anatole 21:150; 22:153,156

ADDENDA

aegeria, Pararge 10:215
alexiares, Papilio 4:64
anthedon borealis, Lethe 25:256-261
aspersa, Borolia 9:57
australis, Pseudaletia 9:57
bore, Oeneis 23:155
inornata ochracea, Coenonympha (IA) 10:81-86
leucaspis, Milanion 21:164
matronula, Pericallia 3:75
mazans ascalaphus, Staphylus 21:164
neamathla, Lerodea 13:71
perenna, Anaea 15:72
ruricola metacomet, Atrytone 23:192,194
sabrina, Catonephele 22:152,156
sublinea, Heliopetes 21:205